DETECTION OF LIGHT

THIRD EDITION

The invention and development of advanced methods to detect light underlies much of modern technology. This fully updated and restructured third edition is unique amongst the literature, providing a comprehensive, uniform discussion of a broad range of detection approaches. The material is accessible to a broad range of readers rather than just highly trained specialists, beginning with first principles and developing the relevant physics as it goes. The book emphasizes physical understanding of detector operation, without being a catalog of current examples. It is self-contained but also provides a bridge to more specialized works on specific approaches; each chapter points readers toward the relevant literature. This will provide a broad and lasting understanding of the methods for detecting light that underpin so much of our technology. The book is suitable for advanced undergraduate and graduate students, and will provide a valuable reference for professionals across physics and engineering disciplines.

GEORGE H. RIEKE is Regents Professor of Astronomy and Planetary Sciences at the University of Arizona. He switched from TeV gamma rays to the infrared just in time to enjoy its growth from primitive beginnings, to culminate with the launch of JWST. He has contributed to many scientific and technical topics along the way, gaining the background for this broad-ranging book. His other books include *The Last of the Great Observatories* and *Measuring the Universe*, which won the American Astronomical Society's Chambliss Astronomical Writing Award.

DETECTION OF LIGHT

THIRD EDITION

GEORGE H. RIEKE
University of Arizona

CAMBRIDGE
UNIVERSITY PRESS

University Printing House, Cambridge CB2 8BS, United Kingdom

One Liberty Plaza, 20th Floor, New York, NY 10006, USA

477 Williamstown Road, Port Melbourne, VIC 3207, Australia

314–321, 3rd Floor, Plot 3, Splendor Forum, Jasola District Centre, New Delhi – 110025, India

79 Anson Road, #06–04/06, Singapore 079906

Cambridge University Press is part of the University of Cambridge.

It furthers the University's mission by disseminating knowledge in the pursuit of education, learning, and research at the highest international levels of excellence.

www.cambridge.org
Information on this title: www.cambridge.org/9781107124141
DOI: 10.1017/9781316407189

First and Second editions © Cambridge University Press 1994, 2003
Third edition © George H. Rieke 2021

First published 1994
Reprinted 1996
First paperback edition 1996
Second edition 2003
Third edition 2021

A catalogue record for this publication is available from the British Library.

ISBN 978-1-107-12414-1 Hardback

Contents

Preface

My goal in writing this book has been to provide a comprehensive overview of the important technologies for photon detection, with emphasis on the underlying physics. The emphasis is always upon the methods of operation and physical limits to detector performance. Brief mention is sometimes made of the currently achieved performance levels, but only to place the broader physical principles in a practical context. This emphasis has required that the book include a lot of derivations (i.e., equations). I have tried to emphasize physical principles, not rigor, but still the strings of equations may put you off. In fact, Stephen Hawking once said with regard to *A Brief History of Time*: "Someone told me that each equation I included in the book would halve the sales." If I rather expansively make the assumption that every human alive would want to buy this book if it had no equations, I conclude that the Press is likely actually to sell only 10^{-132} copies. Since you are now reading the preface, it seems likely that sales have already exceeded projections. I thank you and am more than willing to overlook your skipping as many equations as you wish. Just roll your eyes, say the incantation "Here he goes again," and move on. In general, I have tried to fill in with prose so the qualitative meaning should be clear. You should view this as a book within a book, that is prose explanations of the principles behind a vast array of photon detectors, with more detailed asides interwoven for convenient access should you want more depth.

It is nearly 30 years since the first edition. The perspective is interesting: for a number of detector types, although there has been great progress in capabilities such as array size, read noise, and quantum efficiency, the physical principles have not changed. However, there are a number of new approaches; for example, the role of superconductivity has grown substantially. In addition, detectors used in the X-ray have evolved away from nuclear physics and more into the realm of the approaches used for lower energy photons (but then some of the nuclear physics detectors have moved a bit in this direction also). I have taken advantage of this evolution to describe X-ray detectors wherever feasible, and have dropped the subtitle that implied the book stopped at the ultraviolet. Over this time, a number of detector types have also slid into oblivion. Because I did not want this edition to grow unnecessarily, I have dropped cases where it seemed to me that the type had only a minor role and was not historically important. Another consequence of the passage of 30 years of time is that the physics involved has become more specialized and complex, and now even review articles sometimes seem to be written for the well-initiated, not for general consumption. The descriptions in this book should be helpful in translating the contents of such reviews into something more easily understood. Based on the extensive survey of the literature that accompanied preparation of this edition, these goals have led to a unique book. It combines subject matter from many disciplines that usually have little interaction into a unified treatment.

I have restricted the physics assumed in the book very strictly to the level attainable after only a semester or two of college-level physics with calculus. To supplement this minimal background, the first chapter includes an overview of radiometry, the second introduces solid state physics, and superconductivity principles are discussed in the seventh. Although many readers may want to skim this material, it gives others with less preparation a reasonable chance of understanding the rest of the book. Although the required preparation is modest, the subject matter is carried to a reasonably advanced level from the standpoint of the underlying physics. Because the necessary physics is developed within the discussion of detectors, the book should be self-contained for those who are outside a classroom environment. There are many more specialized books and review articles; each chapter ends with a list of possible sources for more information.

I thank Karen Visnovsky and Karen Swarthout for their assistance with the first edition. Illustrations have also been provided by Shiras Manning and Danny Pagano. Any corrections, suggestions, and comments will be received gratefully.

1

Introduction

We begin with some background material. First, we need to establish the formalism and definitions for the imaginary signals we will be shining on our imaginary detectors. Second, we will describe general detector characteristics so we can judge the merits of the various types as they are discussed. This discussion introduces some common metrics: (1) the quantum efficiency, i.e., the fraction of the incoming photon stream absorbed; (2) noise and the ratio of signal to noise; (3) the fidelity of images produced by a detector array or similar arrangement; and (4) the speed of response of a detector.

1.1 Radiometry

1.1.1 Concepts and Terminology

There are some general aspects of electromagnetic radiation that need to be defined before we discuss how it is detected. Figure 1.1 illustrates schematically a photon of light with terms used to describe it. One should imagine that time has been frozen, but that the photon has been moving at the speed of light in the direction of the arrow. We often discuss the photon in terms of wavefronts, lines marking the surfaces of constant phase and hence separated by one wavelength.

As electromagnetic radiation, a photon has both electric and magnetic components, oscillating in phase perpendicular to each other and perpendicular to the direction of energy propagation. The amplitude of the electric field, its wavelength and phase, and the direction it is moving characterize the photon. The behavior of the electric field can be expressed as

$$E = E_0 cos(\omega t + \phi), \tag{1.1}$$

where E_0 is the amplitude, ω is the angular frequency, and ϕ is the phase. Alternatively, the behavior is conveniently expressed in complex notation as

$$E(t) = E_0 e^{-j\omega t} = E_0 cos(\omega t) - jE_0 sin(\omega t), \tag{1.2}$$

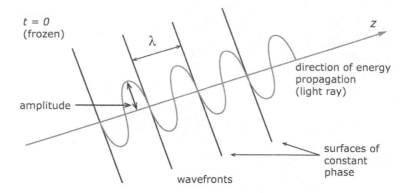

Figure 1.1 Terms describing the propagation of a photon, or a light ray.

where j is the imaginary square root of -1. In this case, the quantity is envisioned as a vector on a two-dimensional diagram with the real component along the usual x-axis and the imaginary one along the usual y-axis. The angle of this vector from the origin and relative to the positive real axis represents the phase.

Most of the time we will treat light as photons of energy; wave aspects will be important only for heterodyne detectors. A photon has an energy of

$$E_{ph} = h\nu = hc/\lambda, \tag{1.3}$$

where h $(= 6.626 \times 10^{-34}$ J s) is Planck's constant, ν and λ are, respectively, the frequency (in hertz $= 1/$seconds) and wavelength (in meters) of the electromagnetic wave, and c $(= 2.998 \times 10^{8}$ m s^{-1}) is the speed of light. In the following discussion, we define a number of expressions for the power output of sources of photons; conversion from power to photons per second can be achieved by dividing by the desired form of equation 1.3.

The spectral radiance per frequency interval, L_ν, is the power (in watts) leaving a unit projected area of the surface of the source (in square meters) into a unit solid angle (in steradians) and unit frequency interval (in hertz). The projected area of a surface element dA onto a plane perpendicular to the direction of observation is $dA \cos\theta$, where θ is the angle between the direction of observation and the outward normal to dA; see Figure 1.2. L_ν has units of W m^{-2} Hz^{-1} ster^{-1}. The spectral radiance per wavelength interval, L_λ, has units of W m^{-3} ster^{-1}. The radiance, L, is the spectral radiance integrated over all frequencies or wavelengths; it has units of W m^{-2} ster^{-1}. The radiant exitance, M, is the integral of the radiance over solid angle, and it is a measure of the total power emitted per unit surface area in units of W m^{-2}.

We will deal only with Lambertian sources; the defining characteristic of such a source is that its radiance is constant regardless of the direction from which it is viewed. A blackbody is one example. The emission of a Lambertian source goes as

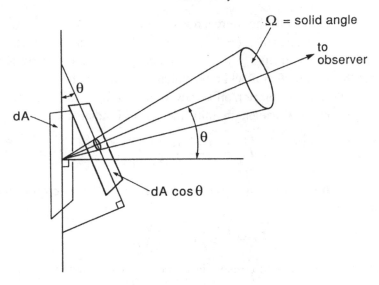

Figure 1.2 Geometry for computing radiance.

the cosine of the angle between the direction of the radiation and the normal to the source surface. From the definition of projected area in the preceding paragraph, it can be seen that this emission pattern exactly compensates for the foreshortening of the surface as it is tilted away from being perpendicular to the line of sight. That is, for the element dA, the projected surface area and the emission decrease by the same cosine factor. Thus, if the entire source has the same temperature and emissivity, every unit area of its projected surface in the plane perpendicular to the observer's line of sight appears to be of the same brightness, independent of its actual angle to the line of sight. Keeping in mind this cosine dependence, and the definition of radiant exitance, the radiance and radiant exitance are related as

$$M = \int L \cos\theta \, d\Omega = 2\pi L \int_0^{\pi/2} \sin\theta \cos\theta \, d\theta = \pi L. \qquad (1.4)$$

The flux emitted by the source, Φ, is the radiant exitance times the total surface area of the source, that is the power emitted by the entire source. For example, for a spherical source of radius R,

$$\Phi = 4\pi R^2 M = 4\pi^2 R^2 L. \qquad (1.5)$$

Although there are other types of Lambertian sources, we will consider only sources that have spectra resembling those of blackbodies, for which the spectral radiance in frequency units is

$$L_v = \frac{\varepsilon \left[2hv^3/(c/n)^2 \right]}{e^{hv/kT} - 1}, \qquad (1.6)$$

where ε is the emissivity of the source, n is the refractive index of the medium into which the source radiates, and k ($= 1.38 \times 10^{-23}$ J K^{-1}) is the Boltzmann constant. The emissivity (ranging from 0 to 1) is the efficiency with which the source radiates compared to that of a perfect blackbody, which by definition has $\varepsilon = 1$. According to Kirchhoff's law, the absorption efficiency, or absorptivity, and the emissivity are equal for any source. In wavelength units, the spectral radiance is

$$L_\lambda = \frac{\varepsilon \left[2h(c/\text{n})^2 \right]}{\lambda^5 \left(e^{hc/\lambda kT} - 1 \right)}. \tag{1.7}$$

It can be easily shown from equations 1.6 and 1.7 that the spectral radiances are related as follows:

$$L_\lambda = \left(\frac{c}{\lambda^2} \right) L_\nu = \left(\frac{\nu}{\lambda} \right) L_\nu. \tag{1.8}$$

According to the Stefan–Boltzmann law, the radiant exitance for a blackbody becomes

$$M = \pi \int\limits_0^\infty L_\nu d\nu = \frac{2\pi k^4 T^4}{c^2 h^3} \int\limits_0^\infty \frac{x^3}{e^x - 1} dx$$

$$= \frac{2\pi^5 k^4}{15 c^2 h^3} T^4 = \sigma T^4, \tag{1.9}$$

where σ ($= 5.67 \times 10^{-8}$ W m^{-2} K^{-4}) is the Stefan–Boltzmann constant.

For Lambertian sources, the optical system feeding a detector will receive a portion of the source power that is determined by a number of geometric factors as illustrated in Figure 1.3. The system will accept radiation from only a limited range of directions determined by the geometry of the optical system as a whole and known as the field of view. The area of the source that is effective in producing a signal is determined by the field of view and the distance from the optical system to the source (or by the size of the source if it all lies within the field of view). This area will emit radiation with some angular dependence. Only the radiation that is emitted in directions where it is intercepted by the optical system can be detected. The range of directions accepted is determined by the solid angle, Ω, that the entrance aperture of the optical system subtends as viewed from the source. In addition, some of the emitted power may be absorbed or scattered by any medium through which it propagates to reach the optical system. For a Lambertian source, the power this system receives is then the radiance in its direction multiplied by the source area within the system field of view, multiplied by the solid angle subtended by the optical system as viewed from the source, and multiplied by the transmittance of the optical path from the source to the system.

Although a general treatment must allow for the field of view to include only a portion of the source, in many cases of interest the entire source lies within the

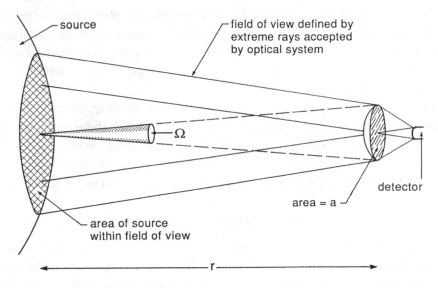

Figure 1.3 Geometry for computing power received by a detector system.

field of view, so the full projected area of the source is used in calculating the signal. For a spherical source of radius R, this area is πR^2. The solid angle subtended by the detector system is

$$\Omega = \frac{a}{r^2}, \tag{1.10}$$

where a is the area of the entrance aperture of the system (strictly speaking, a is the projected area; we have assumed the system is pointing directly at the source) and r is its distance from the source. For a circular aperture,

$$\Omega = 4\pi \, sin^2(\theta/2), \tag{1.11}$$

where θ is the half-angle of the right circular cone whose base is the detector system entrance aperture, and whose vertex lies on a point on the surface of the source; r is the height of this cone.

It is particularly useful when the angular diameter of the source is small compared with the field of view of the detector system to consider the irradiance, E, which is the power in watts per square meter received at a unit surface element at some distance from the source. For the case described in the preceding paragraph, the irradiance is obtained by first multiplying the radiant exitance (from equation 1.4) by the total surface area of the source, A, to get the flux, $A\pi L$. The flux is then divided by the area of a sphere of radius r centered on the source to give

$$E = \frac{AL}{4r^2}, \tag{1.12}$$

where r is the distance of the source from the irradiated surface element. The spectral irradiance, E_ν or E_λ, is the irradiance per unit frequency or wavelength interval. It is also sometimes called the flux density, and is a very commonly used description of the power received from a source. It can be obtained from equation 1.12 by substituting L_ν or L_λ for L.

The radiometric quantities discussed above are summarized in Table 1.1. Equations are provided for illustration only; in some cases, these examples apply only to specific circumstances. The terminology and symbolism vary substantially from one discipline to another; for example, the last two columns of the table translate some of the commonly used radiometric terms into astronomical nomenclature.

1.1.2 The Detection Process

Only a portion of the power received by the optical system is passed on to the detector. The system will have inefficiencies due to both absorption and scattering of energy in its elements, and because of optical aberrations and diffraction. These effects can be combined into a system transmittance term. In addition, the range of frequencies or wavelengths to which the system is sensitive (that is, the spectral bandwidth of the system in frequency or wavelength units) is usually restricted by a spectral filter plus a combination of characteristics of the detector and other elements of the system as well as by any spectral dependence of the transmittance of the optical path from the source to the entrance aperture. A rigorous accounting of the spectral response requires that the spectral radiance of the source be multiplied point-by-point by the spectral transmittances of all the spectrally active elements in the optical path to the detector, and by the detector spectral response, and the resulting function subsequently integrated over frequency or wavelength to determine the total power effective in generating a signal.

In cases where the spectral response is restricted to a range of wavelengths by a bandpass optical filter in the beam, it is generally useful to define the effective wavelength[1] of the system as

$$\lambda_0 = \frac{\int_0^\infty \lambda \, T(\lambda) \, d\lambda}{\int_0^\infty T(\lambda) \, d\lambda}, \tag{1.13}$$

where $T(\lambda)$ is the spectral transmittance of the system. Often the spectral variations of the other transmittance terms can be ignored over the restricted spectral range of the filter. The bandpass of the filter, $\Delta\lambda$, can be taken to be the full width at half maximum (FWHM) of its transmittance function (see Figure 1.4). If the filter cuts

---

[1] We have characterized the response using the mean wavelength; there are a number of other conventions, but for our purposes the differences are minor and unimportant.

Table 1.1 *Definitions of radiometric quantities*

Symbol	Name	Definition	Units	Equation	Alternate name	
L_ν	Spectral radiance (frequency units)	Power leaving unit projected surface area into unit solid angle and unit frequency interval	W m^{-2} Hz^{-1} ster^{-1}	(1.6)	Specific intensity (frequency units)	I_ν
L_λ	Spectral radiance (wavelength units)	Power leaving unit projected surface area into unit solid angle and unit wavelength interval	W m^{-3} ster^{-1}	(1.7)	Specific intensity (wavelength units)	I_λ
L	Radiance	Spectral radiance integrated over frequency or wavelength	W m^{-2} ster^{-1}	$L = \int L_\nu d\nu$	Intensity or specific intensity	I
M	Radiant exitance	Power emitted per unit surface area	W m^{-2}	$M = \int L(\theta)d\Omega$		
Φ	Flux	Total power emitted by source of area A	W	$\Phi = \int M\, dA$	Luminosity	L
E	Irradiance	Power received at unit surface element; equation applies well removed from the source at distance r	W m^{-2}	$E = \dfrac{\int M\, dA}{(4\pi r^2)}$		
E_ν, E_λ	Spectral irradiance	Power received at unit surface element per unit frequency or wavelength interval	W m^{-2} Hz^{-1}, W m^{-3}		Flux density	S_ν, S_λ

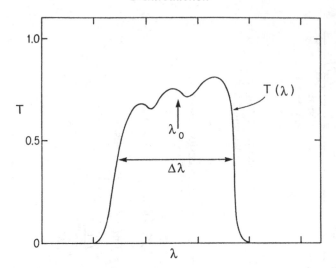

Figure 1.4 Transmittance function $T(\lambda)$ of a filter. The FWHM $\Delta\lambda$ and the effective wavelength λ_0 are indicated.

on and off sharply, its transmittance can be approximated as the average value over the range $\Delta\lambda$:

$$T_F = \frac{\int\limits_{\Delta\lambda} T(\lambda)\, d\lambda}{\Delta\lambda}. \tag{1.14}$$

If $\Delta\lambda/\lambda_0 \leq 0.2$ and the filter cuts on and off sharply, the power effective in generating a signal can usually be estimated in a simplified manner. The signal transmitted by the bandpass filter can be approximated by taking the spectral radiance at λ_0 and multiplying it by $\Delta\lambda$ and the average filter transmittance over the range $\Delta\lambda$. Of course, to obtain the net signal that reaches the detector, this result is multiplied by the various geometric and transmittance terms already discussed for the remainder of the system. However, if λ_0 is substantially shorter than the peak wavelength of the blackbody curve (that is, one is operating in the Wien region of the blackbody) or there is sharp spectral structure within the passband, then this approximation can lead to significant errors, particularly if $\Delta\lambda/\lambda_0$ is relatively large.

Continuing with the approximation just discussed, we can derive a useful expression for estimating the power falling on the detector:

$$P_D \approx \frac{A_{proj}\, a\, T_P(\lambda_0)\, T_O(\lambda_0)\, T_F\, L_\lambda(\lambda_0)\, \Delta\lambda}{r^2}. \tag{1.15}$$

Here A_{proj} is the area of the source projected onto the plane perpendicular to the line of sight from the source to the optical receiver. T_P, T_O, and T_F are the transmittances, respectively, of the optical path from the source to the receiver, of the receiver optics (excluding the bandpass filter), and of the bandpass filter. The area of the receiver entrance aperture is a, and the distance of the receiver from the source is r. An analogous expression holds in frequency units. The major underlying assumptions for equation 1.15 are (a) the field of view of the receiver includes the entire source; (b) the source is a Lambertian emitter; and (c) the spectral response of the detector is limited by a filter with a narrow or moderate bandpass that is sharply defined.

1.2 Detector Types

Nearly all detectors act as transducers that receive photons and produce an electrical response that can be amplified and converted into a form intelligible to suitably conditioned human beings. There are three basic ways that detectors carry out this function:

(a) *Photodetectors* respond directly to individual photons. An absorbed photon releases one or more bound charge carriers in the detector that may (1) modulate the electric current in the material; (2) move directly to an output amplifier; or (3) lead to a chemical change. The most common photodetectors are based on semiconducting materials and are used throughout the X-ray, ultraviolet, visible, and infrared spectral regions. Examples that we will discuss are photoconductors (Chapters 2 and 3), photodiodes (Chapter 3), charge coupled devices (CCDs) (Chapter 5), photographic materials (Chapter 6), photoemissive detectors (Chapter 6), and quantum well detectors (Chapter 6), plus some less common examples scattered about these chapters. The sheer number of types of semiconductor photodetectors provides an indication of their broad application. The unique properties of superconductors enable additional types of photodetector with applications in the submillimeter/millimeter wavelength or with the potential to provide spectral resolution within the detection process. Chapter 7 discusses two examples, microwave kinetic inductance detectors (MKIDs), and superconducting tunnel junctions (STJs).

(b) *Thermal detectors* absorb photons and thermalize their energy. In most cases, this energy changes the electrical properties of the detector material, resulting in a modulation of the electric current passing through it. Thermal detectors have a very broad and nonspecific spectral response, but they are particularly important at infrared and submillimeter wavelengths, and as X-ray detectors. Bolometers and other thermal detectors will be discussed in Chapter 8.

(c) *Coherent detectors* respond to the electric field strength of the signal and can preserve phase information about the incoming photons. They operate by interference of the electric field of the incident photon with the electric field from a local oscillator. These detectors are primarily used in the radio and submillimeter regions but also have specialized applications in the visible and infrared.

Coherent detectors for the visible and infrared are discussed in Chapter 9, and those for the submillimeter are discussed in Chapter 10.

1.3 Performance Characteristics

Good detectors preserve a large proportion of the information contained in the incoming stream of photons. A variety of parameters are relevant to this goal:

(a) *Spectral response* – the total wavelength or frequency range over which photons can be detected with reasonable efficiency.

(b) *Spectral bandwidth* – the wavelength or frequency range over which photons are detected at any one time; some detectors can operate in one or more bands placed within a broader range of spectral response.

(c) *Linearity* – the degree to which the output signal is proportional to the number of incoming photons that were received to produce the signal.

(d) *Dynamic range* – the maximum variation in signal over which the detector output represents the photon flux without losing significant amounts of information.

(e) *Quantum efficiency* – the fraction of the incoming photon stream that is converted into signal.

(f) *Noise* – the uncertainty in the output signal. Ideally, the noise consists only of statistical fluctuations due to the finite number of photons producing the signal.

(g) *Imaging properties* – e.g., the number of detectors ("pixels") in an array. Because signal may blend from one pixel to adjacent ones, the resolution that can be realized may be less, however, than indicated just by the pixel count.

(h) *Time response* – the minimum interval of time over which the detector can distinguish changes in the photon arrival rate.

The first two items in this listing should be clear from our discussion of radiometry, and the next two are more or less self-explanatory. However, the remaining entries include subtleties that call for more discussion.

1.3.1 Quantum Efficiency

To be detected, photons must be absorbed. The absorption coefficient in the detector material is indicated as $a(\lambda)$ and conventionally has units of cm^{-1}. The absorption length is defined as $1/a(\lambda)$. The absorption of a flux of photons, S, passing through a differential thickness element dl is expressed by

$$\frac{dS}{dl} = -a(\lambda)S, \tag{1.16}$$

with the solution for the remaining flux at depth l being

$$S = S_0 e^{-a(\lambda)l}. \tag{1.17}$$

The portion of the flux absorbed by the detector divided by the flux that enters it is

$$\eta_{ab} = \frac{S_0 - S_0 e^{-a(\lambda)d_1}}{S_0} = 1 - e^{-a(\lambda)d_1}, \tag{1.18}$$

where d_1 is the thickness of the detector. The quantity η_{ab} is known as the absorption factor. The quantum efficiency, η, is the flux absorbed in the detector divided by the total flux incident on its surface. Photons are lost by reflection from the surface before they enter the detector volume, leading to a reduction in quantum efficiency below η_{ab}. Minimal reflection occurs for photons striking a nonabsorptive material at normal incidence:

$$r = \frac{(n-1)^2 + \left(\frac{a(\lambda)\,\lambda}{4\pi}\right)^2}{(n+1)^2 + \left(\frac{a(\lambda)\,\lambda}{4\pi}\right)^2}, \tag{1.19}$$

where r is the fraction of the incident flux of photons that is reflected, n is the refractive index of the material, $a(\lambda)$ is the absorption coefficient at wavelength λ, and we have assumed that the photon is incident from air or vacuum, which have a refractive index of n = 1. Reflection from the back of the detector can result in absorption of photons that would otherwise escape. If we ignore this potential gain, the net quantum efficiency is

$$\eta = (1 - r)\eta_{ab}. \tag{1.20}$$

1.3.2 Noise and Signal to Noise

The arriving photons carry a level of information that we want the detector to preserve so far as possible. We now discuss the implications of this requirement.

Ignoring minor corrections having to do with the quantum nature of photons, it can be assumed that the input photon flux follows Poisson statistics,

$$P(m) = \frac{e^{-n}n^m}{m!}, \tag{1.21}$$

where $P(m)$ is the probability of detecting m photons in a given time interval, and n is the average number of photons detected in this time interval if a large number of detection experiments is conducted. The root-mean-square noise in a number of independent events, each of which has an expected noise N, is the square root of the mean, n,

$$N_{rms} = \langle N^2 \rangle^{1/2} = n^{1/2}. \tag{1.22}$$

The errors in the detected number of photons in two experiments can usually be taken to be independent, and hence they add quadratically. That is, the noise in two measurements yielding n_1 and n_2 events, respectively, is

$$N_{rms} = \langle N^2 \rangle^{1/2} = \left[\left(n_1^{1/2} \right)^2 + \left(n_2^{1/2} \right)^2 \right]^{1/2} = (n_1 + n_2)^{1/2}. \qquad (1.23)$$

From the above discussion, the signal-to-noise ratio for Poisson-distributed events is $n/n^{1/2}$, or

$$S/N = n^{1/2}. \qquad (1.24)$$

This result can be taken to be a measure of the information content of the incoming photon stream.[2]

The quantum efficiency is the fraction of incoming photons converted into useful signal in the first stage of detector action. In the simplest form, if the detector converts an individual photon into a single, mobile charge carrier that is collected as the signal, the quantum efficiency is the ratio of the number of charge carriers freed to the number of photons received. For our simple detector example, photons that do not free charge carriers cannot contribute to either signal or noise; they might as well not exist. The portion of information they were carrying is therefore lost. Consequently, for n photons incident on the detector, equation 1.24 shows that the signal-to-noise ratio goes as $\eta n / (\eta n)^{1/2}$, or

$$\left(\frac{S}{N} \right)_d = (\eta n)^{1/2} \qquad (1.25)$$

in the ideal case where both signal and noise are determined only by the photon statistics, where η is the quantum efficiency and the d subscript is to indicate that this value applies just to the detector itself.

Additional steps in the detection process can degrade the information present in the photon stream absorbed by the detector, either by losing signal or by adding noise. The detective quantum efficiency (*DQE*) describes this degradation succinctly in terms of the number of photons that could produce an output signal with an equivalent ratio of signal to noise if no degradation occurred. We define the detective quantum efficiency as

$$DQE = \frac{n_{out}}{n_{in}} = \frac{(S/N)_{out}^2}{(S/N)_{in}^2}, \qquad (1.26)$$

---

[2] A more rigorous description of photon noise takes account of the Bose–Einstein nature of photons, which causes the arrival times of individual particles to be correlated. See the note at the end of the chapter for further discussion, including why this issue can usually be ignored.

where n_{in} is the actual input photon signal, and n_{out} is an imaginary input signal that would produce, with a perfect detector system, the same information content in the output signal as is received from the actual system. Converting to signal to noise, $(S/N)_{out}$ is the observed signal-to-noise ratio, while $(S/N)_{in}$ is the potential signal-to-noise ratio of the incoming photon stream, as given by equation 1.24. By substituting equations 1.24 and 1.25 into equation 1.26, it is easily shown that the *DQE* is just the quantum efficiency defined in equation 1.20 if there is no subsequent degradation of the signal to noise.

1.3.3 Imaging Properties

The resolution of an array of detectors can be most simply measured by exposing it to a pattern of alternating white and black lines (a "bar chart") and determining the minimum spacing of line pairs that can be distinguished, as illustrated in Figure 1.5. The eye can identify such a pattern if the light–dark variation is 4% or greater. The resolution of the detector array is expressed in line pairs per millimeter corresponding to the highest density of lines that produces a pattern at this threshold.

Although it is relatively easy to measure resolution in this way for the detector array alone, a resolution in line pairs per millimeter is difficult to combine with resolution estimates for other components in an optical system used with it. For example, how would one derive the net resolution for a camera with a lens and photographic film whose resolutions are both given in line pairs per millimeter? A second shortcoming is that the performance in different situations can be poorly represented by the line pairs per millimeter specification. For example, one might have two lenses, one of which puts 20% of the light into a sharply defined image core and spreads the remaining 80% widely, whereas the second puts all the light

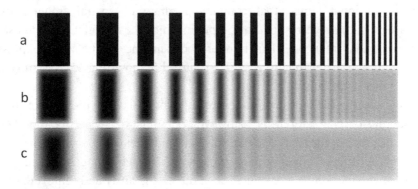

Figure 1.5 Bar chart test of resolution: (a) shows the bar chart with no degradation, while (b) and (c) show the blurring due to the optical system, obviously with lower resolution in (c) than in (b).

into a slightly less well defined core. These systems might achieve identical resolutions in line pairs per millimeter (which requires only 4% modulation), yet they would perform quite differently in other situations.

A more general concept is the modulation transfer function, or *MTF*. Imagine that the detector array is exposed to a field with a sinusoidal spatial variation of the intensity of the input, of period P and amplitude $F(x)$,

$$F(x) = a_0 + a_1 sin(2\pi f x), \tag{1.27}$$

where $f = 1/P$ is the spatial frequency, x is the distance along one axis of the array, a_0 is the mean height (above zero) of the pattern, and a_1 is its amplitude. These terms are indicated in Figure 1.6(a). The modulation of this signal is defined as

$$M_{in} = \frac{F_{max} - F_{min}}{F_{max} + F_{min}} = \frac{a_1}{a_0}, \tag{1.28}$$

where F_{max} and F_{min} are the maximum and minimum values of $F(x)$. Assuming that the resulting image output from the detector is also sinusoidal (which may be only approximately true due to nonlinearities), it can be represented by

$$G(x) = b_0 + b_1 sin(2\pi f x), \tag{1.29}$$

where x and f are the same as in equation 1.27, and b_0 and b_1 are analogous to a_0 and a_1. The modulation in the image will be

$$M_{out} = \frac{b_1}{b_0} M_{in}. \tag{1.30}$$

The modulation transfer factor is

$$MT = \frac{M_{out}}{M_{in}}. \tag{1.31}$$

A separate value of the *MT* will apply at each spatial frequency; Figure 1.6(a) illustrates an input signal that contains a range of spatial frequencies, and Figure 1.6(b) shows a corresponding output in which the modulation decreases with increasing spatial frequency. This frequency dependence of the *MT* is expressed in the modulation transfer function (*MTF*). Figure 1.7 shows the *MTF* corresponding to the response of Figure 1.6(b).

In principle, the *MTF* provides a reasonably complete specification of the imaging properties of a detector array.[3] However, one must be aware that the *MTF* may vary over the face of the array and may have color dependence. In addition, the *MTF* omits time-dependent imaging properties, such as latent images that may persist after the image of a bright source has been put on the array and removed.

---

[3] For optical systems in general, the complete description including phase information is provided by the optical transfer function (*OTF*); the *MTF* is the magnitude of the *OTF*, while the phase is provided by the phase transfer function (*PTF*). For simple photodetectors, the *PTF* can usually be ignored.

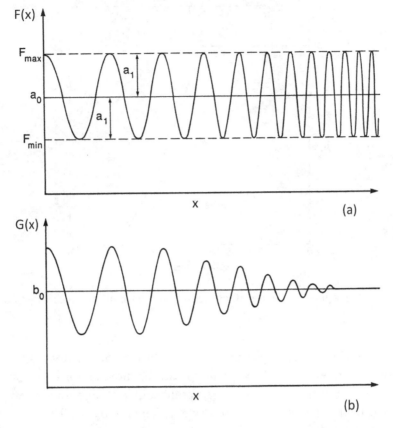

Figure 1.6 Illustration of variation of modulation with spatial frequency: (a) sinusoidal input signal of constant amplitude but varying spatial frequency; (b) how an imaging detector system might respond to this signal.

Computationally, the *MTF* can be determined by taking the absolute value of the Fourier transform, $F(u)$, of the image of a perfect point source. This image is called the point spread function. Fourier transformation is the general mathematical technique used to determine the frequency components of a function $f(x)$ (see, for example, Bracewell 2000; Press et al. 2007). $\mathbf{F}(u)$ is defined as

$$\mathbf{F}(u) = \int_{-\infty}^{\infty} f(x)e^{j2\pi ux}dx, \qquad (1.32)$$

with inverse

$$f(x) = \int_{-\infty}^{\infty} \mathbf{F}(u)e^{-j2\pi xu}du, \qquad (1.33)$$

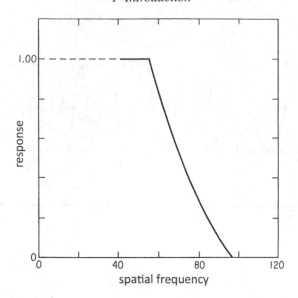

Figure 1.7 The modulation transfer function (*MTF*) for the response illustrated in Figure 1.6(b)

where j is the (imaginary) square root of -1. In the current discussion, $f(x)$ is a functional representation of the point spread function, and u is the spatial frequency. The Fourier transform can be generalized in a straightforward way to two dimensions, but for the sake of simplicity we will not do so here. The absolute value of the transform is

$$|\mathbf{F}(u)| = (\mathbf{F}(u)\mathbf{F}^*(u))^{1/2}, \tag{1.34}$$

where $\mathbf{F}^*(u)$ is the complex conjugate of $\mathbf{F}(u)$; it is obtained by reversing the sign of all imaginary terms in $\mathbf{F}(u)$.

If $f(x)$ is the point spread function, $|\mathbf{F}(u)|/|\mathbf{F}(0)|$ is the *MTF*. This formulation holds because a sharp impulse contains all frequencies equally and hence the Fourier transform of the point spread function gives the spatial frequency response of the detector. The *MTF* is normalized to unity at spatial frequency 0 by this definition. As emphasized in Figure 1.7, the response at zero frequency cannot be measured directly but must be extrapolated from higher frequencies.

Only a relatively small number of functions have Fourier transforms that are easy to manipulate. Table 1.2 contains a short compilation of some of these cases. With the use of computers, however, Fourier transformation is a powerful and very general technique.

The *MTF* of an entire linear optical system can be determined by multiplying together the *MTF*s of its constituent elements. The multiplication occurs on

Table 1.2 *Fourier transforms*

$f(x)$	$F(u)$
$F(x)$	$f(-u)$
$aF(x)$	$aF(u)$
$f(ax)$	$(1/|a|)F(u/a)$
$f(x) + g(x)$	$F(u) + G(u)$
1	$\delta(u)^c$
$e^{-\pi x^2}$	$e^{-\pi u^2}$
$e^{-|x|}$	$2/(1 + (2\pi u)^2)$
$e^{-x}, x > 0$	$(1 - j2\pi u)/(1 + (2\pi u)^2)$
$sech(\pi x)$	$sech(\pi u)$
$|x|^{-1/2}$	$|u|^{-1/2}$
$sgn(x)^a$	$-j/(\pi u)$
$e^{-|x|} sgn(x)$	$-j4\pi u/(1 + (2\pi u)^2)$
$\Pi(x)^b$	$sin(\pi u)/\pi u$

[a] $sgn(x) = -1$ for $x < 0$ and $= 1$ for $x \geq 0$.
[b] $\Pi(x) = 1$ for $|x| < 1/2$ and $= 0$ otherwise.
[c] $\delta(u) = 0$ for $u \neq 0$, $\int \delta(u)du = 1$; that is, $\delta(u)$ is a spike at $u = 0$.

a frequency by frequency basis, that is, if the first system has $MTF_1(f)$ and the second $MTF_2(f)$, the combined system has $MTF(f) = MTF_1(f)MTF_2(f)$. The overall resolution capability of complex optical systems can be easily determined in this way. In addition, the *MTF* gives a complete description of the imaging behavior of detectors and even of many linear optical systems as opposed to single parameter descriptions that may be equivalent for systems having significantly different resolution characteristics.

1.3.4 Frequency Response

The response speed of a detector can be described very generally by specifying the dependence of its output on the frequency of an imaginary photon signal that varies sinusoidally in time. This concept is analogous to the modulation transfer function described just above with regard to imaging.

A variety of factors limit the frequency response. Many of them, however, can be described by an exponential time response, such as that of a resistor/capacitor electrical circuit. To be specific in the following, we will assume that the response is given by the *RC* time constant of such a circuit, although we will find other uses for the identical formalism later. If the capacitor is in parallel with the resistance, charge deposited on the capacitance bleeds off through the resistance with an exponential time constant

$$\tau_{RC} = RC. \tag{1.35}$$

Sometimes a "rise time" is specified rather than the exponential time constant. The rise time is the interval required for the output to change from 10% to 90% of its final value (measured relative to the initial value). For an exponential response, the rise time is 2.20 τ_{RC}.

Let a voltage impulse be deposited on the capacitor,

$$v_{in}(t) = v_0 \delta(t), \tag{1.36}$$

where v_0 is a constant and $\delta(t)$ is the delta function (see footnotes to Table 1.2). We can observe this event in two ways. First, we might observe the voltage across the resistance and capacitance directly, for example with an oscilloscope. It will have the form

$$v_{out}(t) = \left[\begin{array}{cc} 0, & t < 0, \\ \frac{v_0}{\tau_{RC}} e^{-t/\tau_{RC}}, & t \geq 0. \end{array} \right. \tag{1.37}$$

The same event can be analyzed in terms of the effect of the circuit on the input frequencies rather than on the time dependence of the voltage. To do so, we convert the input and output voltages to frequency spectra by taking their Fourier transforms. The delta function contains all frequencies at equal strength, that is, from Table 1.2,

$$V_{in}(f) = v_0 \int_{-\infty}^{\infty} \delta(t) e^{-j2\pi ft} dt = v_0. \tag{1.38}$$

Since the frequency spectrum of the input is flat ($V_{in}(f) = $ constant), any deviations from a flat spectrum in the output must arise from the action of the circuit. That is, the output spectrum gives the frequency response of the circuit directly. Again from Table 1.2, it is

$$V_{out}(f) = \int_{-\infty}^{\infty} v_{out}(t) e^{-j2\pi ft} dt$$

$$= v_0 \left[\frac{1 - j2\pi f \tau_{RC}}{1 + (2\pi f \tau_{RC})^2} \right]. \tag{1.39}$$

The imaginary part of $V_{out}(f)$ represents phase shifts that can occur in the circuit. For a simple discussion, we can ignore the phase and describe the strength

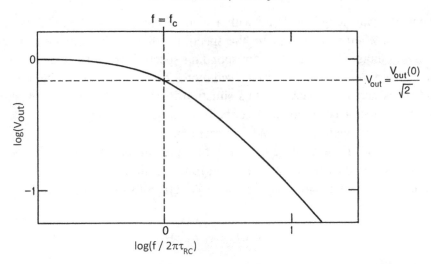

Figure 1.8 Frequency response of an *RC* circuit. The cutoff frequency is illustrated.

of the signal only in terms of the frequency dependence of its amplitude. The amplitude can be determined by taking the absolute value of $V_{out}(f)$:

$$|V_{out}(f)| = \left(V_{out}V^*{}_{out}\right)^{1/2}$$

$$= \frac{v_0}{\left[1 + (2\pi f \tau_{RC})^2\right]^{1/2}}, \qquad (1.40)$$

where V^*_{out} is the complex conjugate of V_{out}. This function is plotted in Figure 1.8. As with the *MTF*, the effects of different circuit elements on the overall frequency response can be determined by multiplying their individual response functions together. The frequency response is often characterized by a cutoff frequency

$$f_c = \frac{1}{2\pi \tau_{RC}}, \qquad (1.41)$$

at which the amplitude drops to $1/\sqrt{2}$ of its value at $f = 0$, or

$$|V_{out}(f_c)| = \frac{1}{\sqrt{2}}|V_{out}(0)|. \qquad (1.42)$$

1.4 Radiometry Example

A 1000 K spherical blackbody source of radius 1 m is viewed in air by a detector system from a distance of 1000 m. The entrance aperture of the system has a radius of 5 cm, and the optical system has a field of view half-angle of 0.1°. The detector operates at a wavelength of 1 μm with a spectral bandpass of 1%, and its optical system is 50% efficient. Compute the spectral radiances in both frequency and wavelength units. Calculate the corresponding spectral irradiances at the detector entrance aperture, and the power received by the detector. Compare the usefulness of radiances and irradiances for this situation. Compute the number of photons hitting the detector per second. Describe how these answers would change if the blackbody source were 10 m in radius rather than 1 m.

The refractive index of air is n $\sim$ 1, so the spectral radiance in frequency units is given by equation 1.6 with ε = n = 1. From equation 1.3, the frequency corresponding to 1 μm is $\nu = c/\lambda = 2.998 \times 10^{14}$ Hz. Substituting into equation 1.6, we find that

$$L_\nu = 2.21 \times 10^{-13}\,\mathrm{W\,m^{-2}Hz^{-1}ster^{-1}}. \tag{1.43}$$

Alternatively, we can substitute the wavelength of 1×10^{-6} m into equation 1.7 to obtain

$$L_\lambda = 6.62 \times 10^{7}\,\mathrm{W\,m^{-3}ster^{-1}}. \tag{1.44}$$

The solid angle subtended by the detector system as viewed from the source is given by equation 1.10. The area of the entrance aperture is 7.854×10^{-3} m^2, so

$$\Omega = 7.854 \times 10^{-9}\,\mathrm{ster}. \tag{1.45}$$

The 1% bandwidth corresponds to $0.01 \times 2.998 \times 10^{14}$ Hz $= 2.998 \times 10^{12}$ Hz, or to $0.01 \times 1 \times 10^{-6}$ m $= 1 \times 10^{-8}$ m. The radius of the area accepted into the beam of the detector system at the distance of the source is 1.745 m, and, since it is larger than the radius of the source, the entire visible area of the source will contribute to the signal. The projected area of the source is 3.14 m^2 (since it is a Lambertian emitter, no further geometric corrections are required for its effective emitting area). Then, computing the power at the entrance aperture of the detector system by multiplying the spectral radiances by the source area (projected), spectral bandwidth, and solid angle received by the system, we obtain $P = 1.63 \times 10^{-8}$ W.

Because the angular diameter of the source is less than the field of view, it is equally convenient to use the irradiance. The surface area of the source is 12.57 m^2. Using equation 1.12 and frequency units, we obtain

$$E_\nu = 6.945 \times 10^{-19}\,\mathrm{W\,m^{-2}\,Hz^{-1}}. \tag{1.46}$$

Similarly for wavelength units,

$$E_\lambda = 2.08 \times 10^{2}\,\mathrm{W\,m^{-3}}. \tag{1.47}$$

Multiplying by the bandpass and entrance aperture area yields a power of 1.63×10^{-8} W, as before.

The power received by the detector is reduced by optical inefficiencies to 50% of the power incident on the entrance aperture, so it is 8.2×10^{-9} W. The energy per photon can be computed from equation 1.3 to be 1.99×10^{-19} J. The detector therefore receives 4.12×10^{10} photons s^{-1}.

If the blackbody source were 10 m in radius, the spectral radiances, L_ν and L_λ, would be unchanged. The irradiances, E_ν and E_λ, would increase in proportion to the surface area of the source, so they would be 100 times larger than computed above. The field of view of the optical system, however, no longer includes the entire source; therefore, the power at the system entrance aperture is most easily computed from the spectral radiances, where the relevant surface area is that within the field of view and hence has a radius of 1.745 m. The power at the entrance aperture therefore increases by a factor of only 3.05, giving $P = 4.97 \times 10^{-8}$ W, as do the power falling on the detector (2.48×10^{-8} W) and the photon rate (1.25×10^{11} photons s^{-1}).

1.5 Problems

1.1 A spherical blackbody source at 300 K and of radius 0.1 m is viewed from a distance of 1000 m by a detector system with an entrance aperture of radius 1 cm and field of view half angle of 0.1 degree.

   (a) Compute the spectral radiances in frequency units at 1 and 10 μm.
   (b) Compute the spectral irradiances at the entrance aperture.
   (c) For spectral bandwidths 1% of the wavelengths of operation and assuming that 50% of the incident photons are absorbed in the optics before they reach the detector, compute the powers received by the detector.
   (d) Compute the numbers of photons hitting the detector per second.

1.2 Consider a detector with an optical receiver of entrance aperture 2 mm diameter, optical transmittance (excluding bandpass filter) of 0.8, and field of view 1° in diameter. This system views a blackbody source of 1000 K with an exit aperture of diameter 1 mm and at a distance of 2 m. The signal out of the blackbody is interrupted by a shutter at a temperature of 300 K. The receiver system is equipped with two bandpass filters, one with $\lambda_0 = 20\,\mu$m and $\Delta\lambda = 1\,\mu$m and the other with $\lambda_0 = 2\,\mu$m and $\Delta\lambda = 0.1\,\mu$m; both have transmittances of 0.8. The transmittance of the air between the source and receiver is 1 at both wavelengths. Compute the net signal at the detector, that is compute the change in power incident on the detector as the shutter is opened and closed.

1.3 For blackbodies, the wavelength of the maximum spectral irradiance times the temperature is a constant, or

$$\lambda_{max} T = C. \tag{1.48}$$

This expression is known as the Wien displacement law; derive it. For wavelength units, show that $C \sim 0.3$ cm K.

1.4 Show that for $h\nu/kT \ll 1$ (setting $\varepsilon = n = 1$),

$$L_\nu = 2kT\nu^2/c^2. \tag{1.49}$$

This expression is the Rayleigh–Jeans law and is a useful approximation at long wavelengths. For a source temperature of 100 K, compute the shortest wavelength for which the Rayleigh–Jeans law is within 20% of the result given by equation 1.6. Compare with λ_{max} from Problem 1.3.

1.5 Derive equation 1.11. Note the particularly simple form for small θ.

1.6 Consider a bandpass filter that has a transmittance of zero outside the passband $\Delta\lambda$ and a transmittance that is the same for all wavelengths within the passband. Compare the estimate of the signal passing through this filter when the signal is determined by integrating the source spectrum over the filter passband with that where only the effective wavelength and FWHM bandpass are used to characterize the filter. Assume a source radiating in the Rayleigh–Jeans regime. Show that the error introduced by the simple effective wavelength approximation is a factor of

$$1 + \frac{5}{6}\left(\frac{\Delta\lambda}{\lambda_0}\right)^2 \tag{1.50}$$

plus terms of order $(\Delta\lambda/\lambda_0)^4$ and higher. Evaluate the statement in the text that the approximate method usually gives acceptable accuracy for $\Delta\lambda/\lambda_0 \leq 0.2$.

1.7 From equation 1.51, show that the Bose–Einstein correction to the rms photon noise $\langle N^2 \rangle^{1/2}$ is less than 10% if $(5\varepsilon\tau\eta kT/h\nu) < 1$. Consider a blackbody source at $T = 1000$ K viewed by a detector system with optical efficiency 50% and quantum efficiency 50%. Calculate the wavelength beyond which the correction to the noise would exceed 10%. Compare this wavelength with that at the peak of the source output.

1.8 Compute the Fourier transform of $f(x) = H(x) + sech(10x)$, where $H(x) = 0$ for $x < 0$ and $= 1$ for $x \geq 0$.

1.6 Note

This note discusses the correction to simple noise estimates due to the boson nature of photons. Equation 1.24 is derived using the assumption that the particles arrive completely independently; the bunching of Bose–Einstein particles increases the noise above this estimate. The full description of photon noise shows it to be

$$\langle N^2 \rangle = n\left[1 + \frac{\varepsilon\tau\eta}{e^{h\nu/kT} - 1}\right], \tag{1.51}$$

where $\langle N^2 \rangle$ is the mean square noise, n is the average number of photons detected, h is Planck's constant, ε is the emissivity of the source of photons, τ is the transmittance of the system optics, η is the detector quantum efficiency, ν is the photon frequency, k is Boltzmann's constant, and T is the absolute temperature of the photon source (see van Vliet 1967). Comparing with equation 1.24, it can be seen that the term in square brackets in equation 1.51 is a correction factor for the increase in noise from the Bose–Einstein behavior. It becomes important only at frequencies much lower than that of the peak emission of the blackbody, and then only for highly efficient detector systems. In most cases of interest, particularly with realistic instrument efficiencies, this correction factor is sufficiently close to unity that it can be ignored. Moreover, the entire theory of this noise behavior is rather complex and the predicted phenomena are yet to be observed (Lee and Talghader 2018).

1.7 Further Reading

Boreman (2001) – good introduction to the *MTF* and its applications
Bukshtab (2019) – advanced and very extensive treatment of photometry and radiometry
Grant (2011) – short and practical guide to practice of radiometry
Grum and Becherer (1979) – classic description of radiometry
McCluney (2014) – a good introduction to radiometry
Palmer and Grant (2009) – excellent introduction to radiometry, starting from first principles
Press et al. (2007) – thorough and practical general description of numerical methods, including Fourier transformation

2

Photodetector Basics

This chapter introduces semiconductor physics, which is the basis of many detector types including those discussed in the following four chapters. As an illustration of the basic operation of semiconductor detectors, it discusses in some detail the simplest form of such devices, in which an incident photon stream is absorbed and modifies the conductivity of a chip of semiconductor material. We complete that discussion with a description of the shortcomings of this approach, to motivate the following chapters describing improved detector types.

2.1 Solid State Physics

Semiconductors are the basis for most electronic devices, including those used for amplification of photoexcited currents as well as those used to detect photons and create such currents. The electrical properties of a semiconductor are altered dramatically by the absorption of an ultraviolet, visible, or infrared photon. A broad variety of detectors is based on this behavior. Metals, on the other hand, have high electrical conductivity that is only insignificantly modified by the absorption of photons, and insulators require more energy to excite electrical changes than is available from individual visible or infrared photons. Pure semiconductors are described as having intrinsic photoconductivity; small amounts of impurities added to them yield extrinsic photoconductivity, which allows detection of photons with lower energy than that required to excite intrinsic photoconductivity. Most detectors operating from the X-ray through the mid-infrared depend on these materials. Because of the broad applications of semiconductors, there is a large infrastructure for their manufacture that is critical for production of sophisticated detector types.

To facilitate our discussion, we will first review some of the properties of semiconductors. The concepts introduced below are used throughout the remaining chapters. The elemental semiconductors are silicon and germanium; they are found in column IVa of the periodic table (Table 2.1). Their outermost electron shells, or

Table 2.1 *Periodic table of the elements*

Ia	IIa	IIIb	IVb	Vb	VIb	VIIb	VIII			Ib	IIb	IIIa	IVa	Va	VIa	VIIa	VIIIa
1 H																	2 He
3 Li	4 Be											5 B	6 C	7 N	8 O	9 F	10 Ne
11 Na	12 Mg											13 Al	14 Si	15 P	16 S	17 Cl	18 Ar
19 K	20 Ca	21 Sc	22 Ti	23 V	24 Cr	25 Mn	26 Fe	27 Co	28 Ni	29 Cu	30 Zn	31 Ga	32 Ge	33 As	34 Se	35 Br	36 Kr
37 Rb	38 Sr	39 Y	40 Zr	41 Nb	42 Mo	43 Tc	44 Ru	45 Rh	46 Pd	47 Ag	48 Cd	49 In	50 Sn	51 Sb	52 Te	53 I	54 Xe
55 Cs	56 Ba	57 La	72 Hf	73 Ta	74 W	75 Re	76 Os	77 Ir	78 Pt	79 Au	80 Hg	81 Tl	82 Pb	83 Bi	84 Po	85 At	86 Rn
87 Fr	88 Ra	89 Ac															

valence states, contain four electrons, half of the total number allowed for these shells. They form crystals with a diamond lattice structure (note that carbon is also in column IVa). In this structure, each atom bonds to its four nearest neighbors; it can therefore share one valence electron with each neighbor, and vice versa. Electrons are fermions and must obey the Pauli exclusion principle, which states that no two particles with half-integral quantum mechanical spin can occupy identical quantum states.[1] Because of the exclusion principle, the electrons shared between neighboring nuclei must have opposite spin (if they had the same spin, they would be identical quantum mechanically), which accounts for the fact that they occur in pairs. By sharing electrons, each atom comes closer to having a filled valence shell, and a quantum mechanical binding force known as a covalent bond is created.

The binding of electrons to an atomic nucleus can be described in terms of a potential energy well around the nucleus. Electrons may be in the ground state or at various higher energy levels called excited states. There is a specific energy difference between these states that can be measured by detecting an absorption or emission line when an electron shifts between energy levels. The sharply defined energy levels of an isolated atom occur because of constructive interference of electron wave functions within the potential well; there is destructive interference at all other energies. When atoms are brought together, the quantum mechanically permitted energy levels of an individual atom split because of the coupling between the potential wells. The "valence states" and "conduction states" in a material are analogous to the ground state and excited states, respectively, in an isolated atom.

---

[1] This exclusion rule does not apply to particles with integral spin, which are called bosons.

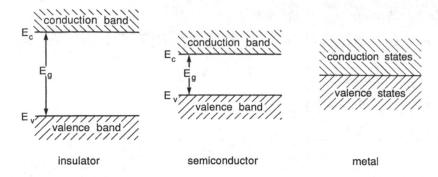

Figure 2.1 Energy band diagrams for insulators, semiconductors, and metals.

In a compact structure such as a crystal, the energy levels split multiply into broad energy zones called bands. This situation can be represented by band diagrams such as those in Figure 2.1. The splitting of energy levels occurs because the Pauli exclusion principle plays an increasing role in the electronic structure of the atoms. If the atoms are close enough to allow the electron wave functions to begin overlapping, then according to the exclusion principle the electrons must distribute themselves so that no two of them are in an identical quantum state.

If the material were at a temperature of absolute zero, all available states in the band would be filled up to some maximum level. The electrical conductivity would be zero because there would be no accessible states into which electrons could move. Conduction becomes possible when electrons are lifted into higher and incompletely filled energy levels, either by thermal excitation or by other means.

There are two distinct possibilities. In a metal, the electrons only partially fill a band so that a very small amount of energy is required to gain access to unfilled energy levels and hence to excite conductivity. In comparison, in a semiconductor or an insulator, the electrons would completely fill a band at absolute zero. To gain access to unfilled levels, an electron must be lifted into a level in the next higher band, resulting in a threshold excitation energy required to initiate electrical conductivity. In this latter case, the filled band is called the valence band and the unfilled one the conduction band. The bandgap, E_g, is the energy between the highest energy level in the valence band, E_v, and the lowest energy level in the conduction band, E_c. It is the minimum energy that must be supplied to excite conductivity in the material. Semiconductors have $0 < E_g < 3.5 \, \text{eV}$.

Despite their differing electrical behavior, the band diagrams for semiconductors and insulators are qualitatively similar. Semiconductors are partially conducting under typically encountered conditions because the thermal excitation at room temperature is adequate to lift some electrons across their modest energy bandgaps. However, their conductivity is a strong function of temperature (going roughly as

$e^{-E_g/2kT}$; $kT \approx 0.025$ eV at room temperature), and near absolute zero they behave as insulators. In such a situation, the charge carriers are said to be "frozen out."

When electrons are elevated into the conduction band of a semiconductor or insulator, they leave empty positions in the valence state. These positions have an effective positive charge provided by the ion in the crystal lattice, and are called holes. As an electron in the valence band hops from one bond position to an adjacent, unoccupied one, the hole is said to migrate (such a positional change does not require that the electron be lifted into a conduction band or receive any appreciable additional energy). Although the holes are not real subatomic particles, they behave in many situations as if they were. It is convenient to discuss them as the positive counterparts to the electrons and to assign to them such attributes as mass, velocity, and charge. The total electric current is the combination of the contributions from the motions of the conduction electrons and the holes.

The key to the usefulness of semiconductors for visible and infrared photon detection is that their bandgaps are in the energy range of a single photon; for example, a visible photon of wavelength 0.55 μm has an energy of 2.26 eV. Absorption of energy greater than E_g photoexcites electrical conductivity in the material. Of course, other forms of energy can also excite conductivity indistinguishable from that due to photoexcitation; a particularly troublesome example is thermal excitation.

In addition to elemental silicon and germanium, many compounds are semiconductors. A typical semiconductor compound is a diatomic molecule comprising atoms that symmetrically span column IVa in the periodic table, for example an atom from column IIIa combined with one from column Va, or one from column IIb combined with one from column VIa. Table 2.2 lists some simple semiconductor compounds and their bandgap energies. Elements that are important in the formation of useful semiconductors are shown in boldface in Table 2.1.

The electrical properties of pure, or intrinsic, photoconductors can be modified dramatically by adding impurities, or doping them, to make extrinsic semiconductors. Consider the structure of silicon, which conveniently lends itself to representation in two dimensions, as shown in Figure 2.2. (The true structure of silicon is three dimensional with tetrahedral symmetry.) The silicon atoms are represented here by open circles connected by bonds consisting of pairs of electrons; each electron is indicated by a single line. In the intrinsic section of the structure, the uniform pattern of line pairs shows that all of the atoms share pairs of electrons to complete their outer electron shells. If an impurity is added from an element in column III of the periodic table (for example, B, Al, Ga, or In, i.e. boron, aluminum, gallium, or indium), it has one too few valence-shell electrons to complete the crystal structure. When the valence shell of the dopant has too few electrons to form all the bonds required for the crystal structure, its atoms tend to capture, or accept, electrons

Table 2.2 *Semiconductors and their bandgap energies*

Columns	Semiconductor	E_g (eV)
IV	Ge	0.67
	Si	1.11
	SiC	2.86
III-V	AlAs	2.16
	AlP	2.45
	AlSb	1.6
	GaP	2.26
	GaSb	0.7
	GaAs	1.43
	InAs	0.36
	InGaAs	variable, 1.43–0.36
	InP	1.35
	InSb	0.18
	AlGaN	variable, 6.2–3.4
II-VI	CdS	2.42
	CdSe	1.73
	CdTe	1.55
	HgCdTe	variable, 1.55–<0
	ZnSe	2.7
	ZnTe	2.25
I-VII	AgBr	2.81[a]
	AgCl	3.33[a]
IV-VI	PbS	0.37
	PbSe	0.27
	PbTe	0.29

[a] Values taken from James (1977). All other values are taken from Streetman and Banerjee (2014), Appendix III. Bandgap energies are temperature dependent, so the tabulated values are only approximate for the purpose of comparison.

from the semiconductor atoms, thus creating holes in the semiconductor valence band. Material in which holes dominate is called p-type. In the band diagram (see Figure 2.3), we add an *acceptor* level at the appropriate excitation energy E_i above E_v. If the impurity is from column V (for example, P, As, or Sb, i.e., phosphorus, arsenic, or antimony), it has sufficient electrons to accommodate the sharing with one electron left over. This "extra" electron is relatively easily detached to enter the conduction band as a negative charge carrier. The impurity is called n-type; the band diagram in Figure 2.3 then has a *donor* level at the appropriate excitation energy, E_i, below E_c. The energy required to move the hole or the extra electron is much smaller than needed to break a bond in the regular silicon crystal, so in either case the material becomes more conductive.

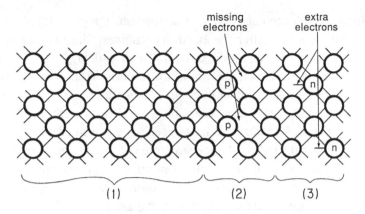

Figure 2.2 Comparison of n-type and p-type doping in silicon. Region (1) shows undoped material, region (2) is doped p-type, and region (3) is n-type.

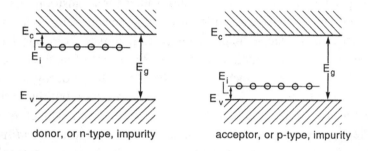

Figure 2.3 Energy band diagrams for semiconductors with n-type and p-type impurities.

Because all semiconductors have residual impurities, weak donor or acceptor levels (or both) are always present. If the electrical behavior is dominated by the effects introduced by the impurities, the material is called extrinsic. In an extrinsic semiconductor, the majority carrier is the type created by the dominant dopant (for example, holes for p-type). The minority carrier is the opposite type and usually has a much smaller concentration.

Figure 2.2 and the discussion above give an overly concrete picture of the semiconductor crystal. Quantum mechanically, the impurity-induced holes and conduction electrons must be treated as collective states of the crystal that are described by probability distribution functions that extend over many atoms. A full theory of the excitation energies required for extrinsic photoconductors requires a quantum mechanical derivation, which is beyond the scope of our discussion.

Impurities can also act as traps or recombination centers. Consider a p-type semiconductor. Thermal excitation, while not always supplying enough energy to raise electrons into the conduction band, may provide sufficient energy to move electrons from the valence band into an intermediate energy state, $E_v < E_t < E_c$, provided by the impurity. After the electron combines with the impurity atom, it will either be released by thermal excitation (after a delay), or the net negative charge will attract holes that will neutralize the impurity atom by recombining with the electron. If the electron is most likely to be released from the atom by thermal excitation, the impurity atom is called a trapping center, or trap; otherwise, it is termed a recombination center. Similar effects can occur at crystal defects, where some of the bonds in the regular lattice structure are broken, providing sites where the intrinsic crystal atoms can attract and combine with charge carriers.

A band diagram can oversimplify the requirements for transitions between valence and conduction bands resulting from the absorption of a photon. In many semiconductors, the energy levels corresponding to the minimum energy in the conduction band and to the maximum energy in the valence band do not have compatible quantum mechanical wave vector values for a transition. In these cases, an electron must either make a transition involving greater energy than the bandgap energy or undergo a change in momentum as it moves from one band to the other. In the latter case, recombination must be by means of an intermediate state, which absorbs the excess momentum and therefore allows decay directly to the valence band. The transition can occur at any crystal atom, but the probability is frequently enhanced at recombination centers or traps. Semiconductors exhibiting this behavior (including silicon and germanium) are said to have *indirect* energy bands. Another class of semiconductors, of which GaAs is one example, allows *direct* energy band transitions; minimum energy electron transitions are permitted without a previous change of electron momentum or the presence of intermediate recombination centers.

The efficiency of photon absorption is parametrized by an absorption coefficient, a, typically with units of cm^{-1}. The difference between these two classes of transition is demonstrated by the behavior of their absorption coefficients (Figure 2.4). The detection process can be reversed by passing a current through a semiconductive device arranged to emit light. In general, semiconductor light emitters such as light emitting diodes (LEDs) and lasers must be based on materials capable of direct band transitions. Detectors made of material allowing only indirect transitions at low energies will tend to have reduced response at the longer wavelengths.

Silicon is the most widely used intrinsic photoconductive detector for the visible, ultraviolet, and X-ray ranges. As shown in Figure 2.4, it has poor absorption at the longest wavelengths (near 1 μm) because absorption requires indirect transitions. In general, as shown in Figure 2.5, the absorption coefficient falls with increasing X-ray photon energy until the energy is sufficient to create electron–hole pairs in a more energetic electron shell. For example, the L shell lies at 0.099 keV and the K shell at 1.84 keV, resulting in the sharp increases in absorption at those energies.

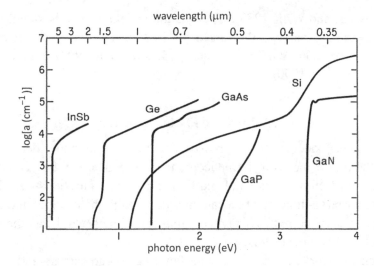

Figure 2.4 Absorption coefficients for various semiconductors. After Stillman and Wolfe (1977) and (for GaN) Muth et al. (1999).

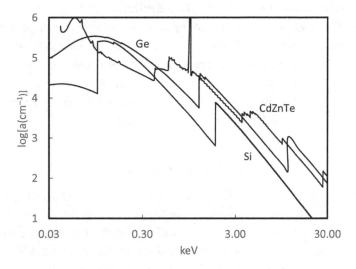

Figure 2.5 Absorption coefficients for various semiconductors in the far ultraviolet and X-ray range. The minimum energy, 0.03 keV, corresponds to a wavelength of 0.0413 μm or 413 Å(angstroms). The absorption coefficient of silicon drops by two orders of magnitude in the factor of 10 wavelength gap between Figure 2.4 and this one. From National Institute of Standards and Technology (2020).

Although we have focused on silicon, the X-ray behavior of other semiconductors is similar. The bandgap of silicon is 1.11 eV; beginning at 3 eV, the photon energy is sufficient to create more than one electron–hole pair, with a "quantum yield" of ~1.25 at 5 eV (i.e., 25% more free electrons than input photons) (Geist and Wang 1983). The quantum yield increases rapidly into the X-ray; for example, between 0.05 and 1.5 keV, an electron–hole pair is created on average for every 3.66 eV of photon energy (Scholze et al. 1998) (e.g., the quantum yield at 3.7 keV is $\sim$1000).

2.2 Intrinsic Photoconductors

Intrinsic photoconductors are the most basic kind of electronic detector, so we will use them as a means to introduce certain general principles. The energy of an absorbed photon breaks a bond and lifts an electron into the conduction band, creating an electron–hole pair that can migrate through the material and conduct a measurable electric current. This process illustrates in a general way the operation of all electronic photodetectors.

All such detectors have a region with few free charge carriers and hence high resistance; an electric field is maintained across this region. Photons are absorbed in semiconductor material and produce free charge carriers, which are driven across the high resistance region by the field. Detection occurs by sensing the result-ing current. The high resistance is essential to allow usefully large signals while minimizing various sources of noise. For the photoconductors discussed in this chapter, the photon absorption occurs within the high resistance region, whereas more sophisticated detectors absorb the photons in a separate region optimized for that purpose. The reasons for separating these two detector zones in modern detectors will be discussed at the end of this section.

2.2.1 Basic Operation

Assume that we arrange a sample of semiconductor as shown in Figure 2.6. The device has transverse contacts on opposite faces; these contacts are usually made of evaporated metal films. The detector is illuminated from a direction parallel to the contacts. Now let us apply an electric field to the detector, and at the same time illuminate it with photons having energy greater than its bandgap energy. If we know the bias voltage, V_b, across the detector and measure the current through it, we can calculate its resistance from Ohm's law, $R_d = V_b/I_d$. By defi-nition, the conductivity, σ (conventionally given in units of (ohm centimeters)$^{-1}$; σ(ohm meters)$^{-1} = 100\sigma$(ohm centimeters)$^{-1}$), is related to the resistance as

$$R_d = \frac{\ell}{\sigma w d},\qquad(2.1)$$

where w, ℓ, and d are the width, length, and depth of the detector, respectively (see Figure 2.6, and note that these quantities are defined with respect to the electrical contacts and not to the incoming photon direction). Charge carriers are generated

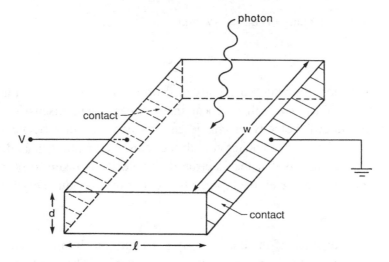

Figure 2.6 Photoconductor with transverse contacts.

both by thermal excitation and by any absorbed photons that have $E \geqslant E_g$; as a result, the conductivity comprises two components:

$$\sigma = \sigma_{th} + \sigma_{ph}. \tag{2.2}$$

The thermally induced component is an example of dark current, that is, signal produced by the detector in the absence of photon illumination. For now, we will assume that the detector is operated at sufficiently low temperature that we can ignore thermally induced conductivity. To understand the detector response, we must derive the dependence of the photoconductivity, σ_{ph}, on the arrival rate of photons at the face of the detector.

Let the coordinate running between the electrodes be x. When an electric field $\mathcal{E}_x = V_b/\ell$ is applied, the resulting current density along the x-axis is the amount of charge passing through a unit area per unit time; it is given by

$$J_x = q_c n_0 \langle v_x \rangle , \tag{2.3}$$

where q_c is the electric charge of the charge carriers, n_0 is their density, and $\langle v_x \rangle$ is their effective drift velocity. From Ohm's law and the definition of conductivity in equation 2.1, we can derive an alternative expression for J_x:

$$J_x = \frac{I_d}{wd} = \frac{V_b}{R_d wd} = \sigma \mathcal{E}_x. \tag{2.4}$$

Rearranging equations 2.3 and 2.4, the electron conductivity is

$$\sigma_n = \frac{-q n_0 \langle v_x \rangle}{\mathcal{E}_x} = q n_0 \mu_n, \tag{2.5}$$

where the electronic charge is $q_c = -q$, and

$$\mu_n = -\frac{\langle v_x \rangle}{\mathcal{E}_x} \tag{2.6}$$

is the electron mobility, which has units of centimeters squared per volt second. The electron mobility, μ_n, will reappear many times in our discussion since it characterizes the drift velocity of charge carriers under an electric field. It is a characteristic of the detector material and operating temperature; it is defined to be positive, which means that the electron drift velocity is taken to be negative. Similar expressions can be derived for the holes. The total conductivity is then

$$\sigma = \sigma_n + \sigma_p. \tag{2.7}$$

The electron component of the conductivity is usually much larger than the hole component, particularly for n-type and intrinsic material, because the electron mobility is higher than that for the holes.

From equation 2.5, it is apparent that the mobility governs the conductivity. Without an applied field, the charge carriers are in rapid motion due to their thermal energy, but this motion is thoroughly randomized in both direction and amount by frequent collisions with crystal atoms or other charge carriers. Application of an electric field adds a slight bias to these motions that results in a current, but the energy transferred to the crystal in collisions imposes a limit on this extra velocity component; $\langle v_x \rangle$ is this terminal velocity. The mobility can then be interpreted as a measurement of the viscosity of the crystal against the motion of the charge carriers.

In general, the mobility is proportional to the mean time between collisions. Impurities in the crystal scatter the conduction electrons and reduce the time between collisions, thus reducing μ_n. As the temperature, T, is decreased, the carriers move more slowly, and the effect of impurity scattering becomes larger, reducing the mobility. If, for example, the impurities are ionized, μ goes as $T^{3/2}$. If the impurities are neutral or frozen out, a much shallower temperature dependence is observed, $\mu \sim$ constant. At higher temperatures, the dominant scattering centers are distortions in the crystal lattice caused by thermal motions of its constituent atoms. The mobility resulting from lattice scattering goes as $T^{-3/2}$. For detector material operating in an optimal temperature regime, the mobility is dominated by neutral impurity or lattice scattering. The mobility of silicon as a function of impurity concentration and temperature is illustrated in Figure 2.7. The samples are dominated by neutral impurity scattering at low temperatures and by lattice scattering at high ones. Typical values for semiconductor material parameters are given in Table 2.3. The values in the table apply at 300 K for high purities (impurity levels at about 10^{12} cm^{-3} for silicon and germanium). Under these conditions, the mobilities are due to lattice scattering.

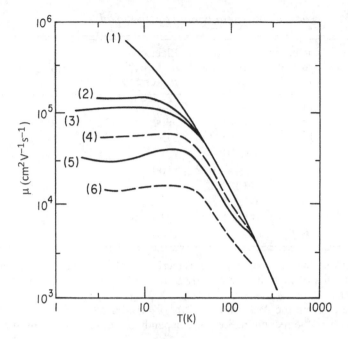

Figure 2.7 Variation of electron mobility in silicon with temperature and impurity concentration. The solid lines are for phosphorus-doped material at concentrations of $\leq 1 \times 10^{12}$, 4×10^{13}, 2.5×10^{14}, and 1×10^{16} cm^{-3}, respectively, for curves (1)–(3) and (5). The dashed lines are for arsenic-doped material at concentrations of 8×10^{15} and 8×10^{16} cm^{-3}, respectively, for curves (4) and (6). After Norton et al. (1973) and Canali et al. (1975).

To derive the dependence of detector conductivity on photon arrival rate, we obtain from equations 2.5 and 2.7,

$$\sigma_{ph} = q(\mu_n n + \mu_p p), \tag{2.8}$$

where n and p are the photoconductive charge carrier concentrations. For intrinsic photoconductivity, we can assume $n = p$ for the following discussion. Let φ photons per second fall on the detector. Some fraction η will create charge carriers; η is the quantum efficiency. The equilibrium number of charge carriers is then $\varphi\eta\tau$, where τ is the mean lifetime of a charge carrier before recombination. Sample values are listed in Table 2.3. The number of charge carriers per unit volume is then

$$n = p = \frac{\varphi\eta\tau}{wd\ell}. \tag{2.9}$$

Using equations 2.8 and 2.9, equation 2.1 becomes

$$R_{ph} = \frac{\ell^2}{q\varphi\eta\tau(\mu_n + \mu_p)}. \tag{2.10}$$

Table 2.3 *Properties of some semiconductor materials*

Material	Dielectric constant,[a] κ_0	Recombination time,[b] τ(s)	Electron mobility,[a] μ_n (cm^2 V^{-1} s^{-1})	Hole mobility,[a] μ_p (cm^2 V^{-1} s^{-1})	E_g (eV)
Si	11.8	1×10^{-4}	1350	480	1.11
Ge	16	1×10^{-2}	3900	1900	0.67
PbS	16.4	2×10^{-5}	575	200	0.37
InSb	17.7	1×10^{-7}	1.0×10^5	1700	0.18
GaAs	13.2	$\geqslant 1 \times 10^{-6}$	8500	400	1.43
InP	12.4	$\sim 10^{-6}$	4000	100	1.35
Hg$_{0.7}$Cd$_{0.3}$Te[c]	16.3	$\sim 10^{-4}$	4.7×10^4	474	0.24
Cd$_{0.9}$Zn$_{0.1}$Te[d]	10	$\sim 5 \times 10^{-6}$	1350	120	1.6
GaN[e]	8.9		1000	30	3.4

[a] Values taken from Streetman and Banerjee (2014), Appendix III. Note that the conventional units of mobility are not MKS; multiply the entries by 10^{-4} (m^2 cm^{-2}) to convert to MKS.
[b] Values taken from the *American Institute of Physics Handbook* (1972). These values are representative; the recombination time depends on impurity concentrations as well as temperature.
[c] Values from Kinch (2008) at 70 K and Itsuno (2012) at 77 K.
[d] Values from Schulman (2006).
[e] Values from Monroy et al. (2003).

We have used the subscript *ph* on *R* to emphasize that we are interested only in the photoinduced effects on the detector resistance. Thus, the detector resistance is equal to the inverse of the quantum efficiency times the photon flux combined with other properties of the detector material and its geometry. With a constant voltage across the detector, the current is proportional to the photon flux, so the detector is linear if its output is measured by sensing the current.

The output of a detector is usually quoted in terms of its responsivity, *S*, which is defined to be the electrical output signal divided by the input photon power. Units of *S* can be amperes per watt or volts per watt depending upon the units in which the output signal is measured. Although the responsivity of a detector is sometimes given only at the wavelength at which it reaches a maximum, it is often convenient to specify the spectral response of the detector in terms of the wavelength dependence of *S*.

From equation 1.3, the power falling on the detector is

$$P_{ph} = \varphi h\nu = \frac{\varphi h c}{\lambda}. \tag{2.11}$$

We define a new parameter, the photoconductive gain, as

$$G = \frac{\tau \mu \mathcal{E}_x}{\ell}. \tag{2.12}$$

A detector can have $G > 1$ if some form of controlled charge multiplication occurs in the detector material. We can then write (using equations 2.1, 2.11, and 2.12)

$$S = \frac{I_{ph}}{P_{ph}} = \frac{V_b}{R_{ph} P_{ph}} = \frac{\eta \lambda q G}{hc}. \tag{2.13}$$

We have also used the definition of the electric field strength in the detector, $\mathcal{E} = V/\ell$. The photocurrent generated by the detector is (substituting equation 2.11 into equation 2.13)

$$I_{ph} = \varphi q \eta G. \tag{2.14}$$

The quantum efficiency of a photoconductor is calculated as in Section 1.3.1. Assuming a constant level of absorption, it is independent of wavelength up to λ_c, the wavelength corresponding to the bandgap energy, which is given by

$$\lambda_c = \frac{hc}{E_g} = \frac{1.24 \ \mu m}{E_g(eV)}; \tag{2.15}$$

beyond λ_c, $\eta = 0$. Assuming each detected photon produces the same effect in the semiconductor (a single electron–hole pair, i.e., unity "quantum yield"), and because the photon energy varies inversely with its wavelength, S is proportional to λ up to λ_c (see equation 2.13). Figure 2.8 illustrates the behavior of the responsivity for an ideal photoconductor. Near λ_c, thermal excitation can affect the production of a conduction electron when a photon is absorbed, so there is a chance that a photon with energy just above the bandgap energy will not produce an electron–hole pair. This effect and others tend to round the quantum efficiency and responsivity curves near λ_c, as shown by the dashed lines. It is also typical for the quantum efficiency to drop at wavelengths well short of λ_c.

2.2.2 Spectral Response

The possible spectral responses for intrinsic photoconductors are as described by Figure 2.8 and equations 2.13 and 2.15. Typical detector materials and their cutoff wavelengths at room temperature are Si ($\lambda_c = 1.1 \ \mu m$), Ge ($\lambda_c = 1.8 \ \mu m$), PbS ($\lambda_c = 3.3 \ \mu m$), PbSe ($\lambda_c = 4.5 \ \mu m$), and the ternary compounds InGaAs, and HgCdTe. For this last material, it is possible to vary the bandgap by altering the proportions of the components. CdTe has $\lambda_c = 0.8 \ \mu m$ (corresponding to $E_g = 1.55$ eV), and HgTe is metallic, so it has $E_g < 0$. The bandgap of $Hg_{1-x}Cd_xTe$

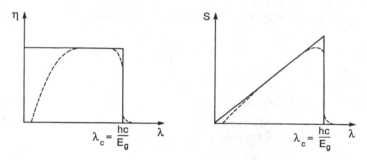

Figure 2.8 Photoconductor response. The solid lines represent the response in the ideal case; the dashed lines show typical departures from the ideal curve.

varies monotonically with x between these values. Intrinsic photoconductors can be made with response to 25 μm with HgCdTe (D'Souza et al. 2000). Adjustable bandgaps are also achieved with $In_{1-x}Ga_xAs$. It is most commonly prepared with $x = 0.47$ giving $\lambda_c = 1.68$ μm. Both of these materials also have far higher electron mobility than hole mobility, important in some applications such as avalanche photodiodes (see Section 3.1.4.2).

The bandgaps and hence cutoff wavelengths of these materials change slightly with operating temperature; for example, at 77 K, the cutoff wavelengths of Si and Ge are 5–10% shorter than at room temperature, whereas the cutoff wavelengths for both PbS and PbSe become significantly longer with reduced temperature and the response extends beyond 4 μm and 7 μm respectively at 77 K. HgCdTe has a relatively complex dependence of bandgap on temperature (Schmit and Stelzer 1969; D'Souza et al., 2000). For example, material with $\lambda_c > 2.25$ μm has a bandgap that decreases with decreasing temperature, resulting in an increase in the cutoff wavelength, whereas material with a $\lambda_c < 2.25$ μm shows a decrease in cutoff wavelength with decreasing temperature.

2.2.3 Frequency Response

The detector must be considered as an electrical element in any readout circuitry, with a resistance as in equation 2.1. Because the detector as shown in Figure 2.6 is a simple parallel plate capacitor filled with a dielectric, its capacitance is

$$C_d = \frac{wd\,\kappa_0\varepsilon_0}{\ell}, \tag{2.16}$$

where κ_0 is the dielectric constant (sample values are included in Table 2.3), and $\varepsilon_0 = 8.854 \times 10^{-12}$ F m^{-1} is the permittivity of free space. This capacitance is in parallel with the detector resistance.

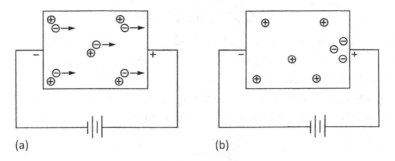

Figure 2.9 Dielectric relaxation in a photoconductor. Charge carriers begin migration from their creation sites in (a); the charge separation that results in (b) reduces the effective field in the detector and decreases its photoconductive gain.

Another factor influencing the speed of response is dielectric relaxation, which arises from space charge effects within the detector. Consider the detector in Figure 2.9(a); assume it is operating at low temperature with $\mu_n \gg \mu_p$ and high photoconductive gain ($G \approx 1$). Let the detector start in the dark at equilibrium and be illuminated by a step function. Under these assumptions, the high electron mobility results in the negative charge carriers being swept out of the detector, leaving a net positive excess charge behind. Such a distribution of excess charge is termed a space charge. The resulting charge separation creates a field that, until neutralized by new charge injected into the sensitive volume, opposes the field set up by the detector electrodes (see Figure 2.9(b)) and reduces the photoconductive gain. The timescale for neutralization of the field is just the RC time constant of the detector, with R from equation 2.10 and C from equation 2.16. The dielectric relaxation time constant is therefore

$$\tau_d = \frac{\varepsilon}{\mu_n n_0 q} = \frac{\varepsilon}{\sigma} \propto \frac{1}{\varphi} ; \qquad (2.17)$$

the final proportionality arises from the relationship between σ and the photon flux (see equations 2.8 and 2.9). Thus, the reduction in frequency response due to dielectric relaxation is inversely proportional to the incident photon rate, and the effect becomes increasingly important at low light levels.

The charge carrier lifetime sets another limit on the speed of the detector, since the conductivity can change only over the timescale for recombination. The corresponding cutoff frequency is

$$f = \frac{1}{2\pi\tau}, \qquad (2.18)$$

where τ is the recombination lifetime (e.g., Table 2.3) and the behavior can be taken to be exponential.

2.2.4 Latent Images

Impurities and crystal defects (including at the interface between the detector material and an electrical contact to it) can hold free charge carriers and release them slowly. This behavior leads to latent images; that is, when a signal is removed from a detector, the detector output may continue to show that signal at a reduced and slowly decaying level. Dielectric relaxation can have a similar effect.

2.2.5 Noise

Along with the generation of an electrical signal, detectors produce various forms of electronic noise that can hide the signal. Thus, in evaluating the use of a detector, we must understand and estimate the various noise sources. As with the photon noise discussed in the previous chapter, we will assume that these noise components are Poisson distributed (equation 1.21) and independent, allowing them to be combined quadratically.

The fundamental noise limitation for any detector is the noise that arises because of the Poisson statistics of the incoming photon stream (see equation 1.25). For a photoconductor, the photons are transformed into free electrons and holes, so this noise appears as a variation in the number of charge carriers. The statistical fluctuations in the number of charge carriers result in generation–recombination, or G–R, noise.

Suppose the detector absorbs N photons in a time t, $N = \eta \varphi t$. For an ideal photoconductor, these photons create N conduction electrons and N holes; we assume $\mu_n \gg \mu_p$ so we need consider the electrons only. Each conduction electron drifts in the electric field and contributes to the conductivity until it recombines. If the detector receives a stream of photons, we can imagine that its net conductivity arises from the superposition of many randomly generated and randomly terminated contributions, one pair for each photoelectron. Since there are then two random events associated with the photoconductivity from each absorbed photon, the root-mean-square (rms) noise associated with them is, in number, $(2N)^{1/2}$. The noise current due to this sequence of generations and recombinations of photoelectrons is then this rms noise in number times the charge of an electron divided by the time interval, and multiplied by the detector gain, or $\langle I_{G-R}^2 \rangle^{1/2} = q(2N)^{1/2}G/t$. Taking the square and using equation 2.14,

$$\langle I_{G-R}^2 \rangle = \left(\frac{2q}{t} \right) \left(\frac{qNG}{t} \right) G = \left(\frac{2q}{t} \right) \langle I_{ph} \rangle G, \qquad (2.19)$$

where $\langle I_{ph} \rangle$ is the detector current averaged over the noise fluctuations. If the detector has a significant number of thermally generated charge carriers, they contribute their own G–R noise component in addition to the one in equation 2.19.

It is noteworthy that the intrinsic $\sqrt{N}$ noise of the photogenerated charge carriers has been degraded by $\sqrt{2}$ because the recombination of these charge carriers occurs in the high resistance detector volume. Alternate detector types discussed in the following chapters avoid this noise increase because for them recombination occurs in relatively high conductivity regions of the detector.

The *G–R* noise is typically measured with an instrument that is sensitive to signals over some range of frequencies. It would be convenient if this response were sharply defined by lower and upper frequencies, f_1 and f_2, with constant response between f_1 and f_2 and no response otherwise. We could then define the frequency bandwidth as $\Delta f = f_2 - f_1$. Unfortunately, it is impossible to build circuitry with the properties just described, and, even if it were possible, the full realization of this concept would require measurements of infinite duration. It is therefore necessary to introduce an equivalent noise bandwidth, which is computed as the sharply defined hypothetical "square" bandwidth $(f_2 - f_1)$ through which the noise power would be the same as the noise power through the frequency response of the actual system, where both have the same maximum gain. Since the power in the electrical signal is proportional to the square of its amplitude, the bandwidth is defined in terms of the square of the response function. Thus,

$$\Delta f = \int_0^\infty |G(\xi)|^2 d\xi, \tag{2.20}$$

where $G(\xi)$ is the electrical response (for example, current or voltage) of the system as a function of frequency ξ, normalized to unity response at the frequency of maximum response.

For a system with exponential response, $G \sim e^{-t/\tau}$, it can be shown that

$$\Delta f = df = \frac{1}{4\tau}. \tag{2.21}$$

Similarly, if a measurement is made by integrating the electrical signal over a time interval T_{int},

$$df = \frac{1}{2T_{int}}. \tag{2.22}$$

Expressions such as equations 2.21 and 2.22 are useful to convert noise measurements taken under one set of conditions to equivalent measurements under different conditions.

The case represented by equation 2.22 is applicable to the assumptions we made in deriving *G–R* noise. Substituting equations 2.14 and 2.22 into the expression for I_{G-R} (equation 2.19),

$$\langle I_{G-R}^2 \rangle = 4q^2 \varphi \eta \, G^2 \, df. \tag{2.23}$$

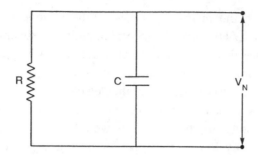

Figure 2.10 Circuit for calculation of Johnson and kTC noise.

We continue to assume that the detector is cold enough that thermally generated conductivity is negligible. Otherwise, the thermally generated charge carriers contribute to I_{G-R} and increase the noise.[2]

There are also noise sources that originate in the detector itself in the absence of external signals. Johnson, or Nyquist, noise is a fundamental form of thermodynamic noise that arises because of thermal motions of charge carriers in any resistive circuit element, such as our photoconductor. Refer to Figure 2.10 for the following discussion. The circuit illustrated has one degree of freedom, V_N. From thermodynamics, if the system is in equilibrium, it has an average energy of $kT/2$ associated with each degree of freedom, or in our case with V_N. Since the energy on a capacitor is $CV^2/2$, we obtain

$$\frac{1}{2}C\left\langle V_N{}^2 \right\rangle = \frac{1}{2}kT. \tag{2.24}$$

The capacitor is a storage device, so the randomly fluctuating potential energy on it must have a corresponding random kinetic energy component, which is the Johnson noise current, I_J. Again, from thermodynamic considerations we expect this energy to be of the form $\langle P \rangle t = kT/2$, where we have expressed the energy as the power, P, times the response time of the circuit, $t = RC$. From elementary circuit theory, the maximum power that can be delivered to a device connected across the terminals of the circuit in Figure 2.10 is $\langle P \rangle = \langle I^2 \rangle R/2$. For a circuit with exponential response (for example, Figure 2.10), the relationship between response time and electrical bandwidth is given by equation 2.21. We therefore obtain

$$\left\langle I_J^2 \right\rangle = \frac{4kT\,df}{R}. \tag{2.25}$$

---

[2] In fact, a number of other simplifying assumptions have been made in the derivation of $G–R$ noise. One of the most significant is that the relaxation time for a charge carrier has been taken to be independent of the density of free carriers. This assumption is not generally true (see equation 2.17). A more rigorous treatment of $G–R$ noise (and other noise mechanisms) can be found in van Vliet (1967), van der Ziel (1976), or Boyd (1983).

While we are discussing the circuit in Figure 2.10, it will be useful to derive the charge on the capacitor. This form of charge noise will be of interest when we discuss circuits that store the signal from a detector on a capacitance before reading it out; it is called kTC or reset noise. From equation 2.24, it is easily shown that

$$\langle Q_N^2 \rangle = kTC, \tag{2.26}$$

where we have made use of the relationship between charge and voltage for a capacitor, $V = Q/C$.

It should be clear from the above discussion that Johnson noise and kTC noise are manifestations of the same thing. On a microscopic scale, they arise because of the random currents generated by the Brownian motion of the charge carriers in a resistor.

In addition to these fundamental noise mechanisms, most electronic devices have increased noise at low frequencies. In many circumstances, this low frequency noise is the irreducible limit to the performance of an electronic system, so it has received substantial attention from both experimental and theoretical viewpoints. Nonetheless, there is no general understanding of it. One of the problems is that many different phenomena appear to have nearly identical low frequency noise behavior:

$$\langle I_{1/f}^2 \rangle = \frac{K I^a \, df}{f^b}, \tag{2.27}$$

where K is a normalization constant and $a \approx 2$ and $b \approx 1$. Since we lack a physical understanding of the cause of this type of noise, it is simply termed "one over f noise."

Formally, equation 2.27 diverges if integrated over all frequencies, but this behavior is more of an academic worry than a practical one. The increase in noise power with increasing bandwidth is mild, since with $b = 1$ the integral of $1/f$ yields a logarithmic result. Given this mild behavior, it is not practical to extend measurements to low enough frequency (even given the life of the Universe to do them!) to yield large powers. Large high frequency noise powers are prevented by the eventual drop of response from all electronic devices at sufficiently high frequencies. In fact, at high frequencies, the noise of most devices is dominated by mechanisms different from those that produce $1/f$ noise and the value of b tends to decrease toward zero. When b is zero, the resulting frequency-independent noise is called white noise. Of course, that would be even more catastrophic if it extended to very high frequencies; this situation just emphasizes the importance of steps to cut off the high frequency response of readout electronics.

Excess $1/f$ noise can be produced by construction flaws, such as bad contacts to a semiconductor device. It is also the typical behavior resulting from temperature variations that affect the outputs of such devices. However, if both of these causes are removed, there is almost always a residual $1/f$ component. This component appears to be associated partially with surface behavior since the amount of $1/f$ noise can be modified by surface treatment of a device, such as by etching away surface damage or changing the ambient atmosphere to which it is exposed (Kogan 1996). In addition, devices that carry signal currents away from their surfaces in the bulk of the semiconductor seem generally to have relatively small $1/f$ noise. An outstanding example is junction field effect transistors (JFETs), discussed in Chapter 5. Although there are still some contrary indications, the favored explanation for this type of noise is that surface traps remove free charge carriers and then return them but with a broad range of trapping times, creating a low frequency noisy modulation of the current. A similar mechanism may account for the noise in the bulk material: the number of charge carriers may fluctuate as they fall into and are released by traps at crystal defects (Kogan 1996). However, none of these hypotheses is firmly established (if one were, we would have a name more physically descriptive than "$1/f$ noise"). In addition, different mechanisms may operate in different situations; for example, the number of free charge carriers in a metal is so high compared with the number of defects and traps that the mechanism suggested above for the bulk of a semiconductor would appear not to be applicable.

Equations 2.23, 2.25, and 2.27 illustrate that the rms noise currents, $\langle I_N^2 \rangle^{1/2}$, are proportional to the square root of the frequency bandwidth, df. The cause of this behavior is illustrated by equation 2.22. Imagine that the signal current is produced by a series of independent, Poisson-distributed events. The number of such events received will increase in proportion to the time of integration, so the accuracy of measurement will improve as the square root of the time, that is, the relative error of measurement goes as the inverse square root of integration time. From equation 2.22, it is equivalent to say the relative error of measurement goes as $(df)^{1/2}$. If the signal frequency falls within the range defined by df, its strength is roughly independent of df and the signal-to-noise ratio improves as $(df)^{1/2}$. Consequently, there is a strong incentive to use electronic filtering to reject frequencies that do not carry significant signal. Such electronic filtering can also be effective in rejecting spurious noise sources that are concentrated at certain frequencies, such as pickup from AC power lines or microphonic electrical signals from mechanical vibrations within the detector system.

The noise mechanisms described above are assumed to be independent, so they are combined quadratically to estimate the total noise of the system:

$$\langle I_N^2 \rangle = \langle I_{G-R}^2 \rangle + \langle I_J^2 \rangle + \langle I_{1/f}^2 \rangle. \tag{2.28}$$

For these mechanisms, the noise excursions can be taken to obey Poisson statistics. Detectors may be subject to other noise mechanisms that are not distributed in this manner. Examples include current spikes that are produced spontaneously in the detector when it has a momentary breakdown or when it is struck by an energetic charged particle such as a cosmic ray. If at all possible, such events should be identified by their departure from Poisson statistics and removed from the detector output either electronically or in later reduction steps. If they remain, they can both dominate the noise of the system and make accurate noise estimation difficult because the normal statistical techniques assume Poissonian noise behavior.

Equation 2.28 has an alternative interpretation that emphasizes the difference between noise and signal. Because noise power goes as $\langle I_N^2 \rangle$, we see that noises add in terms of powers. This behavior holds very generally and occurs because noise currents vary randomly in phase. To summarize succinctly, when the *noise* is to be obtained over a range of frequencies, the *powers* from adjacent frequency intervals are combined, that is, currents are added quadratically. On the other hand, *signal* currents over a range of frequency are correlated in phase and therefore add *linearly*.

2.2.6 Thermal Excitation

Throughout this discussion, we have insisted that the detector be operated at a sufficiently low temperature to suppress thermal excitation. Some of the advantages (besides mathematical simplicity) have been pointed out, and one would expect any low light level detector to be run in this mode. Nonetheless, it is necessary to consider thermal excitation briefly to determine the temperature at which it can be frozen out. The discussion will also demonstrate the steep dependence of thermally excited dark current on detector temperature. In cases where the dark current is appreciable and the detector temperature is not perfectly constant, this dependence introduces $1/f$ noise and can make calibration difficult.

Classically, the electrons in a semiconductor obey Maxwell–Boltzmann statistics, and the relative numbers in two energy levels are related as follows:

$$\frac{n_2}{n_1} = e^{-\frac{(E_2-E_1)}{kT}} = e^{-\frac{E_g}{kT}}. \tag{2.29}$$

More properly, the electrons follow Fermi–Dirac statistics. In this case, the distribution of the electrons over the allowed energy states is governed by the Fermi–Dirac distribution function:

$$f(E) = \frac{1}{1 + e^{(E-E_F)/kT}}, \tag{2.30}$$

where E_F is the energy of the Fermi level. The Fermi level is defined as the energy where an available state would have a 50% likelihood of being occupied by an electron. Equation 2.30 gives, as a function of temperature, the probability that

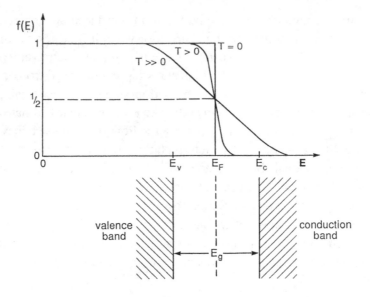

Figure 2.11 Electron probability distribution $f(E)$ as a function of temperature T, compared with an energy band diagram.

any available state at energy E will be occupied by an electron. Moreover, when $T = 0$, $f(E < E_F) = 1$, and $f(E > E_F) = 0$. Thus, at absolute zero, the electrons fill all of the states up to the Fermi level and none above it. At all temperatures, the Fermi level separates the states that are probably occupied ($E < E_F$) from those that are probably empty ($E > E_F$); that is, in accordance with the definition of E_F, $f(E_F) = 0.5$. The relationship of $f(E)$ to the bandgap diagram is shown in Figure 2.11, which illustrates the increasing probability for the electrons to be excited into the conduction band as the temperature is increased.

The concentration of electrons in the conduction band is

$$n_0 = \int_{E_c}^{\infty} f(E)N(E)dE, \qquad (2.31)$$

where $N(E)dE$ is the density of states (conventional units are cm^{-3}) in the energy interval dE. This integral can be simplified by introducing the effective density of states, N_c, located at the conduction band edge, E_c, such that

$$n_0 = N_c f(E_c). \qquad (2.32)$$

This approximation works well because, at modest temperatures, the conduction electrons occupy only the bottom few states in the conduction band; thus, the integrand in equation 2.31 is significantly larger than zero only for energies near E_c, and the integral is well represented by an appropriate average. It is usually the case that $E_c - E_F \gg kT$ (for example, $kT = 0.026$ eV at $T = 300$ K). As a result, equation 2.30 can be simplified to

$$f(E_c) = e^{-\frac{E_c - E_F}{kT}}. \tag{2.33}$$

It can be shown (see, for example, Streetman and Banerjee (2014), Appendix IV) that

$$N_c = 2\left[\frac{2\pi m_n^* kT}{h^2}\right]^{\frac{3}{2}}, \tag{2.34}$$

where m_n^* is the effective mass of a conduction electron, allowing for large-scale effects of the crystal. Similarly, m_p^* is the effective mass of a conduction hole. The values of the effective masses vary slightly with temperature and also depend on the material. For silicon, $m_n^* \approx 1.1 m_e$ and $m_p^* \approx 0.56 m_e$, where m_e is the mass of the electron. For germanium, $m_n^* \approx 0.55 m_e$ and $m_p^* \approx 0.37 m_e$. Combining equations 2.32, 2.33, and 2.34, we obtain for the concentration of electrons in the conduction band:

$$n_0 = 2\left[\frac{2\pi m_n^* kT}{h^2}\right]^{\frac{3}{2}} e^{-\frac{E_c - E_F}{kT}}. \tag{2.35}$$

The applicability of equation 2.35 is subject to the validity of the approximations leading to equations 2.32 and 2.33. Given that most intrinsic semiconductors have relatively large bandgaps and that good detector performance requires temperatures at or below 300 K, equation 2.35 will generally be valid in the following discussions. Analogously, for holes we have

$$N_v = 2\left[\frac{2\pi m_p^* kT}{h^2}\right]^{\frac{3}{2}} \tag{2.36}$$

and for the concentration of holes in the valence band,

$$p_0 = 2\left[\frac{2\pi m_p^* kT}{h^2}\right]^{\frac{3}{2}} e^{-\frac{E_F - E_v}{kT}}. \tag{2.37}$$

For intrinsic semiconductors, $E_c - E_F \sim E_F - E_v \sim E_g/2$; equality will be assumed for these relations unless otherwise noted.

As an example, applying equation 2.37 to silicon at 300 K yields $n_0 \sim 1.38 \times 10^{10}$ cm^{-3}. At 77 K, we get $n_0 \sim 1.8 \times 10^{-18}$ cm^{-3}. These estimates imply that the thermal charge carriers can be frozen out extremely thoroughly without resorting to exceptionally low temperatures (77 K is the temperature of liquid nitrogen, which is readily available). In Chapter 3, however, we will see that this result is no longer valid when we allow for the inevitable impurities in the material.

2.3 Extrinsic Photoconductors

2.3.1 Applications

The spectral response of intrinsic photoconductors is limited to photons that have energies equal to or exceeding the bandgap energy of the detector material. For the high quality semiconductors silicon and germanium, these energies correspond to maximum wavelengths of 1.1 μm and 1.8 μm, InSb and HgCdTe provide response to ∼ 5 μm, and the latter material is now available out to 13–15 μm (next chapter). Beyond ∼15 μm, detector operation must therefore be based on some other mechanism.[3] The addition of impurities to a semiconductor allows conductivity to be induced by freeing the impurity-based charge carriers. This extrinsic process requires smaller energy increments than does intrinsic photoconduction, enabling response at long infrared wavelengths. Because the lower excitation energies also allow for large thermally-excited dark currents, these detectors must be operated at low temperatures. Detailed reviews of the characteristics of such detectors have been written by Bratt (1977) and Sclar (1984).

The electrical conductivity of extrinsic material differs in a fundamental way from that of intrinsic material. In the latter case, charge carriers are created as electron–hole pairs consisting of a conduction electron plus a hole in the valence band from the inter-atomic bond it has vacated. In the former case, individual charge carriers are created whose complementary charge resides in an ionized atom that remains bonded into the crystal structure; this complementary charge is therefore immobile and cannot carry currents. At high temperatures, the intrinsic mechanism dominates in both types of material. At modest and low temperatures, the carrier contributed by the dominant impurity is the most numerous and is called the majority charge carrier. The complementary type, or minority charge carrier, coexists at a lower concentration. It is contributed either by the intrinsic conductivity mechanism or by ionization of impurities of complementary type and lower concentration. The notation *semiconductor:dopant* is used to indicate the semiconductor and majority impurity; for example, Si:As designates silicon with arsenic as the majority impurity.

The dopants do not interfere with the intrinsic absorption process in the material. In fact, because the number of semiconductor atoms is always far larger than the number of impurity atoms, intrinsic absorption always prevails over extrinsic absorption in wavelength regions where the intrinsic mechanism is effective. The response from the intrinsic absorption must be anticipated in any successful use of

---

[3] Although single-element photoconductive detectors of HgCdTe can provide modest performance past 20 μm.

extrinsic photoconductors, for example by using optical filters that strongly block the wavelengths that can excite intrinsic conductivity.

The basic operation of extrinsic photoconductors is similar to that of intrinsic photoconductors except that the excitation energy, E_i, must be substituted for the bandgap energy (E_g). In first-order theory, the response drops to zero at the wavelength corresponding to this energy, $\lambda_c = hc/E_i$. As with intrinsic photoconductors, a number of effects act to round off the dependence of response on wavelength near this cutoff. The absorption coefficient is

$$a(\lambda) = \sigma_i(\lambda)N_I, \qquad (2.38)$$

where $\sigma_i(\lambda)$ is the photoionization cross-section and N_I is the neutral impurity concentration. Table 2.4 gives sample values of the cutoff wavelengths and photoionization cross-sections (at the wavelength of peak response). Detector quantum efficiencies can be calculated as in equations 1.18 and 1.20, given N_I.

It would appear desirable to increase N_I without limit to maximize the quantum efficiency, but upper limits to the impurity concentration arise from two sources. First, there is a limit to the solubility of the impurity atoms in the semiconductor crystal. These limits range from ~10^{16} to 10^{21} cm^{-3} for commonly used dopants (Streetman and Banerjee 2014, Appendix VII). Second, before the solubility limit is reached, the electrical properties of the crystal usually undergo unwanted changes in the form of conductivity modes that cannot be adequately controlled either by operating the detector at low temperature to freeze them out or by other means.

One example of these unwanted conductivity modes is hopping. When impurity atoms are sufficiently close together, the electron wave function from one impurity atom extends at a non-negligible value to a neighboring impurity atom. Under these conditions, conduction can occur directly from one impurity atom to another without supplying the energy necessary to raise an electron into the conduction band.

Given these limitations, typical acceptable impurity concentrations are around 10^{15} to 10^{16} cm^{-3} for silicon and somewhat lower for germanium. Using the photoionization cross-sections in Table 2.4 and equation 2.38, it can be seen that the absorption coefficients are some three orders of magnitude less than those for direct absorption in intrinsic photoconductors. As a result, the active volumes in extrinsic photoconductors must be large to provide adequate quantum efficiency, with dimensions on the order of a millimeter in the direction along the incoming photon beam. These detectors are sometimes termed "bulk photoconductors" to distinguish them from detector types where the absorption takes place in a much smaller volume (see, for example, Section 3.2).

Table 2.4 *Properties of some extrinsic photoconductors*

Impurity	Type	Ge Cutoff wavelength λ_c (μm)	Ge Photoionization cross-section σ_i (cm²)	Si Cutoff wavelength λ_c (μm)	Si Photoionization cross-section σ_i (cm²)
Al	p			18.5[a]	8×10^{-16}[b]
B	p	119[b]	1.0×10^{-14}[c]	28[a]	1.4×10^{-15}[b]
Be	p	52[b]		8.3[a]	5×10^{-18}[d]
Ga	p	115[b]	1.0×10^{-14}[c]	17.2[a]	5×10^{-16}[b]
In	p	111[b]		7.9[a]	3.3×10^{-17}[b]
As	n	98[b]	1.1×10^{-14}[b]	23[a]	2.2×10^{-15}[a]
Cu	p	31[b]	1.0×10^{-15}[b]	5.2[a]	5×10^{-18}[d]
P	n	103[b]	1.5×10^{-14}[b]	27[a]	1.7×10^{-15}[b]
Sb	n	129[b]	1.6×10^{-14}[b]	29[a]	6.2×10^{-15}[a]

[a] Sclar (1984)
[b] Bratt (1977)
[c] Wang et al. (1986)
[d] Sclar (1984)
Photoionization cross-sections apply near peak response; smaller values apply toward shorter wavelengths.

2.3.2 *Stressed Detectors*

The long-wavelength cutoff of a p-type photoconductor can be modified by physically stressing the crystal. This behavior should at least seem plausible because conduction in p-type material occurs through the breaking and remaking of inter-atomic bonds (that is, migration of a hole). An external force places stress on the inter-atomic bonds, so it would be plausible that less additional energy is required to break them.

A particularly dramatic effect can be achieved with diamond lattice crystals stressed along a particular (the [100]) crystal axis. The stress can produce a significant reduction in the acceptor binding energy and can extend the response of Ge:Ga photoconductors from a maximum wavelength of 115 μm (unstressed) to beyond 200 μm (Kazanskii et al. 1977; Haller et al. 1979).

In making practical use of this effect, it is essential to apply and maintain very uniform and controlled pressure to the detector so that the entire detector volume is placed under high stress without exceeding its breaking strength at any point. Detectors stressed in this manner show a wavelength-dependent response similar to that of conventional detectors, with responsivity proportional to wavelength until the wavelength is close to the cutoff.

2.3.3 Limitations

2.3.3.1 Ionizing Radiation Effects

All solid state detectors are very sensitive to energetic particles because these particles are heavily ionizing and hence create large numbers of free charge carriers when they pass through the sensitive volume of the detector. As we have seen, extrinsic detectors must have large volumes to achieve reasonable quantum efficiencies. Consequently, in a high radiation environment, they are struck frequently and generate a high rate of spurious signals. In well-behaved detectors, the duration of such signals is limited by the frequency response of the amplifier. A "radiation hit" then appears as a large voltage spike followed by a relatively quick recovery (that is, limited by the amplifier response time) to quiescent operating conditions.

Unfortunately, detectors operating at low backgrounds and under high radiation conditions can exhibit other effects. The high energy particles create extensive damage in the crystal bonds; the resulting lattice defects can act much as impurities would by introducing additional energy levels that can behave as acceptors and donors. The increase in pseudo-minority impurities can increase the recombination time and result in an increase in the detector responsivity. The behavior is not stable, so the noise also increases, typically much faster than the responsivity. Under low background operating conditions this condition can decay very slowly. Thus, the signal-to-noise ratio can be seriously decreased well after the energetic radiation has been removed. In addition, the slow decay of the responsivity back to its normal value makes the calibration of the detectors time dependent and hence degrades the accuracy of any measurements made with them. These problems are severe with extrinsic germanium detectors, but they also occur in silicon detectors, particularly those that operate at relatively long wavelengths.

The radiation damage can be erased by flooding the detector with free electrons to re-establish thermal equilibrium. The most effective procedure is to heat the detector sufficiently ($\sim$6 K for typical germanium detectors, $\sim$20 K for silicon ones) to re-establish thermal equilibrium, and then to cool it to normal operating temperature ($\sim$2 K for germanium, $\sim$6 K for silicon). Alternatively, the bias voltage can be raised above the detector breakdown level, flooding the active region momentarily with conduction electrons, or the detector can be flooded with a high flux of infrared photons.

2.3.3.2 Transients and Nonequilibrium Response

At low background levels such as encountered in space astronomy, extrinsic detectors can exhibit a variety of undesirable response characteristics, including oscillations, drift, "hook" (so named because of the appearance of the electronic output waveform), and other long time constant behaviors. Some of these effects

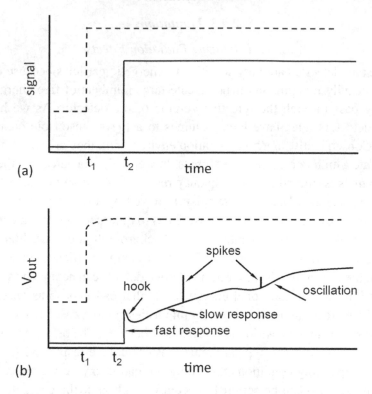

Figure 2.12 Response of a bulk extrinsic photoconductor to two-step input signals shown in (a), one from a moderate (dashed line) and the other from a nearly zero background (solid line). The response of the detector is shown in (b). For the moderate background, the detector follows the signal reasonably well (dashed line). However, starting at nearly zero illumination (solid line), the response has multiple components, including fast and slow response, a hook anomaly, and voltage spikes.

are illustrated in Figure 2.12. An interesting historical overview of the underlying theory, developed largely in Russia and ignored elsewhere, is given by Fouks (1992). Kocherov et al. (1995) give a thorough theoretical discussion.

The initial rapid response is due to the generation and recombination of charges in the bulk region of the detector. The other effects arise at the detector contacts. In discussing photoconductive gain in Section 2.2.1, we assumed that a charge carrier absorbed at one contact is "immediately" replaced by one emitted at the other to maintain electrical equilibrium. However, the necessity to emit a charge carrier to maintain equilibrium is only communicated across the detector at roughly the dielectric relaxation time, so in fact at low backgrounds the detector space charge adjusts to a new configuration at this relatively slow rate. The result of this slow injection of new charge carriers is not just a slow response component, but a slow adjustment of the detector to a new equilibrium under a new level of illumination. Under some circumstances, the response shows oscillatory behavior as the detector adjusts in an underdamped condition (some illustrations are provided by Abergel et al. 2000).

Another example is the hook response. This artifact can result from non-uniform illumination of the detector volume in transverse contact detectors; due to shading by the contacts, the portion of light under them is reduced. When the overall illumination is increased, the resistance of the rest of the detector volume is driven down, leaving a high resistance layer near the contacts that carries most of the electric field. If this area is under the contact that replaces the charge carriers (the "injecting" contact), then the recovery to an equilibrium state requires the time discussed in the preceding paragraph. As a result, after the initial fast response the photoconductive gain in the bulk of the detector is driven down, and the overall response decreases, to recover only slowly (Kocherov et al. 1995; Haegel et al. 2001). In transparent contact germanium detectors, where the entire detector volume including the areas under the contacts is fully illuminated (one of the few advantages of low absorption efficiency!), the hook behavior is reduced or eliminated (Haegel et al. 2001). In other types of transparent contact photoconductor, mild hook behavior is sometimes seen and probably arises through a combination of nonuniform illumination due to high absorption and to other nonequilibrium pseudo-oscillatory effects (Abergel et al. 2000). A thick transparent contact detector with a high absorption coefficient and with the illumination through the non-injecting contact would assume a space charge distribution similar to that in a transverse contact detector. It presumably would have a similar hook response, although it is not clear that such an experiment has been conducted.

2.3.3.3 Spiking

At the contact junction between conductor and semiconductor, large fields are produced by the migration of charges to reconcile the differing bandgap structures. To control the peak fields, it is necessary to produce a "graded contact" by adding dopants through ion implanting to increase the electrical conductivity in a thin layer of semiconductor just below the contact. Nonetheless, as the illumination level of the detector is changed, these fields must adjust. During this process, the field local to a small region near a contact can accelerate charge carriers sufficiently to impact-ionize impurities in the material, producing a mini-avalanche of charge carriers that appears as a spike in the output signal.

2.3.3.4 Background Dependence

High backgrounds produce a steady state concentration of free charge carriers that reduces the dielectric relaxation time constant and hence allows rapid approach to a new equilibrium appropriate for the new signal level (see Figure 2.12). Thus, photo-conductors can be relatively well behaved in high background operation. However, at low backgrounds the signal usually must be extracted while the detector is still in a nonequilibrium state and hence the input photon flux must be deduced from partial response of the detector. A variety of approaches have been used (Coulais et al. 2000). In extreme cases, the fast generation–recombination response may be sufficiently separated in time behavior from the undesirable terms that the signal can be extracted almost entirely from the fast component, which is generally better behaved than the others. However, usually it is not possible to avoid a mixture of fast and slow components in the signal. Empirically fitted corrections have been reasonably successful in restoring good photometric behavior (see, e.g., Church et al. 1996; Abergel et al. 2000). Approximate analytical physical models can also be useful (Schubert et al. 1995; Coulais and Abergel 2000). However, the most successful approach has been to combine modulating the source on and off the detector to put the signal into the "fast" regime, along with frequently flashing a calibration source to track the response variations (Gordon et al. 2007).

2.4 Example: Design of a Photoconductor

Consider an intrinsic silicon photoconductor operating at 1 μm and constructed as shown in Figure 2.6. Let it be 1 mm^2, and operate it at 300 K. Assume the detector breaks down when its bias voltage, V_b, exceeds 60 mV. Determine: (a) a reasonable detector thickness for good quantum efficiency; (b) the responsivity; (c) the dark resistance; (d) the time response; and (e) the Johnson noise.

(a) Since the transverse contacts make it impossible to coat the back of the detector with a reflective metal, the minimum thickness for reasonably good quantum efficiency is one absorption length. From Figure 2.4, at $\lambda = 1$ μm this thickness corresponds to about 80 μm. Allowing for reflection loss at the front face (the refractive index of silicon is n $\sim$ 3.4), the quantum efficiency is 44% (see equation 1.20).

(b) To maintain safely stable operation, we choose to set $V_b = 40$ mV. The photoconductive gain can be calculated from equation 2.12. For simplicity, we will ignore the hole component of photoconductivity. From Table 2.3, we obtain $\tau = 10^{-4}$ s and $\mu = 1350$ cm^2 V^{-1} s^{-1}. The electric field $\mathcal{E}_x = V_b/\ell = (0.04$ V$)/(0.1$ cm$) = 0.4$ V cm^{-1}. These parameters yield $G = 0.54$. Substituting for G and η in equation 2.13, the responsivity will be $S = 0.19$ A W^{-1}.

(c) For the sake of simplicity, we will again ignore the contribution of holes in computing the dark resistance of the detector. The concentration of conduction electrons can be obtained from equation 2.35 with $E_g = 1.11$ eV, $m_n^* = 1.00 \times 10^{-30}$ kg, and $T = 300$ K. We estimate $n_0 = 1.38 \times 10^{16}$ m^{-3}. From equation 2.5, the conductivity of the detector material is $2.98 \times 10^{-6} (\Omega \, \text{cm})^{-1}$. Substituting into equation 2.1, we find that the detector dark resistance is $4.2 \times 10^7 \, \Omega$.

(d) The time response of the detector is determined by the longest of three time constants. The first is the recombination time, which in this case is 10^{-4} s. The second is the *RC* time constant. The capacitance can be calculated from equation 2.16 with the dielectric constant from Table 2.3; the result is $C = 8.4 \times 10^{-15}$ F, leading to $\tau_{RC} = 3.5 \times 10^{-7}$ s. Since we have not specified the circuit in which the detector is operated, the dielectric relaxation time constant will be the same as τ_{RC}. Comparing these three time response limitations, we find that the recombination time is the important limitation.

(e) The Johnson noise current is given by equation 2.25 with $T = 300$ K and $R = 4.2 \times 10^7 \, \Omega$. It is $1.99 \times 10^{-14} df^{1/2}$ A.

2.5 Problems

2.1 Use equations 1.40 and 2.20 to prove equation 2.21.

2.2 A circular photoconductor of diameter 0.5 mm operating at a wavelength of 1 μm has a spectral bandwidth of 1% (0.01 μm). Suppose it views a circular blackbody source of diameter 1 mm at a distance of 1 m and at a temperature of 1500 K. If the detector puts out a signal of 5×10^{-13} A, what is its responsivity and ηG product?

2.3 Suppose the output of the detector in Problem 2.2 is sampled in a series of well-defined one-second intervals. With the view of the source (and other photons) blocked off, it is found that the rms noise current is 2.00×10^{-16} A, which we take to be excess electronics noise. Viewing the source, the rms noise is 4.58×10^{-16} A. Determine the detective quantum efficiency (in the absence of electronics noise) and use it to estimate the photoconductive gain.

2.4 Show that the quantum efficiency of a photoconductor that has reflectivity r at both the front and back faces, absorption coefficient a, and length ℓ from the front to the back is

$$\eta = \frac{(1 - r)(1 - e^{a\ell})}{1 - re^{a\ell}}.$$

Hint: recall the McClaurin series

$$\frac{1}{1 - x} = \sum_0^{\infty} x^n.$$

Comment on the applicability of this result to a detector with a perfectly reflective back face (for example, one with a transparent front contact and a metallic back face).

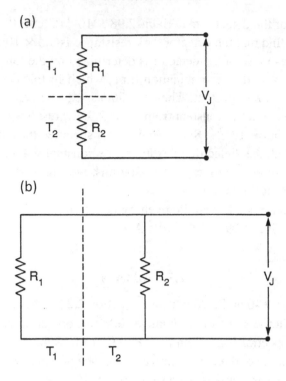

Figure 2.13 Circuits for Problem 2.6.

2.5 Compare the performance of a germanium photoconductor in the same application as the silicon one in the example. Assume the detector has a transparent front contact, sensitive area 1 mm^2, and an absorbing back contact. Assume breakdown occurs at 1 V cm^{-1} and that the detector is biased to half this value. How can the operating conditions be adjusted to make the germanium more competitive and therefore to take advantage of the high absorption efficiency of germanium near 1 μm?

2.6 Consider two resistors at different temperatures. Show that the Johnson noise of the combination when they are connected in series (see Figure 2.13(a)) is

$$\langle V_j^2 \rangle = 4k\,df\,(R_1 T_1 + R_2 T_2),$$

but when they are connected in parallel (see Figure 2.13(b)), it is

$$\langle V_j^2 \rangle = 4k\,df\,R_1 R_2 \left[\frac{R_1 T_2 + R_2 T_1}{(R_1 + R_2)^2} \right].$$

2.7 From the Wien displacement law (Problem 1.3), suggest suitable semiconductors for detectors matched to the peak irradiance from

(a) stars like the Sun ($T = 5800$ K)

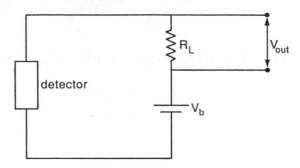

Figure 2.14 Circuit for Problem 2.8.

(b) Mercury ($T = 600$ K)

(c) Jupiter ($T = 140$ K).

2.8 Consider the circuit of Figure 2.14, and assume that the detector is a photoconductor with a resistance of ~1 MΩ that is exposed to a small, varying signal.
Assuming that the system is limited by Johnson noise, derive an expression
for the signal-to-noise ratio as a function of the load and detector resistances
and temperatures (use the results of Problem 2.6). Determine the value of R_L
where S/N is maximum for the case of a detector operating at 4 K, 77 K, and
300 K, always with the load resistor at 300 K. How does the signal-to-noise
ratio vary with detector operating temperature?

2.6 Further Reading

Boyd (1983) – detailed treatment of noise mechanisms in photoconductors and other
    detectors
Buckingham (1983) – overview of noise mechanisms in a broad variety of
    electronic devices
Pierret (1996) – comprehensive text on semiconductor electronics
Rogalski (2010) – review of intrinsic and virtually all other types of infrared
    detectors

3

Infrared (and Optical) Photodetectors

We describe two approaches to overcome the shortcomings of simple photo-conductors as described at the end of the preceding chapter. Both of them deal with these issues by separating the zone of the detector that provides the high electrical impedance from the zone that absorbs the photons and converts them to free electrons. The first example, photodiodes, are fabricated in a number of photoconductive materials, to provide different spectral responses. In the optical, where they are based on silicon, they provide a useful alternative to charge coupled devices discussed in Chapter 5. When constructed in smaller-bandgap semiconductors, they dominate most applications in the near-infrared spectral range. In them, a high impedance is maintained at the thin junction between regions of semiconductor with p-type and n-type impurities, respectively. Photons are converted to photo-electrons typically in the overlying layers away from the junction and through intrinsic absorption, with high absorption cross-sections. The second example, impurity band conduction detectors, are made with extrinsic material, i.e., they can have spectral response extending to longer photon wavelengths (lower energies) than photodiodes and practical examples have response to about 38 μm. Their high impedances are established by a thin layer of intrinsic material, and the photon absorption occurs in a heavily doped layer adjacent to it.

3.1 Photodiodes

A photodiode is based on a junction between two oppositely doped zones in a sample of semiconductor. These adjacent zones create a region depleted of charge carriers, producing a high impedance. In silicon and germanium, this arrange-ment permits construction of detectors that operate at high sensitivity even at room temperature. In semiconductors whose bandgaps permit intrinsic operation in the $1 - \gtrsim 15$ μm region, a junction is often necessary to achieve good performance at any temperature. The spectral response of photodiodes is currently restricted to wavelengths $\sim$15 μm and shorter because of quality and manufacturing issues for intrinsic semiconductors with extremely small bandgaps. Because these detectors operate through intrinsic rather than extrinsic absorption, they can achieve high quantum efficiency in small volumes. Well-established techniques of semiconductor device fabrication allow photodiodes to be constructed in arrays with many thousands, even millions, of pixels.

3.1.1 Basic Operation

As an illustration of the usefulness of photodiodes, we first consider the problems that arise in constructing an intrinsic photoconductor from InSb for the $1-5$ μm region. From Table 2.3, we get the material parameters $\tau \approx 10^{-7}$ s and $\mu \approx 10^5$ cm^2 V^{-1} s^{-1}. We compute the photoconductive gain to be $G \sim 10^{-2}$ V/ℓ^2 (from equation 2.12 and with ℓ, the distance between electrical contacts, in centimeters). The breakdown voltage for InSb is low, so the only way to achieve a reasonably high photoconductive gain is to make ℓ small. From equation 2.1, the detector resistance is $\ell/\sigma wd$, where $\sigma = qn_0\mu$. Because of the large electron mobility in InSb ($\sim$100 times that in silicon), it is impossible to achieve a large resistance with a small value of ℓ (while keeping the other detector dimensions fixed). That is, for this material, it is impossible to achieve simultaneously a high photoconductive gain and a large resistance. If G is small, the signals are small, and the detector system is likely to be limited by amplifier noise. If the resistance is small, the system will be limited by Johnson noise (equation 2.25). The leading alternate material for detectors in the 1–5 μm region, Hg$_{1-x}$Cd$_x$Te, also has very high electron mobility and therefore suffers the same shortcomings as InSb for high sensitivity photoconductor construction.

This dilemma can be overcome by fabricating a diode in the InSb (Rieke et al. 1959). Diodes are made by growing oppositely doped regions adjacent to each other in a single piece of material or by implanting impurity ions of opposite type to the dominant doping of the material with an ion accelerator. The n-type material has a surplus (and the p-type a deficiency) of electrons compared to what are needed for the crystal bonds. As a result, if thermal excitation is adequate to free them, electrons in the vicinity of the junction between the two types of doping diffuse from the n-type into the p-type material where they combine with holes, producing a space charge region with a net negative charge in the p-type material and a net positive charge in the n-type material they left behind. This process is illustrated in Figure 3.1. The region where the charge has diffused from the n-type to the p-type material has nearly all complete bonds and a depletion of potential charge carriers. The high resistance of this depletion region overcomes the dilemma we faced previously in trying to make a photoconductor of intrinsic InSb.

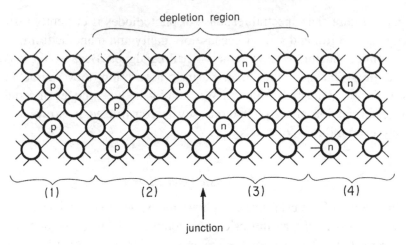

Figure 3.1 Microscopic view of a junction. Region (1) contains neutral p-type material; region (4) contains neutral n-type material; the electrons associated with the impurity n-type atoms in region (3) have diffused into region (2) where they have filled the holes due to the p-type atoms. Regions (2) and (3) have net space charges and are depleted of free charge carriers.

The diffusion of charges is self-limiting because it results in a voltage being set up across the junction; this voltage opposes the diffusion of additional electrons into the p-type material. The voltage at which equilibrium has been reached is called the contact potential, V_0. The material on either side of the depletion region has a relatively small electrical resistance because of its doping; consequently, there is virtually no potential across it; nearly all of the potential across the device appears at the junction. The potential and energy level diagrams for the diode are as shown in Figure 3.2.

The size of V_0 is determined by the Fermi levels on the two sides of the junction. Recall that $f(E_F) = 0.5$ (equation 2.30 and following discussion). The Fermi level in the n-type material prior to contact is at a higher energy than in the p-type, so electrons will flow between the materials until their Fermi levels are the same. Although this behavior might be expected from the definition of the Fermi level, it can also be proven from the formalism we are about to develop (see Problem 3.4). The difference in the Fermi levels before contact is equal to qV_0, as is the difference in the conduction band levels after contact.

We will derive the electrical properties of the diode in detail later, but a qualitative description is useful now. If an external bias is connected such that it adds to the contact potential (that is, the positive voltage is connected to the n-type material), we say that the diode is reverse-biased. Under this condition, the potential across the depletion region is increased by the external voltage, which increases the size

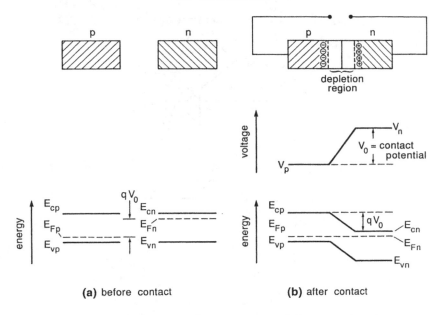

Figure 3.2 Development of a contact potential across a junction.

of the depletion region and thus the resistance of the junction. If the reverse bias is increased, eventually the junction will break down and become highly conducting. At modest reverse biases, current can still be conducted across the junction through processes such as tunneling; the reverse bias brings E_{cn} (the conduction band in the n-type material) below E_{vp} (the valence band in the p-type material; see Figure 3.3), making it energetically favorable for an electron to penetrate the depletion region without first moving into the conduction band of the p-type material, E_{cp}. If the depletion region is narrow enough, the electron wave function can extend across it. As a result, there is a finite probability that the electron will appear on the other side of the depletion region; if it does, it is said to have tunneled through the junction. At high reverse biases, breakdown occurs by avalanching. In this case, the strong electric field can accelerate an electron from the p-type region sufficiently strongly that the electron creates additional conduction electrons when it collides with atoms in the depletion region. This cascade of conduction electrons carries a large current. When the junction is forward-biased, the sign of the applied potential is reversed so it decreases the bias across the depletion region. If the bias voltage is larger than V_0, the junction becomes strongly conducting. The overall behavior of the diode is summarized by the I–V curve (Figure 3.4).

If free charge carriers are generated and recombine in either the n-type or p-type regions they produce little net modulation of the current through the device because the relatively low resistances of these regions allow equilibrium to be

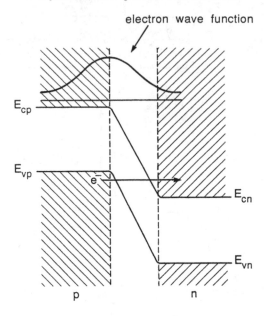

Figure 3.3 Illustration of tunneling through a junction.

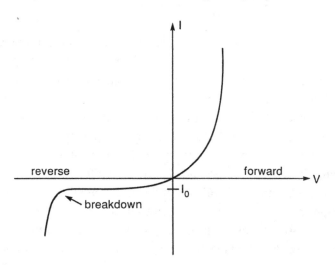

Figure 3.4 Current–voltage curve for a diode. I_0 is the saturation current.

re-established quickly. However, charge carriers produced in or near a reverse-biased or unbiased junction can be driven across it by the junction field and then can recombine on the other side, thus carrying a net current. Charge carriers can be produced thermally or by photoexcitation; as always, we assume that the detector is cold enough that we can ignore the former. Photoexcitation in the p-type material is illustrated in Figure 3.5; a photon is absorbed and excites an electron–hole pair.

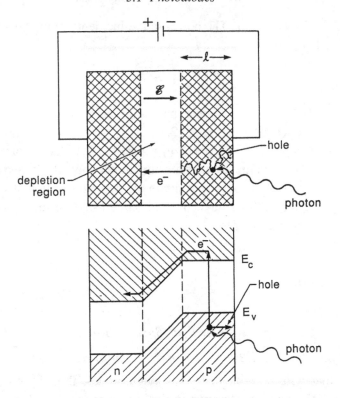

Figure 3.5 Illustration of the detection process in a photodiode. The photogenerated electron diffuses into the depletion region, where the field sweeps it across to the n-type region.

The hole is eventually collected at the negative electrode or recombines and is lost to the detection process. The electron diffuses through the material; if it enters the depletion region, the junction field drives it across, creating a photocurrent. The same process occurs if the n-type material is illuminated, except the roles of the electron and hole are reversed.

As long as a photodiode is designed to allow efficient diffusion of the photo-excited charge carriers into the junction, virtually every absorbed photon will contribute to the photocurrent. Thus,

$$I_{ph} = -\varphi q \eta = -\frac{q\eta}{h\nu} P_{ph},$$ (3.1)

where I_{ph} is the photocurrent, η is the quantum efficiency, φ is the incident photon rate (s^{-1}), P_{ph} is the power in the signal, and the negative sign is consistent with the coordinate system used later in this chapter. Compared with the expression in equation 2.14, equation 3.1 can be interpreted as stating that the photoconductive gain of photodiodes is $G = 1$. The power in the incident photon beam is given by equation 2.11. Thus, the responsivity

$$S = \frac{I_{ph}}{P_{ph}} = \frac{\eta \lambda q}{hc},$$ (3.2)

so long as $\lambda \leq hc/E_g$. This response has a wavelength dependence similar to that for photoconductors (see Figure 2.8). Noise currents are as described in Section 2.2.5, except the G–R noise is reduced by a factor of $\sqrt{2}$ because recombination of the charge carriers occurs in the heavily doped (and hence low resistance) region of the detectors, not in the depletion region. As discussed in the preceding chapter, in this case the noise is called shot noise; for a diode in the photon noise limited regime it is

$$\langle I_S^2 \rangle = 2q^2 \, \varphi \eta \, df$$ (3.3)

(see equation 2.23).

3.1.2 Photodiode Spectral Range

3.1.2.1 Visible and Near Infrared

For reasons that will be explained soon, photodiodes, though constructed with extrinsic material, work only through intrinsic absorption; therefore, the possible spectral responses correspond to the intrinsic bandgaps of the relevant semiconductors. Typical optical/infrared photodiode materials with cutoff wavelengths at room temperature are Si ($\lambda_c = 1.1$ μm), Ge ($\lambda_c = 1.8$ μm), InAs ($\lambda_c = 3.4$ μm), InSb ($\lambda_c = 6.8$ μm), $In_{1-x}Ga_x As$, $Al_x Ga_{1-x} AsSb$, and $Hg_{1-x}Cd_x Te$. The bandgaps of the latter three materials can be tuned by changing the composition; for example, varying the relative amounts of In and Ga in InGaAs adjusts λ_c from 0.9 to >3 μm and varying the amounts of Al and Ga in AlGaAsSb adjusts λ_c from 0.75 to 1.7 μm. Both materials are widely used for optical fiber receivers with a cutoff at ∼1.68 μm, i.e., $x = 0.47$ for InGaAs. High-performance photodiodes with cutoff wavelengths in the 5 μm range can be manufactured in either InSb or $Hg_{0.70}Cd_{0.30}Te$, operating at 30–45 K. The bandgaps and hence cutoff wavelengths of these materials change slightly with operating temperature (for example, the cutoffs of InAs and InSb become 5–10% shorter at 77 K than at room temperature).

3.1.2.2 Ultraviolet

A variety of materials are also being explored for the ultraviolet (UV). Silicon is one possibility but suffers from two significant disadvantages: (1) although the internal quantum efficiency can be high the refractive index fluctuates wildly in the ultraviolet making wide-band antireflection (AR) coatings impossible;

and (2) silicon is *too good* a detector in the visible, making the requirements to reject visible light and avoid contamination of the UV signal difficult to achieve without compromising system efficiency. By manufacturing detectors with cutoff wavelengths in the blue or near ultraviolet, the requirements for blocking longer wavelength radiation are relaxed. These devices are sometimes described as "solar blind" since the complete absorption of UV radiation in the atmosphere at wavelengths < 0.28 μm results in their having very little sensitivity to sunlight. Materials include GaP ($\lambda_c = 0.55$ μm), GaN ($\lambda_c = 0.37$ μm), and diamond films ($\lambda_c = 0.23$ μm). Al$_x$Ga$_{x-1}$N has a tunable bandgap (between $\lambda_c = 0.20$ and 0.37 μm: Lim et al. 1996; Walker et al. 1997; Monroy et al. 2003). It has become the material of choice for most high-performance applications. A variety of detector architectures are employed, including photoconductors (Chapter 2), PIN (Section 3.1.4.1) and Schottky diodes (discussed in a different context in Chapter 10) – see Schühle and Hochedez (2013).

3.1.2.3 HgCdTe Photodiodes

HgCdTe is the most widely used material for near-infrared photodiodes. Detector grade materials with λ_c tuned between ~1 and ~15 μm can be manufactured by adjusting the relative amounts of Hg and Cd (e.g., D'Souza et al. 2000). HgCdTe has the advantages of being able to operate at higher temperatures and being able to tune the bandgap (and hence λ_c) to the application, allowing further optimization of the operating temperature; for example, Hg$_{0.55}$Cd$_{0.45}$Te provides $\lambda_c \sim 2.5$ μm and can operate well at 80–90 K. In addition, this material exhibits very small changes of the lattice constant with x (the Cd fraction), i.e., good crystal structure matching holds for material grown with a range of bandgaps, allowing a broad variety of detector approaches. Superb HgCdTe detectors are now produced by Teledyne using molecular beam epitaxy (MBE – where the detector is built up literally one molecular layer at a time from a molecular beam in vacuum) and by Raytheon using advanced liquid phase epitaxy (where the crystal is grown from a liquid in which the detector material has been introduced at high concentration). There is continued development of alternative detector types with primary goals to improve the performance at elevated operating temperatures, and to drive down costs. However, HgCdTe detectors represent the current state-of-the-art (e.g., Rogalski 2012), so we will concentrate on them.

The control achieved over the internal structure of the photodiodes with MBE (see Figure 3.6) allows adjustment of the molecular composition and hence the bandgap within the detector to improve the performance (Garnett et al. 2004). The junction is at the interface between p- and n-type material. The large bandgap cap layer is grown integral with the smaller bandgap infrared-active layer and reduces the susceptibility to surface-induced excess currents, helping to minimize dark

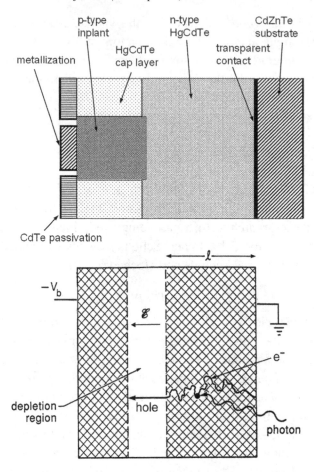

Figure 3.6 Cross-sectional drawing of a HgCdTe photodiode (upper panel). The upper figure is based on the discussion in Garnett et al. (2004), the lower one reproduces Figure 3.5.

current. It also keeps the photo-charge-carriers away from the interface with the metallized contact, helping to reduce trapping and the resulting latent images. The CdZnTe substrate is used to support the array of diodes while they are being grown, but can be removed afterwards. The lower panel shows schematically how a photon is detected, as already discussed. In this case, the process is based on the holes created when photons are absorbed. If the absorbing layer thickness, l, is small enough, the hole falls into the contact potential field before it can recombine and is driven across the depletion region, producing the current that is used for the act of detection. The width of the depletion region is exaggerated for clarity.

Because of its adjustable bandgap, HgCdTe is being developed for devices operating at wavelengths longer than 5 μm. A number of issues have been addressed. First, the bandgap becomes very sensitive to the material composition

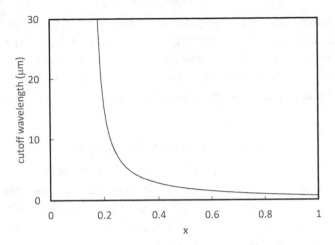

Figure 3.7 Dependence of cutoff wavelength in $Hg_{1-x}Cd_xTe$ detectors with increasing proportions of Hg (decreasing x).

as x is made smaller – i.e., as the proportion of Hg is increased (see Figure 3.7). Process improvements have been necessary to improve the control of the detector composition. In addition, the detectors with response cutoffs $\gtrsim 10~\mu m$ can be optimized by producing them with a gradient of x across the infrared-active region, which can be used to extend the response to longer wavelengths compared with detectors with no gradient (Phillips et al. 2002). There are also some operational and construction concerns. Material with a small bandgap only allows a small contact voltage to maintain the diode depletion region, reducing the well depths achievable with integrating amplifiers (the well depth is the number of charge carriers that can be collected without loss of response). Finally, as the Hg content increases, the detector wafers become increasingly fragile, making it difficult to hybridize them onto readout wafers to manufacture detector arrays.

Nonetheless, excellent detector arrays have been produced with $\lambda_c \sim 10~\mu m$ (McMurtry et al. 2013, 2016a), with dark currents for $> 90\%$ of the pixels less than 1e/s and well depths of up to 75,000 electrons. This effort has also yielded arrays with performance nearly as good out to nearly 13 μm (McMurtry et al. 2016b) and developmental devices with $\lambda_c \sim 15~\mu m$ (Cabrera et al. 2020). These devices have response across the entire 8–13 μm atmospheric window, so mated to readout circuits capable of handling the large background signals they will find wide application in astronomy and other areas.

Development of HgCdTe detectors continues. However, given the challenges already encountered, it seems unlikely that high-performance detector arrays with $\lambda_c \gg 15~\mu m$ will result in the near future. Instead, operation at higher temperatures with low dark current is a more likely development (Lee et al. 2016). Such detectors would reduce the cryogenic infrastructure required for mid-infrared instrumentation, a particularly important direction for space-borne instruments. For the foreseeable future, high-performance detectors for the longer wavelengths will be based on extrinsic photoconductivity (Section 3.2 of this chapter) or will operate on completely different principles, such as bolometers (Chapter 8).

3.1.3 Quantitative Description

From the above description, there are a number of attributes of interest in photodiodes. Once charge carriers are produced in the diode material, they must make their way to the junction by *diffusion*, a process that is also important in understanding the electrical behavior of the diode. Absorption and diffusion combine to control the *quantum efficiency*. The electrical properties of the junction will be described in terms of two important performance aspects: the *impedance* of the junction and the *photoresponse* of the detector. The frequency response of the detector will be controlled by its *capacitance*. These characteristics and their implications for detector performance are discussed in turn in this section, in terms of the relatively simple case where we assume that the junction between p and n type materials is abrupt, i.e., very narrow.

3.1.3.1 Diffusion

Diffusion refers to the tendency of thermal motions to spread a population of particles uniformly over the accessible volume in the absence of confining forces. The operation of a photodiode depends on diffusion of charge carriers into its junction region. After the charge carriers have been created, perhaps near the surface of the diode and away from the junction, diffusion spreads them through the material; those that reach the junction are swept across it by the junction field so that a gradient is maintained against which further diffusion occurs. An efficient photodiode is designed so that virtually all the charge carriers diffuse into the junction.

To understand this process, we need to describe it quantitatively. Refer to Figure 3.8 for the following discussion. Consider a column of semiconductor containing charge carriers at density $n(x)$. To be somewhat more specific, let the carriers be electrons, and let their density decrease with increasing x. (The same general arguments are valid for hole diffusion.) Suppose there is an electric field in the material, $\mathcal{E}_x$, in the direction of increasing x. The electrons thus experience an electrostatic force (per unit volume)

$$F_E = -q\, \mathcal{E}_x\, n(x), \tag{3.4}$$

that drives them in the direction of decreasing x; that is, it will tend to increase the already existing density concentration in this direction. However, the random

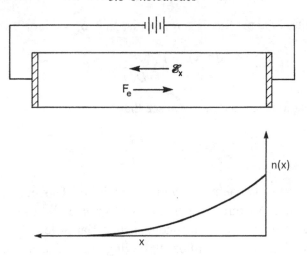

Figure 3.8 Distribution of electrons under joint influence of an electric field and diffusion.

thermal motions of the electrons will tend to spread them out against the effect of the electric field; for example, if the field were removed, we would expect the electrons on average to be spread uniformly through the entire accessible volume. The osmotic pressure that drives this diffusion is

$$P_0(x) = \left(\frac{N}{V}\right) kT = n(x)kT. \tag{3.5}$$

The corresponding osmotic force (again per unit volume) is

$$F_0(x) = -\frac{dP_0(x)}{dx} = -kT\frac{dn(x)}{dx}. \tag{3.6}$$

At equilibrium, $F_0(x) + F_E(x) = 0$, or

$$\frac{dn(x)}{dx} + \frac{q_x n(x)}{kT} = 0. \tag{3.7}$$

The solution of equation 3.7 is

$$n(x) = n(0)e^{-q\mathcal{E}_x\, x/kT}. \tag{3.8}$$

The diffusion coefficient, D (conventionally given in units of cm^2 s^{-1}; D (m^2 s^{-1}) = $10^{-4} D$ (cm^2 s^{-1})), is defined to give the osmotic flux of particles (number cm^{-2} s^{-1}) as

$$S_0 = -D_n\frac{dn(x)}{dx}. \tag{3.9}$$

Using equation 2.6, the flux of particles driven by the electrostatic field is

$$S_E = \langle v_x \rangle\, n\,(x) = -\mu_n \mathcal{E}_x\, n\,(x). \tag{3.10}$$

Under equilibrium, the net flux is zero, that is, $S_0 + S_E = 0$, or

$$\frac{dn(x)}{dx} + \frac{\mu_n \mathcal{E}_x n(x)}{D_n} = 0. \tag{3.11}$$

Comparing equations 3.7 and 3.11, we see that

$$\frac{D_n}{\mu_n} = \frac{kT}{q}, \tag{3.12}$$

connecting the diffusion coefficient with the mobility. This equation is known as the Einstein relation. A similar equation for holes relates D_p to μ_p.

Now consider a volume of semiconductor material of length dx with a flux f_1 of electrons entering at x_1 and a flux f_2 leaving at x_2. The rate of change in the number of conduction electrons in the volume is

$$\frac{f_2 - f_1}{dx} = D_n \frac{d^2 n(x)}{dx^2}. \tag{3.13}$$

If the electrons have a lifetime τ_n before recombination, they will disappear at a rate $n(x)/\tau_n$. They can also be created by thermal- or photoexcitation, which we will represent by a source term g. Then, by continuity, we have

$$\frac{dn}{dt} = D_n \frac{d^2 n}{dx^2} - \frac{n}{\tau_n} + g. \tag{3.14}$$

At equilibrium, $dn/dt = 0$, and for the moment we set $g = 0$, so equation 3.14 becomes

$$\frac{d^2 n}{dx^2} = \frac{n}{D_n \tau_n}. \tag{3.15}$$

The solution of equation 3.15 with the boundary condition that recombination should reduce $n(x)$ to zero for very large x is

$$n(x) = n(0)e^{-x/L_n}, \tag{3.16}$$

where we have defined the diffusion length to be

$$L_n = (D_n \tau_n)^{1/2}. \tag{3.17}$$

The diffusion coefficient and diffusion length play important roles in both the optical and electrical properties of photodiodes, as will be discussed in the following two sections.

3.1.3.2 Quantum Efficiency

To be detected, the photoexcited charge carriers in a photodiode must diffuse into the depletion region out of the relatively low resistance p- or n-type neutral material on either side of the junction. From equation 3.16, this requirement suggests that the layer of neutral material overlying the junction should be no more than a diffusion length thick. The resulting relationship between L and photodiode quantum efficiency leads to important constraints on detector design. For example, following the discussion in Section 2.2, at low temperatures we take μ to be due to neutral impurity scattering and therefore to be independent of T; from the Einstein relation, equation 3.12, D will go as T. Also from Section 2.2, we can take the recombination time, τ, to change approximately in proportion to $T^{1/2}$. The diffusion length then goes as $T^{3/4}$. Obtaining good quantum efficiency with diodes operated at low temperatures requires a compromise between thinning the overlying absorbing layer for good charge collection and making it thick enough for good absorption. Equation 3.17 also demonstrates why it is not feasible to extend the response of a photodiode through extrinsic absorption. The thickness of the overlying absorber must be $l \leqslant (D\tau)^{1/2} = C_1 N_I^{-1/2}$, where C_1 is a constant and N_I is the impurity concentration in the material (taking τ to go roughly as N_I^{-1}). Obtaining good absorption requires that $l \geqslant 1/a(\lambda)$, where the absorption coefficient $a(\lambda) = \sigma_i N_I$, or equivalently $l \geqslant C_2 N_I^{-1}$, where C_2 is a constant. It is generally not possible to meet these two conditions on l simultaneously with extrinsic absorption.

Following Holloway (1986), the above arguments regarding the photodiode quantum efficiency can be made more quantitative by considering the diffusion of charge carriers into the depletion region according to equation 3.14, which can be rewritten (still assuming the steady state case, $dn/dt = 0$):

$$\frac{d^2 n}{dx^2} - \frac{n}{L^2} + \frac{g}{D} = 0. \tag{3.18}$$

We will use g to represent a uniform planar charge source and will adopt the approximation of an infinitely extended diode junction to avoid having to consider edge effects. If g is the number of photogenerated charge carriers per unit time and per unit volume (assuming that thermal excitation is reduced by adequate cooling), then equation 3.18 represents the diffusion of the signal charge carriers from their generation sites to the junction. We let x run from 0 at the junction to c at the diode surface. The general solution of equation 3.18 is

$$n(x) = A \cosh\left(\frac{x}{L}\right) + B \sinh\left(\frac{x}{L}\right) + \frac{gL^2}{D}. \tag{3.19}$$

Equation 3.19 can be simplified if we specify appropriate boundary conditions. We will assume the charge carriers are absorbed with 100% efficiency when they diffuse to the junction,

$$n = 0 \quad \text{at} \quad x = 0. \tag{3.20}$$

We also assume that the absorption by the detector material is very efficient so the photons are absorbed very near the surface of the diode. This assumption means that $g \sim 0$ everywhere but in this thin surface layer. (The situation for inefficient absorption is treated as a limiting case in Problem 3.2). Applying condition 3.20, we set $A = 0$. The flux of charge carriers at $x = c$ can be taken to be

$$S_{in} = \left(D \frac{dn}{dx} \right)_{(x=c)} = b\varphi, \tag{3.21}$$

where φ is the photon flux and b is the fraction of incident photons available for absorption to produce charge carriers (thus accounting for loss mechanisms such as reflection from the surface). The first part of the expression above is identical to equation 3.9 except for a sign change. This change arises because the direction of increasing x has been reversed in the present discussion to allow easier application of the boundary conditions. Equation 3.21 allows us to determine B in 3.19, yielding

$$n(x) = \frac{S_{in} L \sinh \left(\frac{x}{L} \right)}{D \cosh \left(\frac{c}{L} \right)}. \tag{3.22}$$

The quantum efficiency is the flux of charge carriers into the junction divided by the flux of input photons, or (from equations 3.9, 3.21, and 3.22),

$$\eta = \frac{D \left(\frac{dn}{dx} \right)_{(x=0)}}{\varphi} = b \, \text{sech} \left(\frac{c}{L} \right) = \frac{2b}{e^{c/L} + e^{-c/L}}. \tag{3.23}$$

For example, when the material overlying the junction has a thickness of one diffusion length, $c = L$ and $\eta = 0.65b$, but if $c = 2L, \eta = 0.27b$. Our intuitive guess at the beginning of this section that we should have no more than one diffusion length of overlying material proves to be reasonable.

Recalling that L can scale as $T^{3/4}$, equation 3.23 indicates that the quantum efficiency of a photodiode will be low if it is operated at a temperature far below the range for which it was designed. Fortunately, it is found that many diodes operate with higher quantum efficiency at very low temperatures than would be implied by equation 3.23. At more optimum operating temperatures, state-of-the-art photodiodes (e.g., MBE-grown HgCdTe) with suitable multi-layer AR coatings can have quantum efficiencies $\gtrsim 85\%$ between 1 and 5 μm.

3.1.3.3 Current and Impedance

We will next describe the fundamental electrical behavior of a photodiode. A description of this behavior is given by the diode equation, which describes the relationship between the current that flows through the diode and the bias voltage applied across it:

$$I = \left[qA \left(\frac{D_n^p}{L_n^p} p_n + \frac{D_p^n}{L_p^n} n_p \right) \right] (e^{qV_b/kT} - 1)$$

$$= I_0(e^{qV_b/kT} - 1), \tag{3.24}$$

where I_0 is the saturation current, A is the area, V_b is the bias voltage, and T is the temperature. The terms p_p and p_n, i.e., the concentration of p-type impurities in the p-type and n-type regions of the diode, are commonly referred to as majority and minority charge carrier concentrations, respectively. These terms emphasize the difference in location of otherwise identical charge carriers; majority charge carriers are in a region of semiconductor where the doping favors generation of similar carriers, and minority carriers are in semiconductor that is doped to favor generation of carriers of opposite type. The terms n_n and n_p have analogous meanings, while the L's and D's are diffusion lengths and constants. The derivation of equation 3.24 is a bit tedious so it has been put into a Note at the end of the chapter.

As always, an overriding requirement for high sensitivity operation is to cool the detector sufficiently to stop thermally generated carriers from passing through it at a significant rate. With the photodiode, we have an advantage over photoconductors because majority carriers generated outside the depletion region have little effect on the junction current and hence on the effective resistance of the diode. This is demonstrated in equation 3.24, which shows that I_0 depends only on the minority carrier concentrations p_n and n_p, which by definition are much smaller than the majority carrier concentrations.

In principle, the diode resistance (V/I) can be increased by back-biasing the detector close to breakdown. This strategy sometimes produces excess current-related noise, and the best overall performance for low light levels is often obtained at lower bias. A figure of merit for the detector is then the zero bias resistance; it is obtained by rearranging equation 3.24 to solve for V_b and then taking the derivative with respect to I:

$$\left(\frac{dV_b}{dI} \right)_{(V_b=0)} = \frac{d}{dI} \left[\left(\frac{kT}{q} \right) ln \left(1 + \frac{I}{I_0} \right) \right]_{(I=0)}$$

$$= \frac{kT}{qI_0} = R_0. \tag{3.25}$$

As seen in equation 3.24, I_0 depends on diffusion coefficients and lengths and on minority carrier concentrations. D and L go as modest powers of the temperature, as can be seen from equations 3.12, 3.17, and the temperature dependencies of the mobility and recombination time that were discussed in Chapter 2. The behavior of I_0 is dominated by the exponential temperature dependencies of p_n and n_p (see Note). Thus R_0 for an ideal diode increases exponentially with cooling. For real diodes, this improvement occurs only to some limiting value of R_0. For example, tunneling of charge carriers across the junction has relatively little temperature dependence and is likely to place an upper limit on R_0 at very low temperatures. Currents due to generation or recombination of charge carriers thermally in the depletion region also have relatively shallow temperature dependence.

In addition to the dependence on temperature already discussed, I_0 and hence R_0 depend on the level of doping in the diode. This dependence arises from the dependence of p_n and n_p on doping level (see Note). In addition, a second dependence on doping level arises from the diffusion lengths in equation 3.24. Near room temperature, D is roughly independent of impurity concentration, but L is proportional to $N_I^{-1/2}$ through the dependence of recombination time on impurities. We can therefore show that

$$I_0 \approx C_1(N_D)^{-1/2} + C_2(N_A)^{-1/2}, \tag{3.26}$$

where C_1 and C_2 are constants. At low temperatures where μ is due to neutral impurity scattering, D goes inversely as a modest power of N_I, increasing the effect of impurity concentration in reducing I_0 compared with equation 3.26. Thus, R_0 increases at least as fast as the square root of the doping concentration in the diode. All other things being equal, better diode performance should be achieved with higher doping to increase R_0.

We now have to say our mea culpas. Our goal has been only to achieve a first-order understanding of diode behavior, and we have made a number of simplifying assumptions along the way. We have taken the signal frequency to be low enough that the diode could be assumed to be in equilibrium, thus dropping out all time-dependent terms. In the operating regime of interest for high sensitivity detectors, we are justified in considering only diffusion currents, so we have not included terms involving charge carrier drift in the electrostatic field.

With far less justification, we have ignored the generation and recombination of charge carriers in the depletion region. For example, when the diode is moderately forward-biased, some electrons entering from the n-type region recombine with holes entering from the p-type region, which is equivalent to a transfer of a positive charge across the depletion region (because the hole density in the n-type region has been increased and that in the p-type region decreased). When the diode is moderately reverse-biased, the increased field across the depletion region stops this recombination mechanism, but electron–hole pairs can be generated in the depletion region and swept out of it by the junction field. In both of these cases, since there is an additional current in the diode, the equivalent saturation current is larger than the I_0 given in equation 3.24. In the reverse-biased case there may be no clearly defined saturation current; instead, as the width of the depletion region increases with back-bias, so does the volume in which generation of charge carriers can occur and hence the current across the junction. In the forward-biased case, the depletion width decreases with increasing bias, tending to flatten out the inflection in the diode $I-V$ curve as it switches into conduction. An approximate allowance for this latter behavior can be made by taking the $I-V$ curve to be given by

$$I = I_s(e^{qV_b/mkT} - 1),\qquad(3.27)$$

where I_s and m are fitted empirically to the measured curve. The parameter m is called the ideality factor.

3.1.3.4 Response

In the preceding discussion, we derived the electrical properties of an ideal diode in the absence of light. When light falls on this diode, any resulting charge carriers that penetrate to the depletion region will be driven across it and will produce a current. In other words, the photocurrent through the junction is as given in equation 3.1, where the quantum efficiency of the detector can be calculated as shown in equation 3.23 for the case of high absorption efficiency. The total current through an ideal photodiode is given by the sum of the expressions in equations 3.24 and 3.1:

$$I = -\frac{\eta q P}{h\nu} + I_0(e^{qV_b/kT} - 1),\qquad(3.28)$$

where we have converted φ to $P/h\nu$. Figure 3.9 shows the $I-V$ curve for a diode exposed to various levels of radiation.

Figure 3.9 also suggests a variety of photodiode operation modes. For example, the detector can be run by monitoring the output voltage at a fixed current. A simple method might be to use a high input impedance voltmeter so $I \sim 0$. As Figure 3.9 suggests, the detector response may be significantly nonlinear when the device is operated in this way. In the photoconductive mode, the diode is placed in a circuit that holds the voltage across the detector constant (and negative to back-bias it), and the current through the circuit is measured as an indication of the illumination level. As indicated by equation 3.28, the output current in this mode is linear with input power. A particularly attractive operating method is to hold the voltage across the detector at zero; certain types of excess voltage noise are then suppressed (Hall et al. 1975). In Chapter 4, we will illustrate the ability of a simple circuit called a transimpedance amplifier to operate photodiodes in this way. However, this approach proves to be a limitation in photodiode arrays, where it is desirable to accumulate the photocurrent on a capacitance without the cancellation of the resulting change in bias voltage. Fortunately, the nonlinearities are well behaved and can be corrected in data reduction.

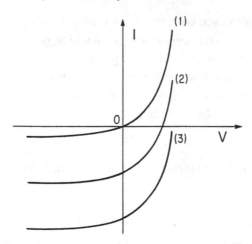

Figure 3.9 Response of a diode to illumination. The illumination increases for curves (1) to (3), starting from zero for curve (1).

3.1.3.5 Capacitance

A photodiode has relatively high capacitance because the distribution of positive and negative charge across its junction forms a parallel plate capacitor with small separation between the plates. The capacitance of the photodiode can control its frequency response and often determines the limiting noise of the amplifier used to read out its signal. Here we show how to estimate the capacitance of the junction and what parameters control it.

Consider an "abrupt junction" at $x = 0$, with N_A the concentration of ionized acceptors for $x < 0$ and N_D the concentration of ionized donors for $x > 0$. Let l_p be the depth of the depletion region into the p-type region and l_n the depth into the n-type region. Applying Poisson's equation in one dimension, we have for the junction

$$\frac{d^2V}{dx^2} = \begin{bmatrix} qN_A/\varepsilon, & -l_p < x < 0, \\ -qN_D/\varepsilon, & 0 < x < l_n, \\ 0, & \text{otherwise.} \end{bmatrix} \tag{3.29}$$

We adopt the boundary condition $dV/dx = 0$ for $x \leq -l_p$ and $x \geq l_n$, and $V(l_n) - V(-l_p) = V_0 - V_b$, where V_0 is the contact potential and V_b is the applied potential. The solution to equation 3.29 is

$$V(x) = \begin{bmatrix} \dfrac{qN_A}{2\varepsilon}(x^2 + 2l_p x), & -l_p < x < 0, \\[3mm] \dfrac{-qN_D}{2\varepsilon}(x^2 - 2l_n x), & 0 < x < l_n. \end{bmatrix} \tag{3.30}$$

Since the electric field in the neutral regions of the diode material is zero, the space charge on either side of the junction must be equal and opposite:

$$N_A l_p = N_D l_n. \tag{3.31}$$

If we express $V_0 - V_b$ from equation 3.30 (and the discussion immediately preceding it) and use equation 3.31, we can solve for lengths of the depletion region into the p-type and n-type material as

$$l_p = \left[\frac{2\varepsilon N_D (V_0 - V_b)}{q N_A (N_A + N_D)} \right]^{1/2},$$

$$l_n = \left[\frac{2\varepsilon N_A (V_0 - V_b)}{q N_D (N_A + N_D)} \right]^{1/2}. \tag{3.32}$$

The width of the depletion region is

$$w = l_p + l_n = \left[\frac{2\varepsilon (N_A + N_D)(V_0 - V_b)}{q N_A N_D} \right]^{1/2}. \tag{3.33}$$

The junction capacitance is

$$C_J = \varepsilon \frac{A}{w} = \kappa_0 \varepsilon_0 \frac{A}{w}, \tag{3.34}$$

where the dielectric constant κ_0 is given in Table 2.3, ϵ_0 is the permittivity of free space, and A is the junction area. Note that C_J goes as the square root of the impurity concentration.

In Chapter 4, we will see that achieving low noise with practical readouts requires that the detector capacitance be kept small, which is achieved in a photodiode by minimizing the impurity concentrations. We already know that a detector with a large resistance is needed to minimize Johnson noise; from the preceding discussion, higher R_0 is achieved with larger impurity concentrations. Thus, the two requirements on the impurity concentrations are in conflict. In practical detectors, some compromise must be made.

The abrupt junction is only one example of a doping density profile that may apply to a diode. Other profiles yield differing relations between capacitance and impurity concentration and bias voltage. Measurements of the dependence of capacitance on voltage can be used to deduce the contact potential and the profile of the impurity concentration (see Problems 3.5 and 3.6).

3.1.4 Photodiode Variations

Both PIN and avalanche diodes, discussed in this section, are used where a simple, rugged, compact, and inexpensive detector is required. For fast time response at extremely low light levels and with minimal dark counts, photoemissive devices (see Chapter 6) are sometimes preferred if their higher complexity and fragility, larger size, and expense can be tolerated. However, continued improvements in photodiodes make them optimum for an increasingly large variety of applications.

3.1.4.1 PIN Diodes

As can be seen from equation 3.34, the capacitance can be reduced and hence the frequency response of a photodiode can be improved by widening the depletion region. If the depletion region is wide enough that most of the photon absorption occurs there, the photoexcited charge carriers will drift under the influence of the junction field rather than having to diffuse into the junction. We have seen that the width of the depletion region can be increased by reducing the doping that creates the junction (see equation 3.33), but with a concomitant undesirable decrease in the detector resistance. Often a better solution is to interpose a high resistivity intrinsic (or nearly intrinsic) layer between the p and n sides of the diode; hence, these detectors are called PIN diodes. Typically, the diode will be illuminated through the p-doped layer, so that photoelectrons are driven across the junction (given the higher mobility of electrons than holes, this choice improves the frequency response). In addition, the p-type layer overlying the intrinsic region is made very thin, so negligible absorption occurs in it. Any photoelectrons produced in this layer would diffuse into the junction and would impose their relatively long diffusion time constant on the response. The time response for a PIN diode is then $w/|\langle v_x \rangle|$, where

$$|\langle v_x \rangle| = \mu \, \mathcal{E}_x = \frac{\mu \, (V_0 + V_b)}{w} \tag{3.35}$$

or

$$\tau_{PIN} = \frac{w^2}{\mu(V_0 + V_b)}. \tag{3.36}$$

The intrinsic region must be thin enough that the charge carriers drift across the junction and are collected before they recombine; in other words, τ_{PIN} in equation 3.36 should be less than the recombination time τ_{rec}. For example, taking parameters for silicon from Table 2.3 and assuming $V_0 + V_b > 1$ V, we can allow $w > 1$ mm, although such a large value would be unusual in a practical detector. Comparing this width to the absorption coefficient for silicon (Figure 2.4), the intrinsic layer can be thin enough for the charge carriers to drift across it and still be thick enough for high absorption efficiency nearly to the bandgap energy. The very high breakdown voltage of PIN diodes also allows them to be used with large biases ($\sim$ 100 V), in which case equation 3.36 shows they will have very fast time response.

3.1.4.2 Avalanche Diodes

If a photodiode is operated at a back-bias so large that it is just short of breakdown, photoexcited charge carriers can be sufficiently accelerated in the depletion region that they produce additional carriers by avalanching. That is, the device has gain: a single absorbed photon can yield multiple charge carriers.

To simplify the discussion, we focus on avalanching by electrons, although in general holes can also trigger this process. The electrons freed by photo-absorption drift toward the pn junction as in equation 3.35. Once a free electron reaches the depletion region, however, it is accelerated to high velocity by the large field maintained by the back-bias on the device. To create an additional free charge carrier, i.e., an ionization event, in a direct transition material such as HgCdTe, the electron must obtain at least enough energy to lift another electron across the bandgap when it collides with a crystal atom, i.e., from the valence band to the conduction band. That is,

$$E_{threshold} = C \times \frac{1}{2} m v_{threshold}^2 = C \times \frac{1}{2} m (\alpha v_d)^2 = C' \times m \, (\alpha \mu \mathcal{E})^2, \quad (3.37)$$

where $E_{threshold}$ is the minimum energy required to free an additional charge carrier, m and $v_{threshold}$ are the mass and threshold velocity of a charge carrier (= electron), α is a factor of order unity, v_d is the drift velocity of the initial charge carrier, we have made use of equation 3.35 for the third relation, $\mathcal{E}$ is the magnitude of the electric field, μ is the mobility, and the C's are constants. Energy losses due to scattering could increase the threshold requirement, but in HgCdTe they are relatively minor. However, even for this material, in practice the required energy for a high probability of ionization is somewhat larger than the bandgap energy, $\sim 1.3 \, E_g$ because of considerations such as the density of available states in the conduction band (Kinch et al. 2004).

Because silicon is an indirect semiconductor, the process is more complex for it (Shockley 1961). With increasing bias on the device, and hence increasing electric field in the depletion region, the drift speed increases until, at about 10^4 V cm^{-1}, it saturates because of collisions by the charge carrier with the crystal lattice; this occurs at a speed of about 10^7 cm s^{-1}. A physical description of the limiting process is that a phonon of heat energy is emitted by the charge carrier and absorbed by the crystal at a typical speed of 1.5×10^7 cm s^{-1}, i.e., at an electron kinetic energy of 0.063 eV. The drift velocity is greatly reduced by this process and the carrier is then re-accelerated by the field; hence the average drift velocity is less than the limiting value. As the electric field is increased further, the interaction becomes more complex and the mobility becomes strongly field-dependent. By the time the field reaches $\sim 2 \times 10^5$ V s^{-1}, the carrier energy is suffcient that its collisions with the lattice can free additional charge carriers; i.e., rather than just raising the new charge carrier across the bandgap, avalanching occurs through energy deposited in the crystal.

Because it relies on stochastic events, this gain mechanism results in additional noise, described by the gain dispersion, $\beta = \langle G^2 \rangle / \langle G \rangle^2$, or by the excess noise factor, F, expressed relative to pure photon noise. A formula derived by McIntyre (1972) is often used for F:

$$F = kG + \left[2 - \frac{1}{G} \right] (1 - k), \tag{3.38}$$

where k is the ratio of the hole impact ionization rate to the electron impact ionization rate and G is the multiplication factor. The avalanche process is general and could be utilized in a normal PIN diode, but the range in position where photons are absorbed would produce large variations in gain and $F \gg 1$. An optimized device has a region about one absorption length in thickness to produce the charge carriers, from which they drift into a region of much larger field where the avalanching occurs. Figure 3.10 shows one such arrangement. Photons pass through the thin p^+ contact into the intrinsic region, where they are absorbed. Ideally, within the intrinsic region the electric field is constant, and causes the photoelectrons to drift into the junction. Much of the voltage drop across the device occurs across this junction, and

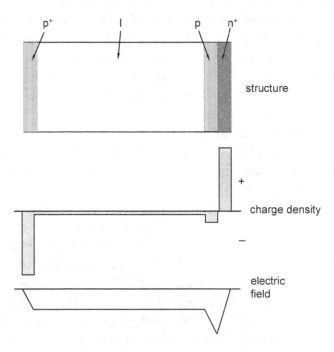

Figure 3.10 Construction of a "reach-through" avalanche photodiode.

the electric field has the characteristic triangular profile. The avalanching occurs only at the large field produced there. The multiplication is therefore similar for nearly all charge carriers and can be controlled by appropriate adjustment of the bias voltage.

Nonetheless, the gain process is relatively noisy in silicon avalanche photodiodes, with F typically 3–6 (Perkin Elmer 2003). Lower excess noise factors for visible-wavelength use can be obtained in multi-layer APDs, making use of different avalanche thresholds in different materials (e.g., Bai et al. 2012). In comparison, HgCdTe avalanche diodes can provide moderate amounts of gain with very little noise penalty, as a result of the large difference in electron and hole mobilities combined with optimized photodiode architectures. The minimum excess noise in equation 3.38 is $F = 2$, for $k = 0$ (as it nearly is for HgCdTe). This formula was derived in the limit of a thick and homogeneous multiplication layer where the ionization rates are independent of carrier history, making the statistics relatively noisy. However, because of their low excitation energy and efficient avalanching, HgCdTe avalanche diodes can have thin multiplication layers and values of F close to 1, independent of the gain (e.g., Finger et al. 2012).

These results apply to operation at modest gains (e.g., up to 100 or so for silicon, somewhat lower for InGaAs or HgCdTe), termed the proportional or linear mode, since the signals remain analog in nature and simply provide an amplified version of the input photon scene with modest (if F is close to 1) degradation. Arrays of such detectors can be hybridized onto conventional two-dimensional arrays of readout amplifiers (Chapter 4). Since the output noise of such amplifiers can be only 10–20 electrons rms, the result can be imaging systems working at high speed

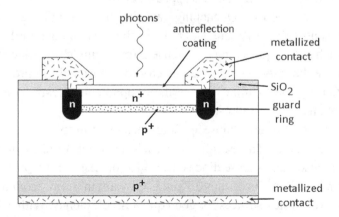

Figure 3.11 Practical layout of a Geiger-mode avalanche photodiode. Unwanted breakdown due to high fields at the edge of the junction is avoided with the guard ring architecture.

with effective read noise as low as 1 electron, i.e., working at the single-photon detection limit. However, compared with staring mode detectors, the equivalent gain-corrected dark currents are large, of order 10 e/s (e.g., Beck et al. 2014). Consequently, for large signals read out quickly the arrays operate with essentially no loss of intrinsic information other than that associated with the avalanche gain itself, i.e., due to $F > 1$, but for long integrations on faint sources conventional detectors are preferred.

3.1.4.3 Geiger-Mode Photodiodes

Where illumination levels are sufficiently modest to allow single-photon counting, the bias on a silicon photodiode can be increased so a single charge carrier creates an avalanche that breaks down the detector (i.e., makes it strongly conducting), producing a pulse with a net gain of a million or more. This type of operation is sometimes called "Geiger mode" after Geiger counters for nuclear particles. The rapid response of an avalanche diode can then allow pulse counting of single-photon events, so long as quench circuitry is used to suppress continuous breakdown and the detector is cooled sufficiently to control dark current (and, indeed, overheating to a destructive level). There are a number of noise (false pulse) mechanisms for these devices in addition to dark current. For example, after-pulses result when charge carriers from an avalanche are held in traps and released after the main event. The avalanche process also yields light (Lacaita et al. 1993), which can produce crosstalk into neighboring pixels in a detector array.

The simplest approach to quenching is to place a resistor between the photodiode and ground, so when the diode becomes strongly conducting due to an avalanche, more of the voltage drop from the bias appears across the resistor leaving less across the diode so it can recover. Quenching can also be accomplished with external circuitry that senses the avalanche and reduces the bias voltage until the diode has recovered. Such active quenching can be used, for example, to reduce the recovery time to improve the duty cycle of the detector, or conversely to provide sufficient dead time after a pulse to allow most of the trapped charge carriers to be released so they do not create after-pulses.

Pulse rise times of a few nanoseconds are achieved in this manner with a dead time of a few hundred nanoseconds after each detection to allow for quenching. The rapid response makes these diodes useful for precise timing. From Figure 2.4, the absorption length in silicon at $\sim$0.7 µm is only about 3 µm, so an intrinsic layer only a few microns thick suffices for good quantum efficiency. With a drift velocity of $\sim$10^7 cm s^{-1}, a charge carrier transits this layer in ≤ 0.1 ns, which represents the fundamental timing uncertainty due to the absorption of photons at different depths in the intrinsic layer. The large and repeatable pulse from the ensuing avalanche provides a strong signal against which to make accurate timing determinations, so this fundamental timing limit is approached in practice.

Such devices essentially saturate on a single photon – once the diode has broken down, additional photons cannot produce any additional effect. For more information on detection rates, arrays of Geiger-mode diodes are used, with the outputs combined (i.e., the arrays are generally not used for areal resolution). The outputs are the sums of pulses from all of the diodes triggered, so there is a distinct pulse level for each number of photons detected within the timing interval. Spatial arrays have also been developed, based on a three-dimensional architecture that allows for time-stamping each individual event (Aull et al. 2018). Because of a general similarity in application to photomultipliers, these devices are called silicon photomultipliers (SiPM). Avalanche photodiodes are discussed in depth by Stillman and Wolfe (1977), Capasso (1985), Kaneda (1985), Pearsall and Pollack (1985), Cova et al. (1996), and, more recently, Hadfield (2009), Eisaman et al. (2011), Singh et al. (2011), and Aull et al. (2018).

3.1.5 Example

We consider a photodiode operating at 300 K, with an area of 1 mm^2, and at a wavelength of 1 μm. We dope the material with arsenic at 10^{15} cm^{-3} to make the n-type side and use a similar concentration of boron to make the p-type side. Compute I_0, R_0, and C_J, and estimate the quantum efficiency at 1 μm.

(a) I_0: From the discussion following equation 2.35, the conduction electron concentration in the intrinsic material is 1.38×10^{10} cm^{-3}. Equations 3.58 and 3.59 (in the Note) then give us $p_n = 1.9 \times 10^{11}$ m^{-3} and $n_p = 2.5 \times 10^{10}$ m^{-3}. From Table 2.3, $\tau \sim 10^{-4}$ s for impurity concentrations $\sim 10^{12}$ cm^{-3}. Assuming τ varies as N_I^{-1}, we get $\tau \sim 10^{-7}$ s. Taking mobilities from this same table and using equations 3.12 and 3.17, the diffusion coefficient for electrons in the p-type material is $D_p^n = 3.5 \times 10^{-3}$ m^2 s^{-1}, and for holes in the n-type material is $D_n^p = 1.24 \times 10^{-3}$ m^2 s^{-1}. These values yield $L_p^n = 1.87 \times 10^{-5}$ m and $L_n^p = 1.11 \times 10^{-5}$ m. From equation 3.55 , I_0 is then $qA(2.6 \times 10^{13}$ m^{-2} s$^{-1}) = 4.1 \times 10^{-12}$ A.

(b) R_0: From equation 3.25, $R_0 = 6.3 \times 10^9$ Ω.

(c) C_J: From equation 3.60 (in the Note), $V_0 = 0.58$ V. From Table 2.3, the dielectric constant for silicon is 11.8. Setting $V_b = 0$, we can use equation 3.33 to compute the width of the depletion region to be 1.23×10^{-6} m. Substituting into equation 3.34, the junction capacitance is found to be 85 pF.

(d) Quantum efficiency: Given a diffusion length of 19 μm for electrons and the absorption length of ~ 80 μm at a wavelength of 1 μm, it is clear that the quantum efficiency will be low, of order 10% (see Problem 3.2). For example, if the absorbing layer is set to one diffusion length, then $\tanh(c/L) = 0.76$. From reflection loss, $b = 0.70$. The generation rate times the absorbing layer thickness is 0.21 (assuming exponential absorption with absorption length 80 μm. These values yield $\eta = 11\%$.

3.2 Extrinsic Doped Silicon Detectors

3.2.1 *Impurity Band Conduction (IBC) or Blocked Impurity Band (BIB) Detectors*

We have seen how photodiodes solve the dilemma that the goals of low electrical conductivity and efficient absorption cannot be combined in a single detector component. However, we showed in Section 3.1.3.2 that photodiodes cannot operate on extrinsic absorption, i.e., for photodetectors with response outside the range accessible with high-quality intrinsic semiconductors. Fortunately, there is another approach that allows separating the detector layers for optimized electrical and optical properties in a detector operating by extrinsic absorption. A device built with this approach is the blocked impurity band (BIB) (or impurity band conduction, IBC – to be used hereafter) detector;[1] it is shown schematically in Figure 3.12 and described in detail by Szmulowicz and Madarsz (1987) (the original papers by Petroff and Stapelbroek from 1980–1985 are difficult to access).

3.2.1.1 *Basic Operation*

In an IBC detector, the absorbing, infrared-active layer is doped heavily to a level where impurity band conductivity through hopping would be completely unacceptable in a conventional extrinsic detector. In the following discussion, we will assume this layer is n-type; in fact, this assumption coincides with the most common type of IBC detectors, which are made of arsenic-doped silicon. Other possible dopants include antimony, used to construct detectors that respond to nearly 40 μm (van Cleve et al. 1994, 1995), phosphorus with response to about 34 μm, and gallium, responding to about 21 μm (Sclar 1984).

Returning to Si:As, we will assume that there is a low, but not negligible, concentration of a p-type dopant in the infrared-active layer. An additional thin, high purity layer, called the blocking layer, is grown over the front of the absorbing layer. This layer provides the large electrical resistance required for low light level operation. One electrical contact is made to the blocking layer, and the second contact is made to the back of the active layer. The details of the arrangement of the contacts

---

[1] Strictly speaking, blocked impurity band and BIB are trademarks of Rockwell International, where this detector type was invented. Therefore, a generic name "impurity band conduction (IBC)" is more frequently used.

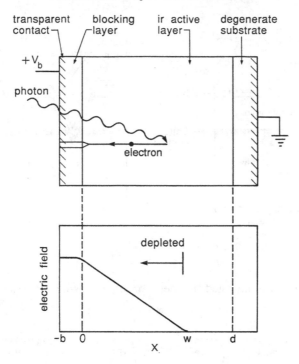

Figure 3.12 Cross-section of an IBC detector (*top*) and electric field within the detector (*bottom*).

depend on whether the detector is to be front illuminated through the first contact and blocking layer or back illuminated through the second contact. In the former case, a transparent contact is implanted into the blocking layer and the second contact is made by growing the detector on a heavily doped, electrically conducting (degenerate) substrate. In the latter case, a thin degenerate but transparent contact layer is placed underneath the active layer on a high purity, transparent substrate. These two geometries are illustrated in Figure 3.13.

For correct operation, a positive bias (relative to that on the active layer) must be placed on the blocking layer of an n-type IBC detector. If a negative bias were placed on the blocking layer, electrons from the contact would drift through the blocking and active layers to be collected on the opposite contact, producing a large dark current. Thus, the device is asymmetric electrically.

Under an appropriate level of positive bias, the detector operates as illustrated in Figure 3.14. Electrons in the impurity band under modest thermal excitation can hop toward the blocking layer, but their progress is stopped there unless they have sufficient energy to rise into the conduction band. When a photon is absorbed in the infrared-active layer, it raises an electron from the impurity band into the conduction band, where it is also attracted to the blocking layer. However, the

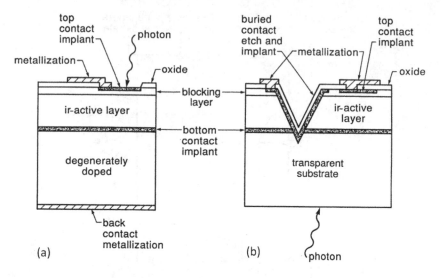

Figure 3.13 Front-illuminated IBC detector (a), compared with back-illuminated detector (b).

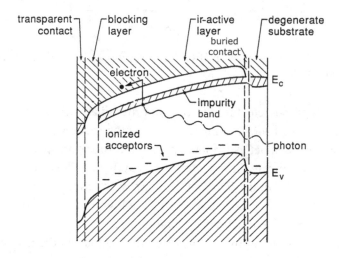

Figure 3.14 Band diagram for Si:As IBC detector.

conduction bands of the active and blocking layers are continuous, so the electron can pass unimpeded through the blocking layer to the contact. At the same time, the holes produced by absorption of a photon migrate to the opposite contact and are collected.

For this detection process to be efficient, there must be an electric field in the infrared-active layer sufficient to drive the photon-produced charge carriers to the blocking layer. Any thermally produced charge carriers must be driven out of the active layer to raise its impedance sufficiently that a significant field develops across it. The positive bias on the intrinsic layer contact drives free negative charge carriers out of the active layer by attracting them to the active/intrinsic layer interface, from where they can traverse the blocking layer and be collected at the electrode. Similarly, it drives any ionized donor sites toward the opposite contact. Consequently, the portion of the infrared-active region near the interface with the intrinsic layer is depleted of charge carriers. Because of the field across this high impedance depletion region, any photogenerated charge carriers within it will drift rapidly to the electrodes. Beyond the depletion region toward the negative electrode, the field is greatly reduced and the collection of free charge carriers is inefficient; some collection still occurs through diffusion from the site where a photoelectron was freed into the depletion region. That is, photons absorbed beyond the depletion region produce relatively little response. Since to first order only the depleted portion of the infrared-active layer is effective in detection, the quantum efficiency depends on the width of the depletion region, which in turn depends on the bias voltage and the density of negative space charge associated with the acceptors.

The width of the depletion region can be determined by calculating where in the detector the electric field becomes zero. To do so, it is necessary to solve Poisson's equation. The p-type impurities in the absorbing layer will be compensated by the arsenic, leaving a concentration of ionized acceptor atoms. These atoms will contribute a negative space charge; thus, Poisson's equation in the infrared-active layer is

$$\frac{d\mathcal{E}_x}{dx} = \frac{\rho}{\kappa_0 \varepsilon_0} = -\frac{q N_A}{\kappa_0 \varepsilon_0}, \tag{3.39}$$

where κ_0 is the dielectric constant and N_A is the density of ionized acceptors. We let x run from the blocking layer/infrared-active layer interface into the detector, assume a blocking layer devoid of impurities (see Figure 3.12), and recall that $dV/dx = -\mathcal{E}_x$. The solution for the thickness of the depletion region is

$$w = \left[\frac{2\kappa_0 \varepsilon_0}{q N_A} |V_b| + t_B{}^2 \right]^{1/2} - t_B, \tag{3.40}$$

where t_B is the thickness of the blocking layer and V_b is the bias voltage. To avoid the problems in reduced response from overcompensated material, it is desirable to keep N_A less than 10^{13} cm^{-3}. Adopting this value, $V_b = 4$ V, and $t_B = 4$ μm, we derive $w = 19.2$ μm.

If the thickness of the infrared-active layer is t_{IR}, the quantum efficiency will increase with increasing V_b until $w \geqslant t_{IR}$. The critical bias level at which $w = t_{IR}$ is termed V_{bC}, which can be shown from equation 3.40 to be

$$V_{bC} = \frac{qN_A}{2\kappa_0\varepsilon_0}\left(t_{IR}^2 + 2t_{IR}t_B\right).$$ (3.41)

If w is the width of the depletion region, the electric field strength in this region is

$$\mathcal{E}_x = \frac{qN_A}{\kappa_0\varepsilon_0}(w - x).$$ (3.42)

For our typical values ($N_A \sim 10^{13}$ cm^{-3}, $w = 19.2$ μm), the field strength near the blocking layer is large; for example, for $x = 1.2$ μm, $w - x = 18$ μm and the field is $\sim 2.8 \times 10^3$ V cm^{-1}, far larger than for the bulk photoconductors we have discussed previously. The field increases linearly across the depletion region, reaching a maximum at the interface with the blocking layer (see Figure 3.12).

The quantum efficiency is obtained by substituting w for d_1 into expressions like equation 1.18. A typical donor concentration is $N_D = 5 \times 10^{17}$ cm^{-3}; comparing with the absorption cross-section in Table 2.4, it can be seen for our example even with $w = 19$ μm that $\sigma_i N_D w = 2.12$, and the peak absorptive quantum efficiency will be about 88% for a single pass, or as high as 98% for a double pass in a detector with a reflective back surface. An inadequate depletion width (or too thin infrared-active layer) will, however, reduce the QE away from the peak toward short wavelengths and narrow the spectral response.

Figure 3.15 shows how these arguments are reflected in the behavior of actual detectors. The theoretical curve of response vs. bias matches the measurements reasonably well, particularly considering that the internal structure of the detector was treated in a simplified manner to avoid introducing too many free parameters.

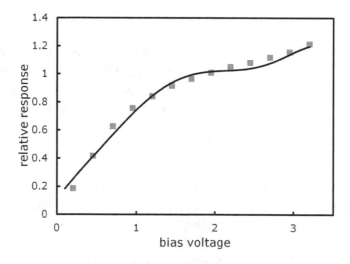

Figure 3.15 Responsivity as a function of bias for a state-of-the-art Si:As IBC detector (Rieke et al. 2015). The dots are measurements and the solid line is a model.

The calculations include the effects of diffusion, which were omitted in the treatment above and in all previous similar calculations for these devices. This process is an essential addition to achieve a good fit to the behavior at low bias voltage. In the model, the infrared-active layer was 35 μm thick with an As concentration of 7×10^{17} cm^{-3}, the acceptor level deduced from the model was 1.5×10^{12} cm^{-3}, and the blocking layer was 4 μm thick. The diffusion length for charge carriers was also deduced from the model to be 2.5 μm. The measured peak quantum efficiency of this detector is ∼75–80% even with an AR coating. The shortfall at the peak is because of losses at the buried contact and is encountered in the back-illuminated geometry used in all arrays of these detectors. At wavelengths shorter than 10 μm, there is a larger shortfall resulting from the drop in absorption coefficient at these wavelengths. However, this effect is partially compensated by a significant improvement in the transmission of the buried contact, and the devices have useful response between 5 and about 27 μm.

From zero to about 1.9 V bias the response steadily grows as an ever larger fraction of the infrared-active region is depleted as the bias increases. From ∼1.9 V to ∼2.5 V, the response is almost independent of bias, as the infrared-active layer is already fully depleted. Above 2.5 V, the responsivity again starts to increase as avalanche gain begins to play a role. The loss of response at low bias due to incomplete depletion of the infrared-active layer is the dominant source of nonlinearity in these detectors as used with conventional integrating amplifiers.

3.2.1.2 Operational Properties of IBC Detectors

Since the recombination in IBC detectors occurs in relatively low resistance material and is not a random process distributed through the high impedance section of the detector, there is only a single random event associated with the detection of a single photon. Consequently, the rms *G–R* noise current is reduced by a factor of $\sqrt{2}$ compared with conventional photoconductors.

A valuable characteristic of IBC detectors is that the infrared-active layer is so heavily doped that it can be made quite thin without compromising the quantum efficiency. Since the surface area of the active volume is small, these detectors can operate efficiently in environments where there is a high density of energetic ionizing particles. Another advantage of IBC detectors is that the infrared-active layers of these detectors need not have extremely high impedance, and hence dielectric relaxation effects are reduced substantially.

Because of the heavy doping of the infrared-active layer, the impurity band increases in width around its nominal energy level and the energy gap decreases between the conduction band (or valence band for p-type material) and the nearest portion of the impurity band. Consequently, the minimum photon energy required to excite photoconduction is lower for an IBC detector than for a conventional bulk

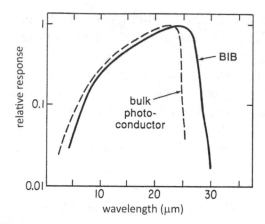

Figure 3.16 Comparison of spectral responses for a bulk Si:As photoconductor and a Si:As IBC detector.

photoconductor with the same dopant, and the spectral response extends to longer wavelengths than that of the conventional detector. This effect is illustrated in Figure 3.16. Because of the high impurity concentrations that can be achieved without degrading the dark current, an IBC detector can also be tailored to provide better quantum efficiency toward short wavelengths than is possible with a conventional detector.

It has been found that cosmic rays can damage the blocking layer crystal structure. If V_b and hence the field across this layer is large, a significant increase in dark current results. To reduce this problem, detectors must be manufactured with $N_A \lesssim 10^{12}$ cm^{-3}, which reduces the required field for a given maximum depletion depth. The use of such ultrapure material allows a number of other detector characteristics, such as temperature of operation, to be controlled in beneficial ways.

At high bias voltages, the photoelectrons in the region of large field can produce avalanche gain in a manner similar to that discussed in Section 3.1.4.2 for silicon avalanche photodiodes. The process can lead to a single photon producing a cascade of electrons entering the blocking layer. Unless the bias voltage is very large, the multiplication stops in the blocking layer. As a result, the charge multiplication is confined and controlled by the setting of the bias voltage and stable photoconductive gains $G > 1$ are produced. The amount of gain increases rapidly with increasing bias voltage up to breakdown. This behavior arises from the strong dependence of α, the impact-ionization coefficient, on the electric field strength. If the width of the gain region is u, the multiplication factor assuming a uniform field in the gain region is

$$M = e^{\alpha u}. \tag{3.43}$$

The field is not uniform, as shown in equation 3.42, so a rigorous solution requires that M be calculated with allowance for the dependence of field on x. It is frequently convenient to do such calculations by numerical simulation (Laviolette and Stapelbroek 1989). Continuing with our example, however, (i.e., $V_b = 4$ V), taking the field to be uniform and evaluating for the position $w - x = 18$ μm, we find $M \sim 2.2$.

Within the detector, the increase of the field near the interface between the infrared-active and blocking layers means that the amplification of the signal by this process occurs preferentially near the blocking layer, leading to designation of this portion of the infrared-active layer as the gain region (although there is no distinct boundary in the detector material; the dimensions of the gain region are set by the bias voltage). The gain process is relatively noisy because of the statistics of the interactions that determine the gain. Additional variations in gain can occur because the gain region is not sharply defined. Let the factor β be the amount by which the noise is increased relative to that for a noiseless gain mechanism. That is, β is the gain dispersion in the device, $\beta = \langle G^2 \rangle / \langle G \rangle^2$. Then the detective quantum efficiency is degraded by this factor:

$$DQE = \frac{\eta}{\beta}. \tag{3.44}$$

The photon-limited noise current then becomes

$$\langle I_S^2 \rangle = 2q^2 \, \varphi \frac{\eta}{\beta} \, (\beta G)^2 \, df. \tag{3.45}$$

The responsivity remains as in equation 2.13.

These results show the importance of (1) high doping in the IR-active layer; (2) making this layer as thick as possible; (3) applying sufficient bias voltage to deplete it completely; but (4) not applying so much voltage as to produce significant gain. These conditions can only be met with optimized processing of the detector material to minimize the minority impurity contamination of the infrared-active layer. Meeting these conditions can provide high sensitivity operation over a broad spectral range: state-of-the-art detectors when suitably AR coated can have quantum efficiency > 50% from ~5 to 25 μm (Ressler et al. 2008), which can simplify system design.

3.2.2 Solid State Photomultiplier

The above discussion suggests the further step of growing the detectors with a distinct and optimized gain region. The result of this concept is the solid state photomultiplier (SSPM) (Petroff et al. 1987). The performance of the SSPM is described by Hays et al. (1989) as follows. The spectral response is similar to that of a conventional Si:As IBC detector. However, a single detected photon produces an output pulse containing ∼40,000 electrons, which is easily distinguished from the electrical noise. These pulses have intrinsic widths of a few nanoseconds, although they are typically broadened by the output electronics to about a microsecond. Under optimum operating conditions, the dark pulse rate is ≤ 1000 s^{-1}; lower dark pulse rates can be obtained at reduced operating temperatures, at the expense of reduced quantum efficiency.

The internal structure of the detector is similar to a Si:As IBC device except that a well-defined gain region is grown between the blocking layer and infrared-absorbing layer with an acceptor concentration of $0.5{-}1 \times 10^{14}$ cm^{-3} and a thickness of about 4 μm. Since the infrared-absorbing region has a lower acceptor concentration, much of the electric field is developed across this gain region, so with an appropriate bias voltage strong avalanching can occur there. At a carefully optimized operating temperature, a controlled dark current is produced by field-assisted thermal ionization (the Poole–Frenkel effect) where the field is near maximum in the gain region. An equivalent current must flow through the infrared-active layer, where by Ohm's law it generates an electric field that pushes any photoexcited electrons into the gain region. Once there, these electrons avalanche to produce the output pulse of the device (this process is discussed by Laviolette and Stapelbroek (1989)). The dark current necessary for correct biasing of the infrared-absorbing region does not contribute to the detector noise (as it would in a conventional IBC detector with modest gain) because it does not produce pulses that could be confused with photon signals.

3.3 Problems

3.1 Show that the voltage across a photodiode held to zero current (that is, measured with a very high impedance voltmeter) is

$$V = \frac{kT}{q} ln \left(1 + \frac{\eta q P}{h \nu I_0} \right).$$

Compute the local voltage responsivity of the diode, dV/dP. For a diode with $I_0 = 10^{-6}$ A, $\eta = 0.5$, and $T = 300$ K, compare the local voltage responsivity for $P = 0$ and $P = 10^{-6}$ W and 10^{-5} W at a wavelength of 0.9 μm.

3.2 Assume that we are operating a photodiode at a wavelength where the absorption is so low that we can assume that the photoexcited charge carriers are produced uniformly throughout the material overlying the junction. The corresponding boundary condition at the surface of the diode is

$$\left(\frac{dn}{dx} \right)_{(x=c)} = 0.$$

Show that the quantum efficiency is

$$\eta = \frac{b \gamma c \, tanh(c/L)}{(c/L)},$$

where γ is interpreted to be the number of charge carriers generated per unit path per photon in the detector material.

3.3 Assume a photodiode with quantum efficiency η is operated at zero bias, receives φ photons s^{-1}, and has no $1/f$ or other excess noise. Derive a constraint on I_0 as a function of φ and η that will result in the diode operating with $DQE \geq 0.5\eta$.

3.4 Use equation 3.11 to demonstrate that the Fermi level must be constant across a junction in equilibrium.

3.5 Consider a junction with $N_A \gg N_D$ and $N_D(x) = Bx^\beta$. The depletion region will then extend from $x \sim 0$ to some distance $x = w$ within the n-type region. Show that the junction capacitance is

$$C_J = A \left[\frac{q B \varepsilon^{\beta+1}}{(\beta + 2)(V_0 - V_b)} \right]^{1/(\beta+2)}.$$

3.6 A one-sided, ideal silicon diode with $N_A \gg N_D$ is square and 100 μm on a side. Its capacitance as a function of bias voltage is measured as in the following table:

V_b (V)	C (pF)
0.500	2.72
0.000	2.00
−1.000	1.49
−2.000	1.26
−3.000	1.12
−4.000	1.025
−5.000	0.95

Determine the contact potential and the doping profile for the diode.

3.7 Redesign the silicon diode of the example for operation at 4 K (assume that freezeout of the junction is of no concern and that the recombination time goes as $T^{1/2}$ and the diffusion coefficient as T). Assume that the doping method gives reliable results only for $n \geq 6 \times 10^{13}$ cm^{-3}. Comment on any improvements and limitations this low operating temperature produces.

3.8 Design an arsenic-doped silicon photoconductor to operate near 20 μm at a temperature of 10 K. Assume transparent contacts, a sensitive area of 1 mm^2, an arsenic concentration of 10^{16} cm^{-3}, a mobility of 6×10^4 cm^2 V^{-1} s^{-1}

below 40 K, a recombination time of 3×10^{-9} s, and a maximum possible electric field before breakdown of 200 V cm^{-1}. Compute the following: the detector geometry that gives a thickness of one absorption length (quantum efficiency of 52% including reflection effects), and (assuming this geometry) the photoconductive gain, the responsivity, the detector resistance (assume $\delta = 1$), and the time response.

3.9 Comment on the expected performance of a Ge:Sb IBC detector. Assume the blocking layer is 4 μm thick, that the infrared-active layer is one absorption length thick, that the detector operates stably at a maximum bias of 30 mV, that the maximum permissible antimony concentration is 2×10^{16} cm^{-3}, and that the mean free path for conduction electrons is similar to that for silicon. Determine an optimum acceptor density, estimate the detector quantum efficiency, and discuss the ability of the detector to provide gain.

3.10 Consider two photoconductors of equal volume and cutoff wavelength, one intrinsic and the other extrinsic (n-type). Assume the intrinsic detector has no impurities and that the minority impurity concentration in the extrinsic detector is zero. Compare dark currents as a function of temperature. Assume $\delta = 1$ for the extrinsic detector, $m_n^* = 1.1\, m_e$ for both materials, and $N_D = 3 \times 10^{15}$ cm^{-3} for the extrinsic detector. Discuss your result.

3.4 Note

This note provides the details to derive the diode equation. Refer to Figure 3.17. We assume that the p-type material contains an adequate supply of conduction holes to supply any current of interest. (When the diode is strongly forward-biased this assumption may be invalid, but this condition is not of interest to us.) To carry a current, these holes must diffuse across the depletion region against the field set

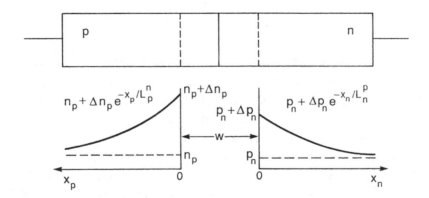

Figure 3.17 Carrier densities in a diode.

up by the contact potential; this situation was discussed in Section 3.1.3.1. In the present case, we can apply equation 3.8 for the positive charge carriers. We place the boundary of the p-type region at $x = 0$ and denote the density of free holes at this point as p_p; the density of free holes at the edge of the n-type region is represented by p_n. The terms p_p and p_n are commonly referred to as majority and minority charge carrier concentrations, respectively. These terms emphasize the difference in location of otherwise identical charge carriers; majority charge carriers are in a region of semiconductor where the doping favors generation of similar carriers, and minority carriers are in semiconductor that is doped to favor generation of carriers of opposite type.

Applying equation 3.8, we find that the ratio of hole concentration at the boundary of the depletion region in the p-type material to that at the boundary in the n-type material is

$$\frac{p_p}{p_n} = e^{qV_0/kT}. \tag{3.46}$$

By the same argument, if a bias V_b is applied to the junction, we obtain

$$\frac{p_p^b}{p_n^b} = e^{q(V_0-V_b)/kT}, \tag{3.47}$$

where the b superscript denotes the carrier concentrations with the bias voltage present. To first order, the bias voltage has little effect on the majority carriers, so $p_p^b = p_p$. Then, using equation 3.46, we get

$$p_n^b = p_n e^{qV_b/kT}, \tag{3.48}$$

and

$$\Delta p_n(0) = p_n^b - p_n = p_n(e^{qV_b/kT} - 1), \tag{3.49}$$

where $\Delta p_n(0)$ is the change in the conduction hole concentration at the boundary of the depletion region that results from applying the bias voltage.

To generate a current, the charge represented by equation 3.49 must diffuse through the remaining n-type material in the diode and reach the contact. From equation 3.16 (and as illustrated in Figure 3.17),

$$\Delta p_n(x_n) = \Delta p_n(0)e^{-x_n/L_n^p}, \tag{3.50}$$

where L_n^p is the diffusion length for holes in the n-type material. From equation 3.9, the osmotic flux of holes in the n-type material is $-D_n^p d(\Delta p_n)/dx$, which produces a current of

$$I_p(x_n) = -qAD_n^p \frac{d[\Delta p_n(x_n)]}{dx_n}, \tag{3.51}$$

where A is the junction area. Using equations 3.49, 3.50, and 3.51,

$$I_p(x_n) = \frac{qAD_n^p}{L_n^p} \, \Delta p_n(x_n), \tag{3.52}$$

and

$$I_p(0) = \frac{qAD_n^p}{L_n^p} \, p_n \, (e^{qV_b/kT} - 1). \tag{3.53}$$

Similarly for the electron component of the current

$$I_n(x_p = 0) = -\frac{qAD_p^n}{L_p^n} \, n_p \, (e^{qV_b/kT} - 1). \tag{3.54}$$

Remembering (Figure 3.17) that x_n and x_p run in opposite directions, the total current through the junction is $I = I_p - I_n$, or

$$I = \left[qA \left(\frac{D_n^p}{L_n^p} \, p_n + \frac{D_p^n}{L_p^n} \, n_p \right) \right] (e^{qV_b/kT} - 1)$$
$$= I_0(e^{qV_b/kT} - 1), \tag{3.55}$$

where I_0 (in brackets) is the saturation current. Equation 3.55 is called the diode equation and gives an approximate quantitative description of the behavior illustrated in Figure 3.4 (except for breakdown).

To make use of this expression, we need to know how to evaluate the various terms in I_0. The diffusion coefficients can be taken from equation 3.12 and the diffusion lengths from equation 3.17. The minority carrier concentrations, that is p_n and n_p, cannot be obtained directly from equations 2.35 and 2.37. They can be derived, however, by first observing that the product of the charge carrier concentrations, $n_0 p_0$, can be written using equations 2.35 and 2.37 as

$$n_0 p_0 = N_c e^{-E_g/2kT} N_v e^{-E_g/2kT} = n_i p_i = n_i{}^2, \tag{3.56}$$

where n_i and p_i are the conduction electron and hole concentrations in the intrinsic material, as indicated by the subscript i. The first equality in equation 3.56 is important because it shows that the dependencies on the Fermi energy in equations (2.35) and (2.37) drop out in the product $n_0 p_0$; that is, this product is independent of any doping of the material that might shift E_F. Therefore, we can associate $n_0 p_0$ with the product of majority and minority carrier densities, that is, $n_n p_n$ or $n_p p_p$. The last equality in equation 3.56 is true because conduction electrons and holes are created in pairs in intrinsic material, allowing us to set $n_i = p_i$. The carrier concentrations in intrinsic material are given by equations 2.35 and 2.37, again with $E_c - E_F = |E_v - E_F| = E_g/2$; they are therefore readily calculated.

Equation 3.56 is referred to as the carrier product equation and is a powerful tool in deriving carrier concentrations. It shows how the minority carrier concentration is suppressed due to the capture of majority carriers by the minority impurity atoms. In particular, the minority carrier concentrations in equation 3.55 can be derived from the majority carrier concentrations by substituting $n_n p_n$ or $n_p p_p$ for $n_0 p_0$, i.e.,

$$p_n = \frac{n_i^{\,2}}{n_n} \quad \text{and} \quad n_p = \frac{n_i^{\,2}}{p_p}. \tag{3.57}$$

Let the dopant concentration in the n-type material be N_D (cm^{-3}) and that in the p-type material N_A (cm^{-3}). Assume the diode is operated at a high enough temperature that all the impurity atoms are ionized. Hence, ignoring compensation by trace impurities, the concentration of majority charge carriers equals the concentration of dopants; for example, $n_n \sim N_D$. In this case we have, from equations 2.35, 3.56, and 3.57,

$$p_n = \frac{\left(N_c \, e^{-E_g/2kT}\right)^2}{N_D}. \tag{3.58}$$

A similar expression can be derived for n_p:

$$n_p = \frac{\left(N_v e^{-E_g/2kT}\right)^2}{N_A}. \tag{3.59}$$

It will also be useful to be able to estimate the contact potential, V_0. From equation 3.46,

$$V_0 = \left(\frac{kT}{q}\right) ln\left(\frac{p_p}{p_n}\right), \tag{3.60}$$

and we have already seen how to obtain all the quantities that are needed to evaluate this expression.

3.5 Further Reading

Capasso (1985), Kaneda (1985), and Pearsall and Pollack (1985) – three articles from a single volume of reviews that emphasize avalanche photodiodes, but cover general principles as well

Eisaman et al. (2011) – review giving overview of single-photon detection approaches

Hadfield (2009) – review of a broad variety of optical single-photon detectors, including APDs

Lei et al. (2015) – review of status and future development of HgCdTe detectors

Rieke (2007) – review of modern infrared detector arrays, still reasonably current

Rogalski, A. (2012) – very broad overview of semiconductor detectors

Schühle and Hochedez (2013) – useful overview of detectors for the ultraviolet based on
　　large bandgap semiconductors
Singh, Srivastav, and Pal (2011) – review of HgCdTe avalanche photodiodes
Szmulowicz and Madarsz (1987) – one of the few detailed descriptions of IBC detector
　　principles in the literature

4

Amplifiers, Readouts, and Arrays

To be useful, the output signal from any of the detectors we have discussed must be processed by external electronics. Conventional electronics, however, are not very well suited for an infinitesimal current emerging from a device with virtually infinite impedance. Nonetheless, highly optimized circuit elements have been developed to receive this type of signal and amplify it. Most of these devices are based on very high input impedance, low noise amplifiers that can be built with field effect transistors (FETs). FETs are used in a variety of circuits that are constructed not only for low noise (usually quoted in terms of the uncertainty in the output in units of electrons into the amplifier), but also to give the desired frequency response and to accommodate the electrical properties and operating temperature of the detector, among other considerations. In the most sensitive circuits, signals of only a few electrons can be sensed reliably. Infrared detector arrays (and some operating in the visible) are built with a FET-based amplifier for each detector. Although the amplifiers are fabricated on silicon wafers, the infrared detectors are usually fabricated on wafers of some other material, so the two must be joined by depositing tiny bumps of indium solder on the detector outputs and amplifier inputs and joining the two wafers by (very carefully) pressing these bumps together. A class of high energy X-ray imager uses a similar approach. An important type of visible-wavelength detector array is closely related, but the silicon-based detectors can be produced together with the FET amplifiers on the same wafer.

4.1 Building Blocks

Amplifiers for the detector signals could be built using two basic kinds of transistor: bipolar junction transistors (BJTs) and field effect transistors (FETs). BJTs are generally unsuitable for directly receiving the signal from the detectors we have been discussing because they have relatively modest input impedances. FETs are used to build first-stage electronics for virtually all high sensitivity detectors.

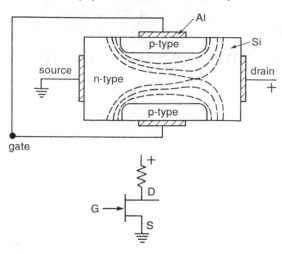

Figure 4.1 Principle of operation of an n-channel junction field effect transistor.

There are two basic classes of FET: the junction field effect transistor (JFET) and the metal–oxide–semiconductor field effect transistor (MOSFET). Although they operate by rather different means, a common terminology is used to discuss their performance. Both FET types have an electrically conducting channel through which current flows from the source terminal to the drain terminal. The current is controlled by an electric field that is established by the voltage on a third terminal placed between the source and drain, the gate.

In the JFET, two diodes are grown back-to-back, and the current is conducted along the common side of the junctions. Figure 4.1 shows a schematic n-channel JFET along with its simplified circuit diagram; a p-channel device reverses the positions and roles of the p- and n-doped materials. The operation of the JFET depends on the very high impedance that is obtained in the depletion region of a diode, a phenomenon that was discussed in the preceding chapter. As the gate voltage is increased to raise the back-bias on the junctions, the depletion regions will grow as shown by the progression of dashed contours in Figure 4.1. When the gate voltage is large enough to cause the depletion regions to join, the current flow is strongly impeded and the transistor is said to be pinched off. If the gate voltage is reduced slightly below this level, a high degree of control can be exercised over the current.

In an enhancement mode n-channel MOSFET, two diodes are formed by implanting an n-type dopant into a p-type substrate, as shown in Figure 4.2. A p-channel MOSFET reverses the positions and roles of the p- and n-doped materials. The metal gate electrode is evaporated onto an insulating layer of SiO_2. Normally, the device is pinched off because the two diodes are back-to-back and

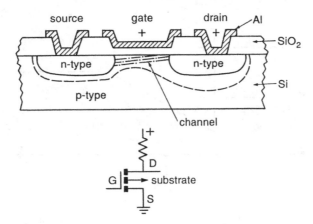

Figure 4.2 Principle of operation of an n-channel enhancement mode metal–oxide–semiconductor transistor.

there is no continuous path of n-type material between the source and the drain. If, however, a positive voltage is put on the gate, it tends to attract negative charge carriers to the underside of the insulator. If the voltage is strong enough, an n-type channel forms, and current is able to flow through the continuous n-type path from the source to the drain. The size of the channel, and therefore the current, is controlled by the size of the voltage (as indicated by the dot-dash lines). There are a number of other types of MOSFET that are variations on this theme. For more detailed discussion of FETs, consult Pierret, (1996) or Sze (2000).

The two basic types of FET have quite different uses. JFETs have very low noise, in part because the current through them does not make contact with an oxide layer where there are many traps due to dangling crystal bonds. The simplest MOSFETs such as in Figure 4.2 convey the current in contact with the insulating oxide layer, i.e., they are "surface channel." MOSFETs can be constructed with the doping adjusted so the current is not in contact with the oxide layer – "buried channel" devices, which have reduced $1/f$ noise compared with surface channel ones. Surface channel MOSFETs are readily constructed in large numbers and complex circuits, since their architecture can be produced with straightforward planar processing. They are central in the readouts for arrays of detectors, but also in nearly every other modern electronic device.

After the detector signal has been amplified by a first-stage FET, it is usually passed on to additional amplifiers. A generic form is the operational amplifier, or op-amp. This designation is left over from the early days of electronic computing, when analog computers were built using amplifiers to simulate various mathematical operations. Although a modern op-amp is a complex integrated circuit, in most situations it can be treated as a single circuit element: a high voltage gain,

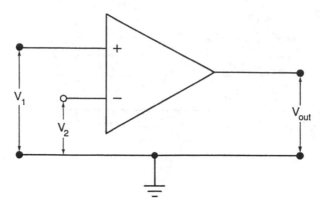

Figure 4.3 Representation of an operational amplifier.

direct current (DC) amplifier with high input impedance. It is convenient to denote these amplifiers by a single symbol (that is, we will not try to show the individual electronic components) as in Figure 4.3. The output voltage, V_{out}, is proportional to the difference between the two input voltages, V_1 and V_2. Note that the "−" (inverting) and "+" (noninverting) symbols on the input terminals designate only that the output voltage changes respectively either in the opposite or in the same direction as the input voltage applied to that terminal. They indicate nothing about the relative voltage difference across the two input terminals. In the uses described below, high input impedance and low noise are achieved by using FETs as the first electronic stage, either as part of the op-amp or as a separate input stage that can be cooled along with the detector and placed very close to it to minimize the length of wiring that must carry high impedance signals. In Figure 4.3, the input FETs have been subsumed into the symbol for the op-amp. Op-amps and other electronic circuit issues are discussed in many places, for example by Horowitz and Hill (2015).

4.2 Load Resistor and Amplifier

The simplest way to measure the signal from a detector is to place the detector in a voltage divider with a resistor, pass a current through it, and measure the voltage developed across it (see Figure 4.4). In this case, the voltmeter impedance is in parallel with the detector impedance; care must be taken that reduced signals and excess noise do not result from use of a low impedance voltmeter. Fortunately, FET input transistors can provide very high input impedances. This arrangement can be satisfactory in situations where only modest performance is required, or for semiconductor bolometers discussed in Chapter 8.

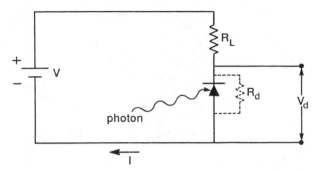

Figure 4.4 Simple readout circuit using a load resistor and sensing the voltage across the detector directly.

However, when we want reasonably good performance with photoconductors and photodiodes without the expense and complication of operating them at very low temperatures, we require (see equation 2.25) a high detector impedance to reduce the Johnson noise. In the circuit of Figure 4.4, it is necessary that $R_L \geqslant R_d$ to get a reasonably high signal ($R_d \gg R_L$ would lead to an output voltage barely varying from the bias voltage, V). The large resistances required can lead to undesirably slow response; as an example, for a photodiode with $R_0 > 10^{12}$ Ω and capacitance > 1 pF, the time response will be > 1 s. A second difficulty is that this circuit does not let us establish and maintain the detector bias independent of the signal. This situation is obviously a problem for a photodiode, where we wish to set the operating point at the optimal position on the I–V curve (Section 3.1.3.4); it can also cause difficulties with other detectors. Finally, for both photodiodes and photoconductors, the detector nonlinearities are passed through the circuit without correction.

4.3 Transimpedance Amplifier (TIA)

4.3.1 Basic Operation

The shortcomings of the simple load resistor circuit for use with high impedance detectors are dealt with simply and elegantly by the transimpedance amplifier (TIA) circuit shown in Figure 4.5. In the TIA, the detector is connected to the inverting ($-$) input terminal of the op-amp, and the output of the op-amp is connected (via R_f) to the same input terminal. The current through R_f opposes the detector current, a situation termed "negative feedback." The output of the op-amp is proportional to the difference of the input voltages, $V_{out} \propto V_+ - V_-$, but with a large gain applied. Therefore, with negative feedback the amplifier drives itself to minimize the voltage difference between its two inputs, $V_+ - V_- \approx 0$. The amplifier output

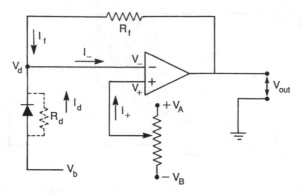

Figure 4.5 Transimpedance amplifier.

required to balance the two inputs provides a measure of the input current from the detector. The noninverting input is used for an adjustable voltage that allows the setting of V_{out} to a desired level in the absence of signal.

Assume for now that V_+ is near ground ($V_+ \approx 0$), in which case $V_- \approx 0$ and $V_d \approx 0$. Because the op-amp has such high gain and input impedance, it can be driven by very small voltages and currents. From Kirchhoff's current law, we know that $I_d + I_f = I_-$, and, because I_- is so small, we can say that $I_d + I_f \approx 0$. If we assume that both R_f and the detector are purely resistive, then it follows from Ohm's law that

$$\frac{V_{out} - V_d}{R_f} = -\frac{V_b - V_d}{R_d},\tag{4.1}$$

which becomes

$$V_{out} = -V_b \frac{R_f}{R_d},\tag{4.2}$$

remembering our assumption that $V_d \approx 0$.

Another way to express the amplifier output is

$$V_{out} = I_f R_f = -I_d R_f.\tag{4.3}$$

This last equation demonstrates one of the key virtues of the TIA: the output voltage depends on the feedback resistance and not on the electrical impedance of the detector. Thus, if the detector current is linear with photon flux, the amplifier output will be linear to the (usually high) degree of linearity that can be achieved for the feedback resistor. The condition on linearity of I_d is met by photodiodes according to equation 3.28, for Si IBC detectors if they are in the realm of nearly constant responsivity as shown around a bias voltage of 2 V in Figure 3.15, and it is also met for photoconductors (see equation 2.14) as long as their photoconductive gains are

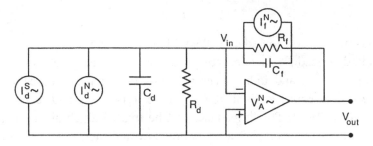

Figure 4.6 Equivalent circuit for the transimpedance amplifier.

independent of the photon flux (but note, for example, the discussion on dielectric relaxation in Section 2.2.3).

The bias on the detector can be set by adjusting V_+ and/or V_b. The feedback of the amplifier automatically maintains V_- at the set point, thus stabilizing the detector bias. This feature can be essential to maintaining the detector current linearity that was assumed above, as well as to minimizing some of the undesirable detector properties we have discussed for extrinsic photoconductors. In particular, the TIA allows operating photodiodes as a current source into approximately zero bias, which both maintains the detector linearity and also suppresses certain types of voltage-dependent noise.

4.3.2 Time Dependencies and Frequency Response

To understand the behavior of the TIA circuit in more detail, we must take into account the nonresistive properties of the circuit elements. To do so, it is convenient to replace the circuit to be analyzed with an electronically equivalent one made up of elementary components such as resistors and capacitors in place of more complex ones such as photon detectors. A Thévenin equivalent circuit is based on voltage sources with series resistances and a Norton equivalent circuit on current sources with parallel resistances. A Norton circuit for the TIA is shown in Figure 4.6.

In this case, a current source produces the detector signal, I_d^S, and a second source produces the equivalent noise current, I_d^N. The capacitance and resistance are represented by the discrete elements C_d and R_d for the detector and C_f and R_f for the feedback resistor, which has associated current noise I_f^N. The intrinsic amplifier noise voltage (that is, the amplifier noise voltage prior to modification by external components) is represented by V_A^N. For the time being, set I_d^N, I_f^N, and V_A^N to zero to compute the response to signals. The input voltage to the op-amp is then

$$V_{in} = I_d{}^S Z_d, \tag{4.4}$$

where Z_d is the detector impedance. We will next determine Z_d.

A standard procedure in alternating current (AC) electrical circuit theory is to treat capacitances and inductances in terms of complex numbers. Thus, the impedance of the capacitor is $1/j\omega C_d$, where j is $\sqrt{-1}$ and ω is the angular frequency ($\omega = 2\pi f$). This impedance can be treated algebraically exactly as if it were a complex resistance. That is, Ohm's law and the rules for computing the net effects of combinations of circuit elements can be applied in exactly the same form as would be used with purely resistive circuits. (Similarly, inductances have an impedance of $j\omega L$; we will not deal with them here but will in later chapters.) Since the detector resistance and capacitance are in parallel,

$$Z_d{}^{-1} = R_d{}^{-1} + \left(\frac{1}{j\omega C_d}\right)^{-1} = R_d{}^{-1} + j\omega C_d, \tag{4.5}$$

or

$$Z_d = \frac{R_d}{1 + j\omega R_d C_d}. \tag{4.6}$$

Since the feedback impedance and detector impedance act as a voltage divider between V_{out} and ground, and since we can assume that $V_{in} \ll V_{out}$, the output voltage is

$$V_{out} = -V_{in}\frac{Z_f}{Z_d}, \tag{4.7}$$

where

$$Z_f = \frac{R_f}{1 + j\omega R_f C_f}. \tag{4.8}$$

Equation 4.7 is the more general form of equation 4.2. Substituting into it from equations 4.4 and 4.8, and multiplying it by $(1 - j\,\omega\tau_f)/(1 - j\,\omega\tau_f)$ to put it into a convenient form, we obtain

$$V_{out} = -I_d^S Z_f = -\frac{I_d^S R_f}{1 + j\omega\tau_f}$$

$$= -\frac{I_d^S R_f}{1 + \omega^2\tau_f^2} + j\,\frac{\omega\tau_f I_d^S R_f}{1 + \omega^2\tau_f^2}, \tag{4.9}$$

where $\tau_f = R_f C_f$. The amplitude of V_{out} is its absolute value,

$$|V_{out}| = (V_{out}V^*{}_{out})^{1/2} = \frac{|I_d^S|R_f}{\left(1 + \omega^2\tau_f^2\right)^{1/2}}. \tag{4.10}$$

The signal cutoff frequency f_f is the frequency at which the signal has decreased by a factor of $\sqrt{2}$, or from equation 4.10,

$$f_f = \frac{\omega_f}{2\pi} = \frac{1}{2\pi R_f C_f} = \frac{1}{2\pi \tau_f}. \tag{4.11}$$

The imaginary part of equation 4.9 gives the phase of V_{out}. This parameter can be ignored for our purposes, although it is quite important in a more complete treatment of the circuit. For example, excessive phase shifts can cause the TIA output to be unstable.

Notice that the output voltage is independent of R_d and C_d. As we found for the simpler analysis, as long as R_f is linear and I_d^S is a linear function of the photon flux, φ, the detector/amplifier system is linear. Moreover, even if $R_d C_d = \tau_d \gg \tau_f$, the frequency response of the system is determined by τ_f, the limit imposed by the feedback resistor. As the detector impedance decreases with increasing frequency, the TIA gain increases to compensate (equation 4.7). The high frequency response is improved substantially for photodiodes because the capacitance of the feedback resistor can be made much smaller than that of the detector.

Now we set I_d^S to zero and use I_d^N, V_A^N, and V_f^N to compute the noise, where V_f^N is the voltage noise of the feedback resistor. The equivalent noise of the amplifier, V_A^N, is subject to the amplifier gain, That is, from equation 4.7,

$$\left\langle \left(V_{out,A}^N \right)^2 \right\rangle^{1/2} = \left\langle \left[V_A^N (f) \right]^2 \right\rangle^{1/2} \left(\frac{R_f}{R_d} \right) \left[\frac{1 + j\omega\tau_d}{1 + j\omega\tau_f} \right], \tag{4.12}$$

where we indicate explicitly that V_A^N is likely to have its own frequency dependence (for example, $1/f$ noise), which is modified by the external time constants τ_d and τ_f. The noise voltage amplitude is

$$\left| \left\langle \left(V_{out,A}^N \right)^2 \right\rangle^{1/2} \right| = \left\langle \left[V_A^N (f) \right]^2 \right\rangle^{1/2} \left(\frac{R_f}{R_d} \right) \left(\frac{1 + \omega^2\tau_d^2}{1 + \omega^2\tau_f^2} \right)^{1/2}. \tag{4.13}$$

Equation 4.13 shows that, although the time constant for signals is reduced by the TIA compared with that of the detector alone, in the interval of time response, $\tau = 1/\omega$, with $\tau_f < \tau < \tau_d$, the amplifier noise is boosted with increasing frequency. τ_d is the frequency at which amplifier noise has been boosted by $\sqrt{2}$ (assuming that $\tau_f \ll \tau_d$); that is,

$$f_d = \frac{1}{2\pi R_d C_d} = \frac{1}{2\pi \tau_d}. \tag{4.14}$$

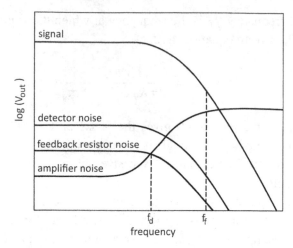

Figure 4.7 Frequency dependence of signal and noise for the TIA.

The current noise from the detector is amplified in exactly the same way as the signal current, so, following equation 4.10, the output voltage due to it is

$$\left|\left\langle\left(V_{out,d}^N\right)^2\right\rangle^{1/2}\right| = \left\langle\left[I_d^N\left(f\right)\right]^2\right\rangle^{1/2} \frac{R_f}{\left(1+\omega^2\tau_f^2\right)^{1/2}}. \tag{4.15}$$

Again, I_d^N contains its own frequency dependencies such as $1/f$ noise. For an ideal feedback resistor, the current noise I_f^N is Johnson noise; converting from current to voltage noise,

$$\left\langle\left(V_f^N\right)^2\right\rangle^{1/2} = \frac{\left\langle\left(I_f^N\right)^2\right\rangle^{1/2}R_f}{\left(1+\omega^2\tau_f^2\right)^{1/2}} = \left(\frac{4kT R_f\, df}{1+\omega^2\tau_f^2}\right)^{1/2}. \tag{4.16}$$

The total output voltage noise is given by the square root of the quadratic sum of the contributions from three sources: amplifier, detector, and feedback resistor. Thus

$$\left\langle\left(V_{out}^N\right)^2\right\rangle^{1/2} = \left[\left\langle\left(V_{out,A}^N\right)^2\right\rangle + \left\langle\left(V_{out,d}^N\right)^2\right\rangle + \left\langle\left(V_f^N\right)^2\right\rangle\right]^{1/2}. \tag{4.17}$$

The dependencies of signal and noise on frequency are shown in Figure 4.7 for the case where V_A^N and I_d^N are independent of frequency, and the feedback resistor contributes only Johnson noise. The boost in high frequency response applies to the noise as well as the signal; it is useful in maintaining fidelity to the input signal but usually does not improve the intrinsic signal-to-noise ratio. This figure illustrates the importance of limiting the frequency response of the circuitry to the region where a high signal-to-noise ratio is achieved. If high frequency noise is admitted

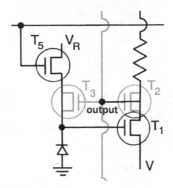

Figure 4.8 Simple source follower integrating amplifier. The source follower FET (T_1) and reset switch (T_5) are the basic amplifier, with the detector shown as a diode. The FET bias is V, and V_R is the reset level, i.e., the bias voltage put on the detector (give or take small offset voltages in the circuit). When used in a detector array, additional switching FETs shown in grey are used to address the amplifier to read its output and enable resetting it.

into the signal chain, it can dominate the signal. Electronic filtering can be used to avoid this problem.

Further discussion of TIAs can be found in Wyatt, Baker, and Frodsham (1974), Hall et al. (1975), Rieke et al. (1981), and Graeme (1996).

4.4 Integrating Amplifiers

A description of the operation of a FET that is equivalent to the one at the beginning of this chapter is that an electric charge, Q, deposited on the capacitance at the gate, C_g, establishes a voltage, $V = Q/C_g$, that controls the current through the channel of the FET. The value of the voltage, and hence the charge on the gate, can be read out at will by monitoring this current. This way of thinking about FETs is appropriate when they are used in integrating amplifiers.

4.4.1 Simple Source Follower Integrators

A simple source follower integrating amplifier (Figure 4.8) is a direct implementation of this concept. It can provide better signal-to-noise ratios than can a TIA when observing signals with long time constants. This benefit comes at the price of a relatively complex set of electronic circuitry (and/or software) to convert the signals to an easily interpreted output format.

Referring to Figure 4.9, in the operation of this amplifier charge leaks from the bias supply through the detector to be deposited on a storage capacitor, C_s.

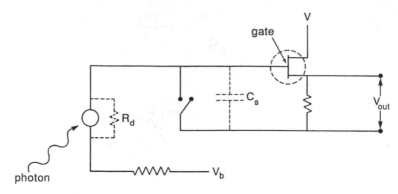

Figure 4.9 Schematic operation of a simple source follower integrating amplifier.

In practice, rather than a discrete capacitor as shown, the storage is on the combined capacitance of the MOSFET gate and the detector. Minimizing the integrating capacitance is usually a goal to produce more voltage for a given charge and hence to maximize the signal to noise. The leakage rate is modulated by the detector impedance, which is a function of the photon flux, φ. Charge can be accumulated noiselessly on the capacitor and read out and reset, ideally whenever enough signal has been collected that it stands out above the amplifier noise. Another limit, however, is when the accumulated charge puts the FET into saturation or some other non-optimum mode. This amount of charge is called the well depth, and a reset must be carried out before this number of charges is exceeded.

The source follower FET both collects the charge and produces a relatively low impedance output signal that matches well to the following electronics. It can have large enough gate (input) impedance, particularly when operated at low temperatures, that its primary effect on the detector side of the circuit is to add the gate capacitance in parallel to the capacitance of the detector and reset switch. For convenience, we define C_s to be this total capacitance. The current in the channel of the FET is a measure of the charge collected on C_s. The uncertainty in the measurement of the charge is typically expressed in electrons and is called the read noise, Q_N.

This type of amplifier is used in the majority of detector arrays because it has a minimum number of transistors (important, for example, if the amplifier has to hide behind the detector area) and minimum power dissipation. This latter advantage arises because the integration process is passive – it does not require that any of the FETs be operational. This enables operating a detector in long integrations with virtually no power, then applying power to the circuit and reading the accumulated charge, after which either the reset switch can be closed to start a new integration,

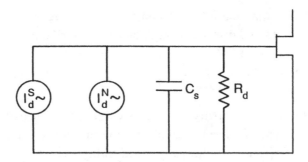

Figure 4.10 Equivalent circuit for a simple source follower integrating amplifier.

or power to the circuit can be turned off to continue the integration (This latter approach is described as a "nondestructive read.")

The bug that accompanies this feature is that the detector bias voltage decreases as signal is accumulated, leading to nonlinear response. For photodiode detectors, as the bias voltage decreases the diode capacitance rises (equations 3.33 and 3.34), reducing the incremental voltage change for additional collected charge. For a Si:As IBC detector, the reduction in detector bias voltage reduces the depletion depth in the infrared-active layer, resulting in reduced quantum efficiency and responsivity (the region below $\sim$1.7 V in Figure 3.15). However, these nonlinearities are simple in character and can be corrected accurately in post-detection software, so this issue is not crippling. The maximum signal where these corrections are still acceptably accurate defines the total charge that can be accumulated before there should be a reset. A more fundamental limitation arises with detectors that can tolerate only a small bias voltage; for them, there is a potentially serious limit in the total charge that can be collected before it is necessary to close the reset switch and restore the bias to its nominal value.

We would like this collected charge to be proportional to the time-integrated signal current from the detector. To derive the conditions required for this proportionality to hold, refer to the equivalent circuit in Figure 4.10, where we show the input to the FET in terms of the equivalent signal generator I_d^S in the detector, the detector resistance R_d, and the storage capacitor C_s. Assume that the detector produces a constant current, I, beginning at time $t = 0$. By integrating the time response of the RC circuit,

$$V_g = \int_0^t V_g = \int_0^t V_0 \, e^{-\frac{t'}{\tau_{RC}}} \, d\left(\frac{t'}{\tau_{RC}}\right), \tag{4.18}$$

we find that the voltage on the gate of the FET will be

$$V_g = V_0(1 - e^{-t/R_d C_s}). \tag{4.19}$$

For times much less than the RC time constant of the circuit, we can expand the exponential in a series and discard the high order terms to find

$$V_g \approx V_0 \left[\left(\frac{t}{R_d C_s} \right) - \frac{1}{2} \left(\frac{t}{R_d C_s} \right)^2 + \cdots \right], \quad t \ll R_d C_s. \qquad (4.20)$$

The circuit has the desired property of linearity within a tolerance of $(1/2)(t/R_d C_s)^2$ (recall that the last term retained in an alternating infinite series is larger than the net contribution of all the remaining discarded terms). The form of this tolerance shows why a high detector impedance is generally required to achieve successful operation in an integrating amplifier. When R_d is very large, the gate voltage is linear within a tolerance that goes as R_d^{-2}.

A modification to the integrating amplifier, called direct injection, can reduce the bias changes. A MOSFET is placed between the detector and integrating capacitor; the photocurrent passes through the channel of the MOSFET. The FET isolates the capacitor and its voltage swings from the detector. Direct injection amplifiers work best at medium to high signal levels but not at low ones because of a number of detrimental effects of the FET on the noise and bias stability at low signal currents. There are a number of other possible integrating amplifier designs, but they are not used in the low signal applications to be discussed here.

4.4.2 Capacitive Transimpedance Amplifier (CTIA)

There is a class of circuit that integrates but still maintains the detector bias as well as providing intrinsically linear output: the capacitive transimpedance amplifier, or CTIA, shown in Figure 4.11. For practical reasons, a simplified version of the op-amp is often employed with only two FETs, i.e., requiring one more than the simple source follower. The CTIA is clearly the readout circuit born with a silver spoon in its mouth. Here, however, as in many other books with mediocre plot lines, one should not be surprised to discover that the CTIA may not entirely live up to our expectations when turned out into the real world. For example, power must be supplied continuously to a CTIA to maintain the feedback that maintains the detector bias. Powered-up FETs can emit low levels of light that are seen by the detectors, creating a false signal and the attendant noise. The overall result is that the CTIA takes a bit more real-estate (three FETs instead of two), requires significantly more power than the source follower integrator, and is challenging to implement while shielding the detectors from its emission. However, it can provide an advantage in the specific situation where the detector requires a small and stable bias voltage.

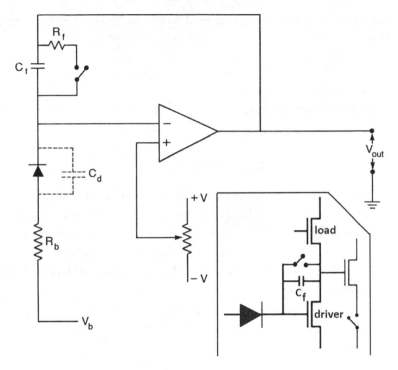

Figure 4.11 Capacitive transimpedance amplifier. The inset shows how the CTIA might be implemented for a detector array to minimize the transistor count. Inset after Hoffman et al. (2005).

We will use the full diagram in Figure 4.11 (rather than the simplified two-FET form) to analyze the circuit. Referring to this figure, a current can be generated from the feedback capacitor to the input of the amplifier by varying the output voltage of the amplifier appropriately,

$$I_f = C_f \frac{dV}{dt}. \tag{4.21}$$

As a result, the amplifier can balance itself by varying its output to produce an I_f that equals the output current of the detector, I_d. The output voltage of the amplifier is then proportional to the total charge generated by the detector, while the input voltage of the amplifier, that is, the detector bias, is maintained. The system is reset by closing a switch around the feedback capacitor, thus discharging it and (momentarily) establishing a conventional TIA such as we analyzed earlier. We already know that the TIA will drive itself to a stable set point, so the circuit will re-establish itself for a repeat integration.

The gain of the CTIA can be computed from equation 4.7 and the expression $Z_f = 1/j \, \omega \, C_f$ to be

$$V_{out} = -V_{in} \left(\frac{C_d}{C_f} \right). \tag{4.22}$$

When the conditions for linearity as in equation 4.20 are satisfied, $V_{in} = Q/C_d$, giving

$$V_{out} = -\frac{Q}{C_f}. \tag{4.23}$$

If the capacitance of the feedback can be made independent of voltage, the output of the CTIA will be linear with collected charge. Equation 4.23 also suggests that C_f must be kept relatively small to provide a large output signal and to minimize the influence of the electronic noise of the amplifier on the achievable ratio of signal to noise. The well depth in the CTIA is set by the voltage differential across the feedback capacitor, giving

$$N_{max} = \frac{C_f V_{max}}{q}. \tag{4.24}$$

4.4.3 Integrating Amplifier Readout Strategies

From the above discussion, returning to the source follower integrator, and assuming the linearity conditions are met, the output waveform for a constant rate of

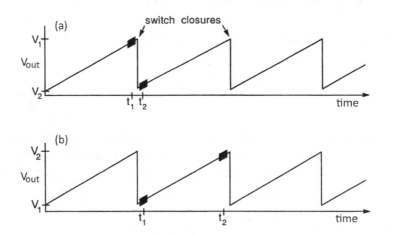

Figure 4.12 Sampling strategies for integrating amplifiers: (a) illustrates double correlated sampling around a reset; and (b) is nondestructive reading at the beginning of the integration ramp with a second read at the end; in this case, it has been elected to reset after the second read.

charge flow onto C_s is shown in Figure 4.12(a). We assume the FET output is a linear function of its gate voltage and that the gain of the FET is unity. Then the signal can be measured by determining $V_{out} = V_1$ at t_1, resetting, and determining $V_{out} = V_2$ at t_2. The accumulated charge is

$$Q = (V_1 - V_2)C_s, \tag{4.25}$$

the usual expression for the charge on a capacitor. This technique is called double correlated sampling.

The minimal approach is similar to Figure 4.12(a) but omits the second read at t_2. In general, it has significantly higher noise than double correlated sampling, but it can be useful when speed is of the essence since it requires only a single read cycle per reset.

The charge Q includes any contributions from the detector dark current and leakage through the FET onto its gate, as well as the photocurrent. The limiting noise, assuming other sources of read noise are negligible, is the square root of the total number of charges collected from these currents. There are at least two other potential sources of noise. One excess noise source is the FET, which will be discussed later. The other is the charge on C_s, which, when sampled in the way described above, is subject to kTC noise (equation 2.26) that arises in the resistor/capacitor circuit formed when the reset switch is closed. In fact, if the detector, reset switch, and FET gate are replaced by an equivalent circuit with a resistor and capacitor in parallel (consult Figure 4.10), the simple source follower integrating amplifier circuit is an accurate reproduction of the circuit in Figure 2.10, which we used to derive the expression for kTC noise. Upon reset, we therefore have a minimum charge uncertainty (in electrons) on C_s of

$$\langle Q_N^2 \rangle = \frac{kTC_s}{q^2}, \tag{4.26}$$

where q is the charge on the electron.

For example, assume all noise sources except kTC noise are negligible. A relatively small gate capacitance with a MOSFET/detector combination is of the order of 10^{-13} F; at $T = 4$ K, the resulting noise is $\langle Q_N^2 \rangle^{1/2} = 15$. Such a circuit is said to have 15 electrons read noise.

Returning to the CTIA, it is also subject to reset noise associated with the capacitance C_f. Without affecting our conclusion, we can simplify the following derivation by assuming that the output of the amplifier with the reset switch closed has been set to ground by adjusting the voltage to the noninverting input, and that $V_b = 0$, as is often the case with a photodiode. When the switch is opened, there will be a charge uncertainty of

$$\langle Q_f^2 \rangle^{1/2} = \left(kTC_f \right)^{1/2} \tag{4.27}$$

on C_f. However, the bottom plate of C_f and the upper plate of C_d must be at the same voltage, giving us

$$V_{in} = \frac{Q_f}{C_f} = \frac{Q_d}{C_d}. \tag{4.28}$$

If $\langle Q_d^2 \rangle^{1/2}$ represents the charge uncertainty on the detector capacitance corresponding to $\langle Q_f^2 \rangle^{1/2}$, we have

$$\langle Q_d^2 \rangle^{1/2} = \left(\frac{C_d}{C_f} \right) \langle Q_f^2 \rangle^{1/2} = \left[kT \left[C_d \left(\frac{C_d}{C_f} \right) \right] \right]^{1/2}, \tag{4.29}$$

where we have substituted for $\langle Q_f^2 \rangle^{1/2}$ from equation 4.27. Comparing with equation 4.27, the reset noise is equivalent to the noise from simple integrators except that the effective capacitance is the detector capacitance multiplied by the ratio of detector to feedback capacitances. Like taxes, reset noise cannot be avoided with the simple sampling strategy in Figure 4.12(a).

Therefore, the sampling strategy in Figure 4.12(b) is usually employed when minimum read noise is needed. The signal is measured by differencing measurements at the beginning and end of the integration, that is, $V(t_2) - V(t_1)$, with no intervening reset. Assuming the open reset switch and the gate of the FET have high impedance, the virtue of this strategy can be seen from the time constant for changes of charge on C_s,

$$\tau_s = R_d C_s. \tag{4.30}$$

For example, if $C_s = 10^{-11}$ F and $R_d = 10^{15}$ Ω, then $\tau_s = 10^4$ s. If $t_2 - t_1 \ll \tau_s$, the kTC noise is "frozen" on C_s during the integration. It is a bit like paying taxes so slooowwwly that the government never gets the check. More quantitatively, the noise is reduced by the factor for response of an RC filter (the factor multiplying V_0 in equation 4.19). The disadvantage of this strategy is that it depends both on the FET and on the electronics that follow it to have extremely small low frequency noise. At some level, nearly every electronic device has $1/f$ noise, which can limit the read noise in this situation.

Nonetheless, in many applications the detector readouts and support electronics are sufficiently stable and the integration times sufficiently short that $1/f$ noise is not a serious issue. Where this is not quite the case, means to estimate and compensate for the $1/f$ noise can be provided with "reference pixels." This term refers to amplifiers identical to those that read out the detectors, but with no detectors connected to them. Reference pixels are often fabricated along the edges of an array so that they are addressed in the overall cadence by which the live pixels are interrogated. Much of the $1/f$ noise is common across many pixels in the array, or even the entire array; therefore, the reference pixels reflect the effects on the live pixels. As a result, the strategy in Figure 4.12(b) is not strongly affected by low frequency noise if care has been taken to minimize it in the basic circuit and any residual issues are removed with reference pixels.

We now consider the contribution to read noise from the electronic noise of the FET. As can be seen from equation 4.20, the capacitance on the gate of the FET can set a limitation on the circuit performance because the output voltage for a given collected charge is inversely proportional to C_s. Because of electronic noise in the FET (and possibly in the electronics that follow it), there is a minimum voltage difference that can be sensed. As C_s increases, the charge that must be accumulated to produce this minimum voltage difference increases proportionally. As a result, for a FET with given intrinsic voltage noise, the read noise in electrons increases in proportion to the effective gate capacitance. On the other hand, the intrinsic voltage noise of a FET can be reduced by manufacturing it with a larger gate. To first order, the noise goes as the square root of the gate area. Therefore, the read noise of a detector/FET system decreases with increasing gate size but only so long as the gate capacitance does not dominate C_s.

In most high-performance detector systems, the amplifier dominates the noise if an appropriate sampling strategy is adopted. It is, consequently, very important to minimize the amount of amplifier noise that is added to the signal. In general, for integrating amplifiers, the response of the FET and following circuitry must extend to much higher frequencies than implied by the integration time between resets. This requirement arises because it is necessary for the amplifier to follow the reset waveform, and it is particularly important when such amplifiers are used in arrays in which electronic switching is used to route the output signals from a large number of detectors through a single output line (as will be discussed below). Because of the relatively high frequencies involved, in the following discussion we will ignore the $1/f$ noise of the FET and take its noise output to be white; that is, equal noise power at all frequencies. As a result, the FET noise goes as the square root of the frequency bandwidth of the circuitry (as discussed in Section 2.2.5). Optimum signal to noise therefore requires that this bandwidth be limited to the smallest value possible that admits adequate signal power. These arguments can be illustrated qualitatively by considering an example based on a simple integrating amplifier being read out twice, once before and once after reset. We will largely ignore normalization constants (and signs) because we are interested only in the frequency behavior of the signal components. If the time coordinate is greatly magnified, the signal can be represented by a step function, such as $sgn(t)$. The signal voltage amplitude as a function of frequency is proportional to the absolute value of the Fourier transform of $sgn(t)$; referring to Table 1.2 and setting the normalization constant to unity, we have

$$V_S(f) = \left[\left(\frac{j}{\pi f} \right) \left(\frac{-j}{\pi f} \right) \right]^{1/2} = \frac{1}{\pi f}. \tag{4.31}$$

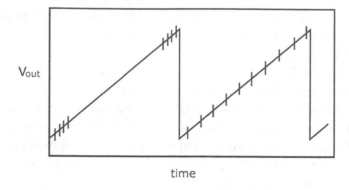

Figure 4.13 "Fowler sampling" (left) and multiaccum sampling (right).

Assuming white noise, the frequency dependence of the noise is

$$\left\langle \left[V_N \left(f \right) \right]^2 \right\rangle = C, \tag{4.32}$$

where C is a constant.

Without frequency filtering, the net output for the signal is just the integral of equation 4.31 over frequency (this integral may be modified slightly by the manner in which the signal is actually measured). Therefore, the signal grows only logarithmically with increasing frequency bandwidth. The net noise output (again with no further frequency filtering) is the square root of the integral of equation 4.32 over frequency and grows as the square root of the frequency bandwidth. Without some reduction in noise toward high frequencies, the signal-to-noise ratio in the measurement will tend toward zero as the bandwidth increases. To maximize the signal-to-noise ratio, the electronic frequency response must be adjusted to reject as much noise as possible while still passing the frequencies that contain the signal.

Strong filtering to limit the higher frequencies can minimize the amplifier noise, but can result in some signal being carried over into the next read (called "crosstalk"). To avoid crosstalk, the equivalent noise reduction can be achieved with nondestructive readout strategies. For example, as illustrated in Figure 4.13, multiple amplifier readings can be taken at each of t_1 and t_2 and two averages computed; this strategy is frequently called "Fowler sampling" after Al Fowler, an early advocate. These averages can be treated just as we have discussed for single readings at t_1 and t_2. Moderate electronic filtering must be used to eliminate noise at frequencies above the maximum multiple sampling frequency. Similar gains can be achieved by sampling the output repeatedly between t_1 and t_2 (Figure 4.13 – "sampling up the ramp": SUR) and computing a linear fit to the readings.[1]

---

[1] This strategy is often called "multiaccum."

The slope of this fit is proportional to the current deposited by the detector on the gate of the integrating FET. Sampling up the ramp can have additional advantages. For example, a bright source that would saturate the amplifier in the full integration can be measured from the integration slope computed from the first few samples, before saturation. If the detector is operating in an environment where it is occasionally struck by ionizing particles (e.g., cosmic rays), the data prior to a particle hit can be recovered if this readout strategy is used, and often the data after a hit are also recoverable. In addition, since the reads are distributed over the entire integration time, the data *rate* that must be handled by the support electronics is reduced.

In any of these cases, to avoid having amplifier noise limit the detector performance, we need to collect enough charges on the integrating capacitor so that the intrinsic statistical noise exceeds the amplifier noise. That is, the square root of the number of collected charges should be greater than the read noise. In general, the FET will have a maximum output voltage, V_{max}, above which extreme nonlinearity or saturation will occur. The maximum number of charges that can be collected without driving the output beyond V_{max} is known as the well depth. Continuing to assume unity gain, the well depth is

$$N_{max} = \frac{C_s V_{max}}{q}. \tag{4.33}$$

Although this relationship suggests that the well depth could be increased by increasing C_s, we have seen a variety of negative effects that would ensue from large values of C_s.

It is possible for the statistical noise to dominate the amplifier noise only if N_{max} is larger than the square of the read noise in electrons. If the amplifier satisfies this requirement, the integration time is long enough to collect sufficient charge, and dark currents and other such phenomena do not provide significant charge compared to the signal, then the detector/amplifier system can be photon noise limited. Assuming negligible dark current, the photon-noise-limited dynamic range of the system can be described in terms of the well depth and read noise as

$$R = \frac{N_{max} - Q_N^2}{Q_N^2}. \tag{4.34}$$

It is the factor in signal strength over which the system can operate near photon-noise-limited sensitivity. For example, a system with $N_{max} = 10^6$ and $Q_N = 100$ can detect signals varying in strength by up to a factor of 99 and still remain nearly photon noise limited.

Another specification of dynamic range refers to the ratio of N_{max} to the minimum signal that can be detected at $S/N = 1$, that is, to the read noise. For the parameters just given, the dynamic range measured in this way is 10^4; the circuit will be read noise limited over most of this range.

4.5 Detector Arrays

Photoconductors and photodiodes are combined with electronic readouts to make detector arrays. The progress in integrated circuit design and fabrication techniques has resulted in continued rapid growth in the size and performance of these solid state arrays. In the infrared, these devices are based on a (conceptually) simple combination of components that have already been discussed; an array of integrating amplifiers is connected to an array of detectors. These devices can provide many thousands to millions of detector elements, each of which performs near the fundamental photon noise limit for most applications between wavelengths of 1 and 40 μm. For visible and near infrared detection, monolithic structures can be built in silicon, forming arrays with millions of high-performance pixels. One highly developed example of these visible detectors is the charge coupled device, or CCD, discussed in the next chapter. So-called CMOS imaging arrays are an alternative to CCDs. They combine a grid of silicon detectors with individual readout amplifiers into an array, usually with somewhat lower per-pixel performance than a CCD but increased flexibility in use, potentially lower overall imager cost, and higher speed at low noise.

4.5.1 Infrared Detector Arrays

A rudimentary detector array could be produced by building a supply of detectors and amplifiers and wiring them together. For more than a few pixels, however, this procedure is a bit tedious. A much more streamlined construction method is possible using integrated circuit technology to miniaturize the amplifiers so that each one is no bigger than the detector it serves. The amplifiers are produced in a grid (or other suitable pattern), each with an exposed contact for the input signal. The detectors are produced in a mirror-image grid and have exposed output contacts. Bumps of indium solder are deposited on both sets of contacts, and the detector and amplifier pieces are "flip chipped" by carefully aligning them and squeezing them together. When the indium bumps deform under this pressure, their naturally occurring oxide coatings tear and expose the bare metal. The indium metal on the two contacts welds, forming the necessary electrical connections between the detectors and amplifiers as well as providing the mechanical strength to hold them together. In some cases, epoxy is flowed into the space between the readout

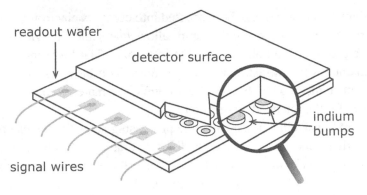

Figure 4.14 Hybrid indium bump bonded infrared detector array. The readout wafer is manufactured on a silicon wafer and the detector wafer can be of any suitable semiconductor material appropriate for the wavelength of response and other characteristics.

and the detectors to increase the bonding strength. The resulting device is called a bump bonded direct hybrid array; it is illustrated in Figure 4.14.[2]

Our breezy discussion of indium bump bonded arrays glosses over many difficulties. Fundamentally, the success probability of each individual processing step is less than unity, sometimes much less. The probability of obtaining a satisfactory array at the end of all the steps is the product of all the intermediate probabilities, so achieving high yields can be difficult. A low yield means many arrays are thrown out along the way; consequently the price of a finished device is high.

The salient advantage of the technique, however, is that the readouts and detectors can be optimized separately in terms of choice of materials and processing. Readouts are almost always made in silicon, for which there is an extremely well-developed technology for fabrication of high-performance ultra-miniature electronic circuits. The robust oxide that can be grown on silicon (or forms naturally)

---

[2] A variety of restrictive conditions must be met by both the readout and the detector array to permit direct hybridization. Where these conditions cannot be met, an alternative type of array can be built using "Z-plane technology." In this approach, the readouts are built in modules that are mounted perpendicular to the sensitive face of the detectors (rather than parallel as in the direct hybrid approach). The readout modules can extend a significant distance behind the detectors. However, these modules are made thin enough that the detector/readout modules can be mounted in a mosaic to provide a large sensitive area. These arrays allow bulkier readouts than bump bonded hybrids, they help isolate the detectors from luminescence and heat produced by the readout, and with mosaicking they can be built in larger formats than bump bonded hybrids. This construction also enables more sophisticated circuitry, for example with avalanche photodiodes to time tag the events (e.g., Aull et al. 2018). However, the added construction complexity makes them unpopular unless these attributes have a high priority.

is the fundamental reason it can be processed into complex electronics. In addition, the high purity and good crystallography allow fine-scale lithography to miniaturize the electronics, and also to provide low noise. The huge integrated circuit industry maintains the infrastructure needed for silicon-based array readouts. The detectors can be made of InSb, HgCdTe, or any other material that gives optimum performance for the intended application.

Even for extrinsic silicon detectors, the processing steps for the electronics and the infrared detectors can be sufficiently different that it is advantageous to produce them separately and bump bond them together – for example, high temperatures used in producing readouts may damage the detectors. However, it is feasible with silicon photodiodes to produce the detectors and readouts together on the same wafer. These devices will be discussed in Section 4.5.2.

Limitations arise in hybrid array construction because of the necessity for very accurate dimensional control during the construction of both the readout and the detector arrays. For detectors of material other than silicon, this difficulty is exacerbated because the devices are constructed and stored at room temperature but must be cooled for operation. Dimensions must be held within tolerance limits over a large temperature range in spite of the differing thermal contraction and expansion properties of the detector material and its silicon readout. With commonly used detector materials, it is possible to attain device sizes up to about 1–2 cm square while still retaining adequate dimensional control to survive cooling. Arrays with very large numbers of pixels depend on very small unit cells and on thinning the detector or readout layers so that they can stretch to accommodate the contraction of the other material, allowing arrays a few times larger than this limit. In addition, the bonding forces must be controlled so that the indium bumps are sufficiently compressed without breaking the thin wafers of material that carry the detectors and readouts or having the wafers slip relative to each other causing adjacent bumps to touch. Despite these difficulties, direct hybrid infrared arrays of 2048×2048 to 4096×4096 pixels (with pixel sizes of 15 to 30 μm) have been constructed successfully. A 4096×4096 array with 15 μm pixels is more than 6 cm on a side, and if the array is cooled by $200\,^\circ$C, the corner-to-corner dimensional mismatch significantly exceeds the size of a pixel, so stretching of the detector or readout layers is essential.

There is another important restriction placed on the detectors in a hybrid array. These devices and their contacts are constructed on one side of a wafer of suitable semiconductor, and this side is subsequently placed in contact with the readout. Therefore, the photons must enter from the other side, an arrangement called backside illumination. The detector wafer must allow the photo charge carriers to migrate from their generation sites to the inputs of the readout amplifiers. Since photodiodes operate by intrinsic absorption and hence the entire wafer is absorptive, not just the diode alone but the entire wafer carrying the diode must be thin enough to have good collection efficiency (as expressed in equation 3.23). The required thickness may be 10 μm or less and must be held to the correct value over the entire array; for example, there may be a tolerance of 1 μm in thickness uniformity across a 1 cm array. Two approaches have been taken to meet this requirement. In the first, the detector array is produced on a wafer that is sufficiently thick to be strong enough for bump bonding. After the detector array has been bonded to the readout, the resulting sandwich is mechanically thinned to the final detector thickness. In the second, a thin layer of detector material is grown onto a substrate of some other material that is both mechanically rugged and transparent at the wavelengths of interest. The detectors are grown with their contacts on the exposed surface, which is bump bonded to the readout. The detectors can then be backside illuminated through the transparent substrate, or this substrate can be removed chemically after bump bonding since the readout then provides the necessary mechanical strength.

Of course, none of these measures are required for Si:As IBC arrays, since the silicon detector wafer matches the silicon readout thermal contraction perfectly and the backside illumination allows the photons to reach the infrared-active layer through the substrate on which the detector layers are grown, so long as their energies are less than the silicon bandgap so they are not absorbed there.

There are a number of additional restrictions placed on the readout circuits that are used with infrared arrays. The necessity of having small unit cells requires that the circuits have a minimal number of components. A second concern is electroluminescence of the circuits. As mentioned in Chapter 2, solid state devices (for example, LEDs) can *emit* light if a current is passed through them. Even though silicon is an indirect semiconductor and this process is suppressed, low-level self-emission by the readout can be a deadly problem in a high-performance detector array. Electroluminescence is reduced by operating the readouts at the minimum possible drive voltages, and residual emission can be shielded locally on the readout.

As we have already mentioned, simple source follower integrator circuits based on MOSFETs can collect charge on their gate capacitors equally well when they are off as when they are on. Turning these circuits off causes electric charges to be introduced at various points within the circuit through unavoidable point-to-point capacitances; however, when power is restored, the circuit recovers its previous "on" state so accurately that its read noise is not degraded. By leaving the MOSFET amplifiers off while the circuit is accumulating a signal, the power requirements for these arrays can be kept very small. Low power consumption is an important consideration in the design of cooling apparatus, and it is also helpful in keeping the detector array at the proper operating temperature. Moreover, the problems with

electroluminescence can be minimized if power is provided to the readout only
for the brief instant when its output is actually being measured. This operating
mode cannot be used with a CTIA because this circuit must be on continuously if
it is to remain in balance, so reducing its power requirements and controlling its
self-emissions can be more difficult than with other types of readout.

The readout is given the final job of reducing the number of array output lines
from one per pixel to a manageable number (for example, one, two, or four for the
entire array of many thousands of pixels). The signals appear sequentially on these
output lines, where they can be processed serially by the following electronics. It is
said that the signals have been "multiplexed" through a single set of electronics, and
the entire readout is frequently referred to as a multiplexer, or MUX, even though
it has a number of additional functions.

A particularly simple implementation of the multiplex switching action addresses
a column of unit cells by turning on the power to their amplifiers; the cells in
other columns that are attached to the same output line but which have their
power off will not contribute signal. This arrangement is illustrated in Figure 4.15.

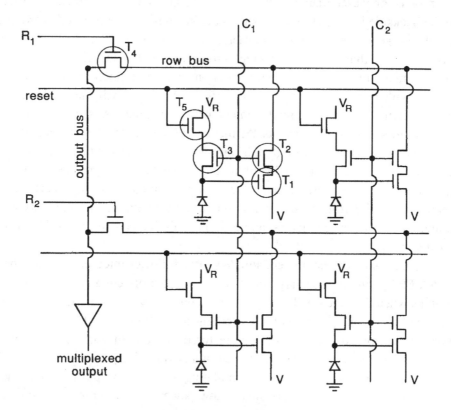

Figure 4.15 Four cells of the readout for an infrared array. The upper left cell is
as in Figure 4.8. The other cells are identical but are shown in less detail.

The operation can be understood in terms of the upper left cell. Signal is collected from the photodiode on the gate of T_1. Readout of the signal is enabled by using R_1 and C_1 to close the switching transistors T_2, T_3, and T_4. They apply power to T_1 and connect it to the output bus so a reading can be taken through the output amplifier. Pulsing the reset line connects V_R to the photodiode through T_5 and T_3 to re-establish the detector bias. A second reading can be taken, if desired, with the reset switch, T_5, closed. This switch is then opened and another reading taken of the output of T_1, after which the cell is turned off by opening T_2, T_3, and T_4. Note that no other cell in the array has been disturbed by this sequence of events.

An important aspect of some multiplexing schemes is that appropriate control electronics allow the MUX to address any unit cell or combination of unit cells on the array. This feature is called random access, and it allows sub-units of the array to be used whenever it is beneficial (for example, when time response is desired that is faster than can be achieved by switching through the entire array). The array in Figure 4.15 operates in this manner.

Indium bump bonded direct hybrid arrays have been manufactured for wavelengths from the visible to nearly 40 μm. High-performance PIN silicon diodes have excellent visible and red response, extending to or even past 1 μm because the diodes can have sufficiently thick absorbing layers to mitigate the poor absorption efficiency of silicon. However, such detectors are not compatible with conventional integrated circuit processing and are generally integrated with readouts through hybridization. Between 1 and ~15 μm, arrays can be based on well-developed photodiode technology in HgCdTe, with InSb as an alternative material out to 5 μm, and InGaAs at short wavelengths, e.g., to 1.7 μm. Between roughly 5 and 40 μm, arrays can be based on extrinsic silicon photoconductors or silicon IBC detectors. At wavelengths longer than 40 μm, although photodiodes and photoconductors are no longer the detector type of choice, a very successful bolometer array has been built along similar lines (Billot et al. 2006). Other bolometer arrays, to be discussed in Chapter 8, use indium bump bonding hybridization in a variety of ways, including onto different types of readout.

4.5.2 CMOS Arrays

A related approach to the infrared arrays is used for low cost silicon-based arrays for the visible wavelength range. Charge coupled devices (CCDs), to be discussed in the next chapter, are virtually perfect detectors for this spectral region, but even perfection can have its flaws. CCDs operated to achieve the lowest noise are unavoidably slow to be read out. The charge transfer, and hence many desirable performance attributes, can be degraded by exposure to energetic ionizing radiation. Perhaps most importantly, their production has become a relatively specialized sideshow in integrated circuit fabrication, removed from the center stage of cell phone and computer chip processing. A useful alternative to CCDs is monolithic arrays with photodiodes next to their readout amplifiers, all on the same silicon wafer and all part of the same processing steps. As with the arrays discussed in the preceding section, an amplifier is provided for every pixel with the signals multiplexed to the array output by on-chip electronics rather than charge transfer. They have become ubiquitous in this form.

These devices are conventionally called CMOS (for complementary metal–oxide–semiconductor) arrays, although this term applies to the readouts for the direct hybrid arrays also. Early CMOS arrays required substantial area for the amplifiers, and this competition for real estate meant that their photodiodes fell far short of covering the entire area (they had low "fill factors") leading to low quantum efficiency. However, as transistors have been made smaller to allow increasing densities in commercial electronics, their claims on array real estate have shrunk sufficiently to change the situation. A further improvement is to place a grid of tiny microlenses over the array, one per pixel, to help improve the fill factor. Microlenses can boost the effective quantum efficiency to $\sim$70% at its peak, but they absorb in the ultraviolet eliminating any response there, and their efficiency is reduced if the array is illuminated off-axis. In addition, the photodiodes need to be very thin to remain compatible with standard processing, so the red and particularly infrared sensitivity of these arrays is modest. Finally, if the array is to return color information, filters are placed on individual pixels (such as in the widely used Bayer pattern with each 2×2 set of pixels containing one red, one blue, and two green ones). In such a case, the area can at best be filled only to the $\sim$25–50% level in any one color.

Scientific applications therefore generally center on high-performance, single-color monolithic arrays. The drive in the commercial processing is largely toward arrays with tiny pixels that are less useful for most scientific applications such as astronomy but are great for cell phone cameras. Thus, science-grade arrays share in standard commercial processing but are usually purpose-designed for large pixels and low noise among other attributes. Some of their advantages over CCDs are (1) they are much more tolerant of energetic radiation, which can degrade the charge transfer in CCDs; (2) they can be combined with complex circuitry on the same wafer, for example to expand the entire readout process to provide a digital output, not just an analog one; (3) they generally have much lower power requirements; (4) they readily allow for random access to individual pixels and to sub-arrays of pixels; (5) they can support faster readout cadences; and (6) they are resistant to blooming and smearing of signals from bright sources. The advantages of CCDs over CMOS arrays include (1) 100% of the sensor area is

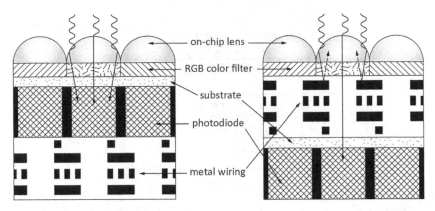

Figure 4.16 Comparison of back illuminated (left) with front illuminated (right) CMOS pixels. The "metal wiring" is the layer with the FET readout amplifiers, with metal traces laid down by photolithography.

active pixels – although back illuminated CMOS arrays have nearly closed this gap; (2) low fixed pattern noise – patterns imposed on an image due to systematic variations in the properties of the photodiodes and their readouts; and (3) lower dark current. Thus, each array type has a distinct range of applicability.

The simplest form of CMOS imager has a readout transistor for each pixel that can switch that pixel to an output line, with no amplification. These "passive pixel sensors" achieve fill factors of ~70% with read noises less than 50 electrons, in some cases significantly less. Lower read noises can be achieved by providing gain in a multi-FET amplifier (e.g., Figure 4.8) for each pixel. These "active pixel sensors" have been demonstrated with noise of a couple of electrons at normal video frame rates (30 frames/second). Similar format CCDs have read noises of 10–20 electrons at these frame rates.

A further improvement in CMOS array performance is offered by backside illumination, in which the silicon wafer is thinned sufficiently to allow the photons to enter the photodiodes from the side opposite the readout amplifiers and their traces, as illustrated in Figure 4.16. Initially this approach was applied to cell phone camera sensors, because their tiny pixels made the competition for real estate between sensor and amplifier extreme. However, science-grade arrays are now available, e.g., a 2048×2048 array with 11 μm pixels from Princeton Scientific.

4.6 X-ray Detection with DEPFETs

Depleted p-channel field effect transistors (DEPFETs) are devices that combine the detector and the readout FET in a single unit (Kemmer and Lutz 1987). The basic

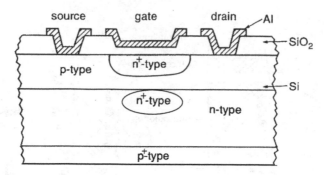

Figure 4.17 Schematic of a DEPFET. The unit cell consists of a MOSFET (upper part) merged with a junction (lower part). We show the MOSFET as in Figure 4.2 except with a p-channel rather than n-channel polarity.

architecture resembles that of a back illuminated CMOS pixel, with an amplifier integral with each pixel and, when fabricated in arrays, the capability for full random access. The FET (Figure 4.17) is grown on an n-doped wafer with a p^+-doped contact on the backside. The field regulating the current in the channel is generated through the combination of the conventional gate (top) and the n^+-doped input that forms a virtual gate on the opposite side (in the center). The device is designed to extend the depletion region over the entire volume, not just under the contacts – this feature is called "sideways depletion." The bottom junction portion can be made thick, e.g., 300 μm, for efficient absorption of X-rays. When photons free holes and electrons in the depleted bulk wafer, the holes are collected in the back p^+ contact and the electrons collect in the potential minimum established by the n^+ virtual gate. These electrons create a field that modulates the channel current just as the field from electrons on the traditional gate does. Because the virtual gate has very low capacitance, the noise in terms of electrons is low, potentially $\lesssim 2$ electrons. The conventional gate is used to set the optimum operating point for the FET. A separate electrode (not shown) is used to clear the collected electrons and start a new integration.

The DEPFET has the advantage of providing a long absorption path (~ 300 μm) for efficient detection of X-rays up to and beyond 10 keV (e.g., Müller-Seidlitz et al. 2018). Compared with X-ray detection with CCDs, the DEPFET avoids the need to transfer the charges so it can be read out faster. A modification of the simple unit DEPFET can use CCD principles to transfer charge to additional amplifiers to reduce the effective readout noise further (Lutz et al. 2016). Although we have discussed DEPFETs in terms of X-ray detection, the DEPFET concept is versatile and they have potential uses ranging from their excellent near-infrared response to high-performance energetic particle detectors.

4.7 High Energy X-ray Detectors

The detectors of choice for moderate energy X-rays are typically based on silicon. However, as shown in Figure 2.5, the absorption coefficient of silicon is low at high X-ray energies, with an absorption length of ~ 100 μm at 10 keV and falling steeply beyond. Focal planes with a general resemblance in construction to the direct hybrid infrared arrays have been used for higher energy X-rays, such as those for the NuSTAR imaging X-ray telescope. In this example, they are based on the much higher absorption in CdZnTe (see Figure 2.5). The inner electron shell binding energy for cadmium is 26.71 keV; for tellurium it is 31.81 keV, and for zinc it is 9.66 keV, giving these materials a significant advantage over silicon for photoelectric absorption. Furthermore, their much higher atomic numbers (48 for Cd, 30 for Zn, 52 for Te) increase the interaction by Thompson and Compton scattering. The bandgap for CdZnTe is 1.57 eV, so the detectors can operate at room temperature without a significant rate of dark events.

Fabricating these detectors is challenging because two-component semiconductors (such as CdTe) tend to have a high rate of crystal lattice imperfections that can trap the free charge carriers and degrade the detector performance. The introduction of Zn to replace some of the Cd reduces the concentration of large-scale imperfections, but traps remain as a problem. In addition, the free electron lifetimes in these detectors are short and the electron mobilities are small, making the diffusion lengths short (see equations 3.12 and 3.17). As a result, the detectors must be relatively thin and operated with a significant electric field for good charge collection efficiency, as shown by Hecht's equation for the ratio of collected charge to the charge generated by photon absorption as a function of the mobility and the distance the charge needs to travel to be collected:

$$\frac{Q}{qN_0} = \frac{\mathcal{E}}{L}(\mu_e\tau_e(1 - e^{(L-x)/\mathcal{E}\mu_e\tau_e}) + \mu_h\tau_h(1 - e^{-x/\mathcal{E}\mu_h\tau_h})), \tag{4.35}$$

where Q is the charge collected at the positive electrode, N_0 is the number of charge carriers created when a photon is absorbed, x is the interaction depth, L is the detector thickness, $\mathcal{E} = V/L$ is the electric field and V is the voltage across the negative and positive electrodes, and μ_e, μ_h, τ_e, and τ_h are the electron and hole mobilities and lifetimes respectively, available from Table 2.3 plus $\tau_h \sim 2 \times 10^{-7}$ s. The number of charge carriers freed per absorbed photon is the quantum yield, typically one for the optical and infrared detectors we have discussed to this point but far larger for semiconductor-based X-ray detectors.

The NuSTAR detectors are in a 32×32 array of CdZnTe with total dimensions of 2 cm square by 0.2 cm thick with a pixel pitch of 605 μm. The detector wafer has a monolithic negative contact on one side, with an individual positive contact for each pixel on the opposing side; a typical bias voltage is 300 V. The pixels are read out by an application-specific integrated circuit (ASIC) as described by Rana et al. (2009). In a procedure analogous to indium bump bonding direct hybridization, the individual positive contacts are bonded with conductive epoxy to the inputs of the ASIC, which provides 32×32 data-handling circuits on a single wafer with the same pitch as the detector pixels. Each of these circuits has a preamplifier, shaping amplifier, discriminator, sample-and-hold circuit, and analog-to-digital converter, whose output is delivered to the higher-level electronics.

In principle, the energy resolution on a detected X-ray should be the inverse square root of the number of charge carriers it frees. The high quantum yield for CdZnTe (Takahashi and Watanabe 2001) would imply high energy resolution, but the low mobility and short charge-carrier lifetimes undermine this potential since many free charge carriers do not survive to contribute to the signal (equation 4.35). In addition, the outputs of CdZnTe detectors have prominent low energy tails. With care to correct for various forms of non-ideal behavior, they can achieve energy resolution of 1 to 2% (at energies approaching 100 keV).

4.8 Example: Readout Performance

Assume a photodiode is operated near zero bias and at a temperature of ~ 170 K, where its zero bias resistance $R_0 = 2 \times 10^{12}$ Ω and $C = 36$ pF. Compare the performance of this detector combined with the following readout circuits: (a) a TIA operating at 1 Hz bandwidth with $R_f = 10^{11}$ Ω and $C_f = 1$ pF; (b) a simple source follower integrating amplifier operating with the signal determined by sampling, resetting, and sampling again with 0.5 s between resets; and (c) a simple integrating amplifier operating as in (b) but with the signal derived from samples at the beginning and end of the integration ramp (without resetting). In all cases, assume that $1/f$ noise is negligible and that none of the systems are shot noise limited. The diode properties (other than capacitance and resistance) will cancel in the comparisons. Just to be specific, though, we will assume $\eta = 0.6$ and that we are operating at 0.9 μm.

(a) For the TIA, the limiting noise is Johnson noise, and from equation 4.16 with $\omega = 2\pi f = 6.28$ Hz and $\tau_f = R_f C_f = 0.1$ s, we get

$$\frac{\left\langle \left(V_f^N \right)^2 \right\rangle^{1/2}}{df^{1/2}} = 2.59 \times 10^{-5} \text{ V Hz}^{-1/2}. \tag{4.36}$$

Similarly, from equation 2.25 with $R = R_0$,

$$\frac{\left\langle \left(I_d^N \right)^2 \right\rangle^{1/2}}{df^{1/2}} = 6.85 \times 10^{-17} \text{ A Hz}^{-1/2}. \tag{4.37}$$

Taking this value as the current noise from the detector, from equation 4.15 we get

$$\frac{\left\langle \left(V_{out,d}^N \right)^2 \right\rangle^{1/2}}{df^{1/2}} = 5.80 \times 10^{-6} \text{ V Hz}^{-1/2}. \tag{4.38}$$

Combining the two noise contributions quadratically, the total noise is

$$\frac{\left\langle \left(V^N \right)^2 \right\rangle^{1/2}}{df^{1/2}} = 2.65 \times 10^{-5} \text{ V Hz}^{-1/2}. \tag{4.39}$$

From equations 3.1 and 4.10, the signal voltage is

$$|V_{out}^S| = 8.14 \times 10^{-9} \varphi \text{ V s}, \tag{4.40}$$

where φ is the photon arrival rate. The power incident on the detector is $P = hc\,\varphi/\lambda$, so the voltage responsivity of the circuit is

$$S = \frac{|V_{out}^S|}{P} = 4.10 \times 10^{16} \lambda(\text{m}) \text{V W}^{-1}, \tag{4.41}$$

or for $\lambda = 0.9 \times 10^{-6}$ m,

$$S = 3.69 \times 10^{10} \text{ V W}^{-1}. \tag{4.42}$$

If we take the output of the TIA to a circuit with an equivalent electronic bandwidth of 1 Hz, this result implies that the detector/amplifier would detect a flux of $\varphi = 3255$ s^{-1} at a signal to noise of unity at the implied net integration time. However, this is an optimistic estimate, since it does not allow for the comparison of the signal viewing the source with that not viewing the source, a step that is virtually always necessary for detection at low levels. To include this effect, we assume a specific measurement strategy: that the signal is chopped in a square wave manner and that the detector output is measured as the rms voltage. The response is then reduced by the factor 0.4502, with no change in the noise. Hence, a signal level of $\varphi = 7230$ s^{-1} is required for a signal to noise of unity.

(b) Assuming that $\varphi = 7230$ s^{-1}, we need to compute the signal to noise that would be achieved by the integrating amplifier. First calculate the signal current in electrons per second:

$$I^S = \eta \varphi = 4338 \text{ electrons s}^{-1}. \tag{4.43}$$

The output noise during one readout is given by equation 4.26 to be

$$\langle Q_N^2 \rangle^{1/2} = \frac{(kTC)^{1/2}}{q} = 1816 \text{ electrons}. \tag{4.44}$$

In analogy with the measurement strategy adopted for the TIA, a measurement is made by spending 0.5 s "on source" and 0.5 s on background. In the 0.5 s on source, we accumulate (0.5 s)(4338 electrons s^{-1}) = 2169 electrons. The noise from the two reads is $\sqrt{2}$(1816 electrons) = 2568 electrons. Thus, S/N = 0.84, so this detector/amplifier arrangement is 1.0/0.84 = 1.2 times *less* sensitive than the TIA system in part (a).

(c) The arguments involving signal calculation are the same as in case (b). In this readout scheme, the noise per reset is diminished by a factor $(1 - e^{-t/RC})$. The time constant of this detector is $\tau = R_0C = 72$ s. The kTC noise in 0.5 s is then

$$(1 - e^{-t/RC})(1816 \text{ electrons}) \approx 13 \text{ electrons}. \tag{4.45}$$

In two reads, the total noise is then $\sqrt{2}$(13 electrons) = 18 electrons. We then get S/N = 2169/18 $\sim$ 120, so this readout scheme gives a S/N about 120 times greater than the one using the TIA.

4.9 Problems

4.1 Repeat the example for a net integration of 100 s. Assume the integrating amplifiers perform two integrations of 50 s each.

4.2 Evaluate the nonlinearity of a simple integrator with the photodiode designed in the example at the end of Chapter 3. Compute the response in volts per charge for diode biases of 0, −1, and −2 V.

4.3 Plot the signal and noise as a function of frequency from 1 to 100 Hz for the TIA and detector in the example just above. Assume the amplifier electronic noise is $V_A^N = 1 \times 10^{-7}$ V Hz$^{-1/2}$ independent of frequency. Take the detector noise to be $I_D^N = (1 \times 10^{-15}$ A Hz$^{-1/2})(1/f)$, where f is in hertz.

4.4 Given the following data set for a simple integrating amplifier, derive the node capacitance and the amplifier noise:

Signal (mV)	RMS noise (μV)
0.32	32
0.96	35
3.2	44
9.6	63
32	106
96	178
320	321

4.5 The signal from an integrating amplifier is extracted by a destructive read, with a sample followed by resetting followed by the second sample. The amplifier noise is white. The output is taken to a simple RC low-pass filter, with time constant $\tau = RC$ adjusted to maximize the ratio of signal to noise. If Δt is the time interval between samples, derive the value of τ in terms of Δt that gives the maximum signal to noise.

4.6 Compare the signal to noise achievable through two means of multiple sampling on the output of an integrating amplifier. In the first place, assume that eight samples are taken at the beginning and eight at the end of the integrating ramp (in a time interval short compared with the rate of change in the output), that each set is averaged, and that the slope of the ramp is computed from the difference of these averages divided by the time interval between them. For the second case, assume that the 16 samples are distributed uniformly over the integration ramp and the slope is determined by a least squares fit to them.

4.10 Further Reading

Bielecki (2004) – review of approaches to read out optical detectors

Driggers et al. (2012) – general description of detection systems with emphasis on analysis methods

Fossum and Pain (1993) – classic overview of infrared array readout electronics

Graeme (1996) – technical and detailed description of amplifiers to read out photodiodes

Horowitz and Hill (2015) – practical discussion of electronic circuitry

Janesick (2001) – includes thorough discusion of CCD output amplifiers and test procedures, which is generally applicable to other types of solid state detector array also

Rogalski (2012) – up to date general review of array photodetectors

Vampola (1993) – general discussion of readout amplifiers

5

Charge Coupled Devices

Charge coupled devices (CCDs) were the first high-performance semiconductor detector arrays. They take advantage of a number of unique properties of silicon to allow transferring the collected photo-charge out of the sensitive area. An amplifier at the edge of this area processes the collected charge packets sequentially and outputs them to external circuitry where they can be digitized, stored, and manipulated. There need be no competition for real estate between detectors and amplifiers, and therefore nearly perfect fill factors can be achieved. The integration of detector and readout onto a single monolithic piece of silicon reduces the number of processing steps (compared with hybrid arrays), and therefore increases yields and reduces costs. This approach depends on the high purity and crystal quality obtainable for silicon and its robust and easily grown oxide. Attempts to extend the CCD principle to other materials were relatively unsuccessful because they did not share these advantages (Rogalski and Piotrowski 1988). Historically important articles on the development of the CCD concept are available in Melen and Buss (1977), and Durini (2015) provides a summary.

5.1 Basic Operation

5.1.1 Operation of a Single CCD Pixel

Refer to Figure 5.1 for the following discussion. A thin oxide layer has been grown onto a piece of silicon. A metal (or heavily doped semiconductor) electrode has been evaporated onto the oxide, which as an insulator blocks the passage of the electrons from the silicon onto the electrode (see Figure 5.1(a)).[1] This structure is called a metal–oxide–semiconductor, or MOS, capacitor; we have already encountered it as the key element in a MOSFET (Chapter 4). The electrical properties

---

[1] We will discuss CCDs with electrons as the operational photo-charge carrier; they can also be manufactured to operate with holes in this role.

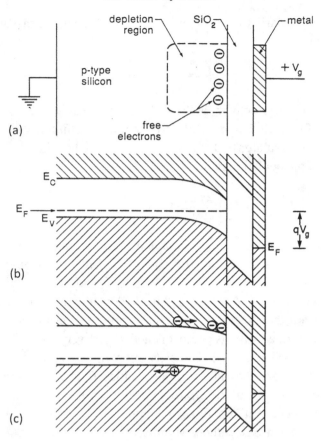

Figure 5.1 Charge collection at a single pixel of a CCD: (a) illustrates the collection of free charge carriers at the silicon–oxide interface; (b) shows the band diagram immediately after application of a voltage V_g to the gate to form a well; and (c) shows the filling of the well as it collects charge carriers.

of the silicon have been controlled in this example by doping it p-type; because extrinsic absorption is relatively weak compared with intrinsic absorption, this doping has little effect on the photoresponse.

Suppose the bulk semiconductor is grounded, and at time $t = 0$ a positive voltage V_g is put on the electrode (see Figure 5.1(b)). The voltage provides the bias for an intrinsic photoconductor in the silicon. Any photons absorbed by the silicon will produce a photocurrent that will flow toward the contacts. Because of the p-type doping, there will be virtually no conduction electrons in the absence of illumination. The voltage will cause any holes in the immediate vicinity of the electrode to drift away, creating a depletion region. This process is virtually identical to the creation of a depletion region in an IBC detector. It can be described mathematically similarly to equations 3.39 and 3.40. That is, Poisson's equation in the bulk silicon is

$$\frac{d\mathcal{E}_x}{dx} = \frac{\rho}{\kappa_0 \varepsilon_0} = -\frac{q N_D}{\kappa_0 \varepsilon_0}, \tag{5.1}$$

where $\mathcal{E}_x$ is the electric field, κ_0 is the dielectric constant of silicon, 11.8, and N_D is the density of ionized donors, which are the minority impurity in the silicon. We let x run from the insulator/bulk silicon interface into the silicon. The solution for the thickness of the depletion region is

$$w = \left[\frac{2\kappa_0 \varepsilon_0}{q N_D} |V_g| + t_I^2 \right]^{1/2} - t_I, \tag{5.2}$$

where t_I is the thickness of the insulator and V_g is the gate voltage. For example, with $N_D = 3 \times 10^{14}$ cm^{-3}, $|V_g| = 10$ V, and $t_I = 0.1$ μm, the width of the depletion region is 6.5 μm.

The free electrons that are created within the depletion region or that diffuse into it will drift toward the electrode and collect opposite it at the Si–SiO$_2$ interface. The detector is operated with its photoconductive gain very close to unity, so nearly every photoelectron is collected. The electrons will accumulate against the oxide layer because there are virtually no opportunities for them to recombine in the depletion region. The role of the electrode positive voltage in driving all the holes away is critical because it suppresses recombination and allows the electrons to be stored for long periods without significant losses. However, a few electrons may get trapped in imperfect crystal bonds arising because the oxide and pure silicon do not match well in crystal structure. This is a general issue with "surface channel" CCDs; we will eventually discuss an alternative architecture that keeps the electrons away from the oxide layer. Assuming that all the electrons generated in the photoconductor diffuse into the depletion region and the trapping in the oxide interface is negligible, the electrons collected against the insulating layer in a long exposure will accurately reflect the number of photons absorbed by the silicon photoconductor during this time.

Free electrons can also be created by thermal excitation; at room temperature, the thermal dark current can dominate the photocurrent and may limit the maximum permissible integration time. In low light level applications, the dark current must be suppressed by cooling the array to 130–160 K, where dark currents in the best devices become virtually unmeasurable. In the following discussion, we make our usual simplifying assumption that the device is operated at a low enough temperature that thermal generation can be ignored.

Although we have assumed p-type bulk semiconductor and positive V_g, the same arguments would hold with n-type semiconductor and a negative V_g; in such a case, holes from any photocurrent would collect at the storage well. Similar structures can be built on other semiconductors besides silicon, in which case they are termed metal–insulator–semiconductor (MIS) devices. Infrared arrays can utilize MIS structures grown on a small-bandgap semiconductor. However, these devices tend to have undesirably small breakdown voltages (leading to small well capacity) and poor charge transfer (leading to high noise and compromised imaging). These difficulties have led to the general adoption of hybrid arrays for the infrared with the readout fabricated on a wafer of silicon (see preceding chapter).

5.1.2 Charge Transfer Process

The CCD provides a detector architecture in which the surrounding electrodes are electrically isolated from each other except while charge is transferred from one to another. If the transfer is complete, it injects no statistical noise and the electrical isolation results in no contribution from kTC noise. In the CCD, the collected charges are passed from one electrode to another along columns to one edge of the array and then passed along the edge to an output amplifier. As in the discussion of a single pixel, we will assume that the photo-charge carriers are electrons and that the silicon has been doped p-type. Refer to Figure 5.2 for the following discussion,

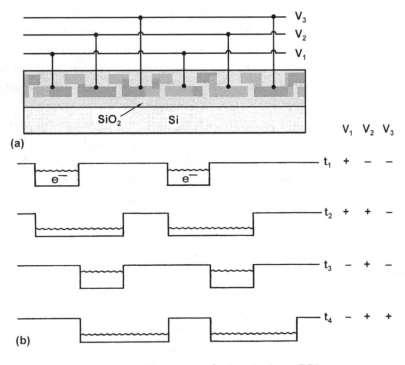

Figure 5.2 Charge transfer in a 3-phase CCD.

and assume that we want to move the collection of electrons under the left electrode toward the right.

The process is usually described in terms of electrons flowing between the storage wells under the electrodes. Figure 5.2 shows a row of MOS capacitors and beneath them four configurations of their storage wells in a time sequence. Assume that we start with V_1 positive at time t_1 and hold it at that voltage for an extended period of time to collect signal electrons in the resulting wells. At t_2, we begin the transfer of this signal by placing a positive bias on V_2 and enlarging the storage well. At t_3, we set V_1 negative, contracting the well and driving the signal electrons under the electrode adjacent to the one that initially collected them, i.e., advanced one gate width. At t_4, we set V_3 positive to create a condition similar to that at t_2 but with the charge advanced one gate width, thus continuing to pass the charge down the column of MOS capacitors. After t_4, V_2 is set negative, collecting the charge entirely under the electrode at V_3, so it will have been advanced two gate widths and is completely isolated from its starting point. That is, the pattern of biases on the electrodes acts both to isolate electrically the successive packets of signal charge that are being transferred and to disconnect the successive storage wells from the charge packet as it is passed along.

Although for simplicity the CCD electrodes are sometimes shown as simple plates deposited on a plane oxide layer, in practice such construction only works well if the gaps between electrode edges are kept extremely small. It is easier to construct CCDs that have electrodes on two or three different levels separated by thin layers of insulating material; this construction allows some degree of overlap at the electrode edges and facilitates efficient transfer of charge by eliminating unwanted potential minima between the electrodes. Silicon nitride is used as an insulator between overlapping sections of electrodes.

Figure 5.2 illustrates a 3-phase CCD with three sets of electrodes and voltage lines. The description above, however, applies equally well to a 4-phase CCD. Much more flexibility in CCD design is achieved by making use of the dependence of well depth on the doping of the silicon and the thickness of the SiO_2 insulator layer. This behavior follows from equation 5.2, which shows explicitly how the width of the depletion region, w, depends on the insulator thickness below the electrode, t_I. For example, if the electrodes in a 4-phase CCD are placed at different depths in the oxide and connected in pairs such that the well generated by one member of the pair for a given voltage is deeper than the well for the other member, the CCD method of operation can be applied to a 2-phase device, as shown in Figure 5.3. Other modes of operation are possible if (1) the electrodes of a CCD are driven with a more complex waveform than a simple, abrupt change of voltage level; or (2) more complex systems of electrodes are used than the ones we have described (as indicated by equation 5.2); or (3) additional doping is introduced under some electrodes, for example by ion implantation.

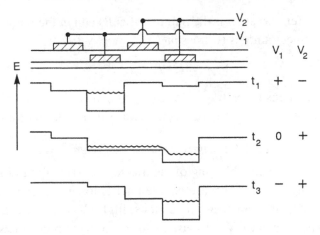

Figure 5.3 Charge transfer in a 2-phase CCD.

5.2 CCD Performance at Faint Light Levels

5.2.1 Charge Transfer and Noise

5.2.1.1 Mechanisms Affecting Charge Transfer

The charge collection and transfer process in CCDs is subject to a number of noise mechanisms. As with all detectors, CCDs are subject to shot noise; because they do not allow recombination, this noise is just the square root of the number of collected charges. There are two additional noise mechanisms associated with the charge transfer. The first arises from poor charge transfer efficiency (*CTE*). If the transfer inefficiency $\varepsilon = 1 - CTE$ is the fraction of N_0 charges that are not transferred, a total of εN_0 charges is left behind. If the *CTE* were perfect ($= 1$), then no charges would be left and there would be no noise associated with the transfer process. However, if the transfer inefficiency is greater than zero and is Poisson distributed, there is an uncertainty of $(\varepsilon N_0)^{1/2}$ in the number of charges left behind and hence an equal uncertainty in the number transferred out of the unit cell. Assuming an identical cell precedes the one under discussion, there will be a similar uncertainty in the number transferred into the cell. Thus, for n transfers, the net uncertainty is

$$N_{n,TL} = (2\varepsilon n N_0)^{1/2}. \tag{5.3}$$

The second mechanism arises from the trapping of charge carriers. In a surface channel device, charge is collected and transferred right at the Si–SiO$_2$ interface, as indicated in Figure 5.1. Since this interface represents a discontinuity in the Si crystal, it contains incomplete bonds that act as traps. On average, these traps will reach some equilibrium occupancy level and will not affect the total charge transferred. This occupancy level, however, is subject to statistical fluctuations, causing charges to be added to and subtracted from a charge packet randomly.

For $l = mn$ transfers, where n is the number of cells and m the number of transfers per cell, the net uncertainty is (Carnes and Kosonocky 1972)

$$N_{n,T} = (2kTlN_{ss}A)^{1/2}, \tag{5.4}$$

where N_{ss} is the density of traps (usually given in units of $(cm^2\ eV)^{-1}$), k is Boltzmann's constant, T is the temperature, and A is the interface area.

5.2.1.2 Charge Transfer Efficiency

Since poor *CTE* results in a blurring of the image due to charge trailing behind, as well as degrading the noise because charge packet losses occur in processes that are subject to the usual statistical fluctuations, high *CTE* is an overriding priority for good CCD performance. We discuss in this section the procedures to maximize the *CTE*, and in the following one a modification of the CCD architecture to reduce or even eliminate interface trapping and its degradation of charge transfer.

A number of mechanisms influence the *CTE*. Referring to Figure 5.2, if there is a large amount of charge in the well at t_1, electrostatic repulsion will cause the charge carriers to spread into the wider well at t_2. This self-induced drift mechanism is similar to the drift under external fields discussed in Chapter 3. We can derive an approximate time constant for the process as follows. Assume we have N_0 electrons in one well and none in the other; the wells are separated by the electrode spacing L_e. Then the voltage difference between these wells is $V_{21} = N_0q/C_0$, where C_0 is the capacitance,

$$C_0 = A\kappa_0\epsilon_0/X_0, \tag{5.5}$$

with X_0 the thickness of the oxide layer and A the gate area. Thus, $\mathcal{E}_{21} = N_0q/L_eC_0$. From equation 2.6, the average drift velocity $\langle v_{21} \rangle = -\mu\ \mathcal{E}_{21}$. We define the self-induced drift time constant to be $\tau_{SI} = -L_e/\langle v_{21} \rangle$. Substituting for $\langle v_{21} \rangle$ and $\mathcal{E}_{21}$ gives

$$\tau_{SI} \sim \frac{L_e}{\mu\mathcal{E}_{21}} = \frac{L_e^2C_0}{\mu N_0q}. \tag{5.6}$$

More rigorously (Carnes et al. 1972),

$$\tau_{SI} = \frac{2L_e^2C_0}{\pi\mu N_0q}. \tag{5.7}$$

For example, taking the electrode to be 15 μm square, $X_0 = 0.1$ μm, and $N_0 = 3 \times 10^5$, we find $\tau_{SI} \sim 0.002$ μs.

For small amounts of charge, thermal diffusion tends to drive the electrons across the storage well. Sufficient time must elapse to allow virtually all the charges to diffuse into the next well to obtain high *CTE*. Drawing on the discussion of diffusion in Chapter 3, but redefining symbols such that they are applicable to the CCD transfer geometry, the exponential time constant for thermal transfer of charges is (from equation 3.17)

$$\tau_{th} \sim \frac{L_e^2}{D},$$ (5.8)

where τ_{th} is analogous to the recombination time, the electrode spacing, L_e, is analogous to the diffusion length, and D is the diffusion coefficient. More rigorously, it can be shown (Carnes et al. 1972) that

$$\tau_{th} = \frac{4L_e^2}{\pi^2 D}.$$ (5.9)

Calculating D from the information in Table 2.3 and the Einstein relation (equation 3.12), we find that $\tau_{th} \sim 0.026$ μs for an electrode 15 μm square on an oxide layer 0.1 μm thick, and with $T = 300$ K.

Charge transfer can also be driven by a third mechanism: the fringing fields between the electrodes. Contrary to our illustrations, the well produced under one electrode is not completely independent of the voltages on neighboring electrodes. The resulting "fringing fields" enhance transfer efficiency both by producing appropriate slopes in the floors of the potential wells and by rounding off their edges. The effect of fringing fields depends on the CCD construction details and on the characteristics of the voltage signals used to transfer charge. Fringing fields usually need to be calculated through numerical rather than analytic techniques. To understand their effects, though, we can consider a position at the center of an electrode but at variable depth below it. If the position is very close to the electrode, then the field can be calculated approximately by assuming the electrode is an infinite charge plane and the neighboring electrodes can have little effect. As the position is moved further from the electrode it is centered under, we depart from the simple infinite plane geometry so the influence of neighboring electrodes grows and the fringing fields they produce can assist the charge transfer. However, as the distance continues to increase, the fields produced by all the electrodes decrease and the influence of fringing fields is reduced. Thus, the effect of fringing fields on charge transfer is a complex function of the electrode design and the placement of the charge collection wells. They are most effective for CCD geometries with small electrodes (≤ 15 μm) and in which the depletion regions are relatively far below the electrodes (i.e., buried channel, to be discussed below).

An approximate time constant for fringing field transfer is (Janesick 2001)

$$\tau_{ff} = \frac{L_e}{2\mu\mathcal{E}_{min}},$$ (5.10)

where $\mathcal{E}_{min}$ is the minimum fringing electric field strength under the gate,

$$\mathcal{E}_{min} \sim \frac{2\Delta V_g \,\kappa_0\epsilon_0}{C_{eff}}, \tag{5.11}$$

where ΔV_g is the potential difference between gates, κ_0 is the dielectric constant (11.8) for silicon, and C_{eff} is the series combination of the capacitances of the gate $(A\kappa_0\epsilon_0/X_0)$ and in the channel. For 15 μm gates a typical value of $C_{eff} \sim 5 \times 10^{-14}$ F. With $\Delta V_g = 3$ V, and nearly empty wells, τ_{ff} is ~0.004 μs.

For suitably designed CCDs with partially filled wells, $\tau_{th} \gg \tau_{SI}$ and τ_{ff}, so initially the electrostatic-driven and fringing field transfer will dominate. The time constant of the electrostatic process goes as N_0^{-1}, so eventually it will slow down. We can define a critical number of charge carriers in the well, N_0^{crit}; if the number of charges in the well is below this number, thermal diffusion dominates over the electrostatic transfer process. We calculate N_0^{crit} by setting $\tau_{th} = \tau_{SI}$:

$$N_0^{crit} = \frac{\pi D C_0}{2\mu q}. \tag{5.12}$$

For nearly filled wells, screening by the collected charge substantially reduces ΔV_g, the fringing field time constant becomes very long, and the transfer must be completed by diffusion. Diffusion may also be required to complete the transfer under large (>15 μm) electrodes. For operation at very low light levels, a very high *CTE* is required. Thus, the time constants derived for thermal diffusion can be taken as conservative upper limits. More detailed discussions, including other charge transfer mechanisms, can be found in Séquin and Tompsett (1975) and Janesick (2001). The example at the end of this chapter uses these results to illustrate the timing involved in obtaining good *CTE*.

The *CTE* normally refers to the unit imaging cell. For a CCD that operates with m phases ($m = 2, 3,$ or 4 for the examples we have discussed) and a charge transfer mechanism that is characterized by a simple exponential, such as thermal diffusion, the *CTE* takes the form

$$CTE = (1 - e^{-t/\tau})^m, \tag{5.13}$$

where τ is the time constant (for example, equation 5.9).

The *CTE* can be affected by imperfections in the construction of the CCD. For example, a flaw in an electrode can lead to a minimum in its potential well that can trap residual charge, or a potential minimum between electrodes may obstruct charge transfer between them. Traps that degrade the *CTE* can be produced either by design errors or by flaws in the processing such as lifting of the edges of the electrodes, diffusion of implanted dopants, lattice defects in the silicon, or unwanted impurities. In such cases, complete transfer may not be possible in any time interval, no matter how long. This sort of problem can be reduced by adjusting electrode voltages, time constants, etc., as well as by slowing down the overall read cycle. Nonetheless, the readout times calculated from electrostatic repulsion, thermal diffusion, and fringing fields usually represent only lower limits.

Fringing fields also have the "potential" to cause trouble. If the columns are not adequately shielded from each other, the fringing fields can allow electrodes in one column to affect charge transfer in a neighboring column; electrons may even be transferred across columns. To prevent such an occurrence, CCDs are manufactured with strongly doped (and hence conductive) barriers between their columns, which are held at ground potential to shield adjacent columns. The electric fields do not extend into these "channel stops," ensuring that the columns are electrically isolated from each other so that charge can be transferred efficiently along the intended direction.

Since poor charge transfer efficiency results in a portion of the signal being trailed behind when an image is transferred out of the CCD, the *CTE* can be measured by imposing a sharply confined image onto the array and measuring carefully the amount of charge that is deferred to later pixels upon readout. One technique is based on illuminating the CCD with monoenergetic X-rays to inject large and repeatable amounts of charge into single pixels. From knowledge of the X-ray energy and the quantum yield and measurements of the collected charge, the *CTE* can be calculated accurately. The *CTE* can also be measured by observing the resulting image spreading; this latter technique, however, is not sensitive to *CTE* problems that spread the lost charge over many pixels.

5.2.1.3 Buried Channel

Interface trapping noise can be reduced in two ways. One approach is to maintain a low level of charge in the wells virtually continuously so that the traps are always filled. Doing so requires that a small, uniform charge level be introduced immediately after each readout; this charge is called a fat zero. Although this procedure can significantly reduce interface trapping noise, it is not completely effective because the well tends to spread as more charge is accumulated, leading to the exposure of additional traps. In addition, a fat zero increases the net noise by the factor $(1 + N_{fat}/N)^{1/2}$, where N_{fat} is the average number of charges placed in a well by the fat zero and N is the average number collected from other sources (and we assume the read noise is negligible). With a low-noise CCD operated at low signal levels, the noise added by a fat zero can degrade the performance, and it is seldom used with modern, high-performance CCDs.

A more satisfactory way to reduce interface trapping noise is to modify the structure of the CCD so that the charges are kept away from the oxide layer. Such CCDs are said to have a buried channel. A thin (typically ~ 1 μm thick) layer of

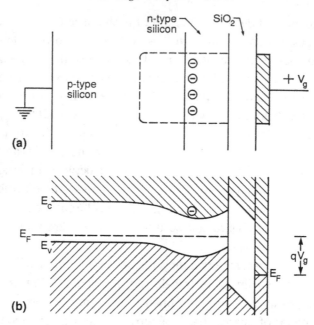

Figure 5.4 Physical arrangement (a), and band diagram (b) for a buried channel CCD.

silicon with the opposite dopant type of that in the bulk silicon of the detector substrate is grown under the oxide layer. Such a construction is shown in Figure 5.4, where we continue the example of a CCD with a p-type substrate and electrons as the active charge carriers, but now with a thin n-type layer between the SiO_2 and the p-type substrate. The two layers, n-type and p-type, form a junction. At the edge of the CCD, a heavily doped n-type region makes contact from the surface through a hole in the SiO_2 to the thin n-type layer. A positive voltage is applied to this contact until the junction is sufficiently strongly back-biased that the n-type region in it has been completely depleted. The depletion region of the CCD then consists of two zones: one just under the SiO_2 layer has a net excess hole density and the second, in the p-type region, has an excess negative ion density. When the CCD electrode is pulsed positive, the space charge in these zones acts to create a potential well within the n-type material. With appropriate selection of layer depths and doping, and of the voltages used to operate the device, this well forms away from the Si–SiO_2 interface. Since the trap density is much smaller in bulk silicon than at the crystal surface, interface trapping noise in this well is very small.

The second important advantage of a buried channel is that the distance between the electrode and well can be increased from ~0.1 μm for surface channel to a large enough value so that fringing fields from neighboring electrodes can assist the charge transfer, as discussed previously in this section. Typically, the wells are

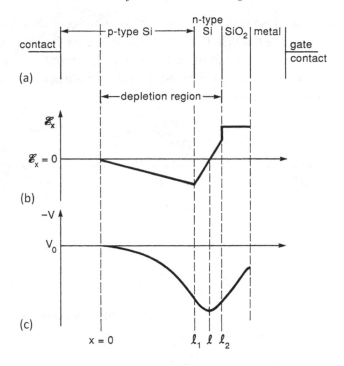

Figure 5.5 Principle of operation of a buried channel: (a) physical arrangement; (b) electric field; and (c) negative voltage, proportional to electron energy level.

placed ∼0.5 μm below the electrodes, leading to faster charge transfer and better *CTE* compared with typical surface channel devices. A disadvantage is that the well capacity is reduced because the distance between the electrode and the well is greater (see equation 5.2, and imagine that the distance of the buried channel from the oxide thickness is added to t_1).

To understand these arguments more quantitatively, we refer to Figure 5.5. Figure 5.5(a) shows a cross-section through the CCD; Figure 5.5(b) shows the electric field in various zones of the CCD; and Figure 5.5(c) shows the negative of the corresponding electrostatic potential. In the case of p-type semiconductor, the potential energy of the minority carriers (electrons) is proportional to the negative of the electrostatic potential (if the potential is in units of electronvolts, the constant of proportionality is unity). Therefore, Figure 5.5(c) can equivalently be taken as a diagram of the electron potential energy. We note that the potential reaches a minimum at $x = \ell$, which is removed from the Si–SiO$_2$ interface that lies at $x = \ell_2$. We will use Poisson's equation to derive and quantify this result, and will simplify the derivation by assuming the CCD electrode is large compared with the depth of field into the silicon we have to consider, so the problem can be studied in one dimension.

We use a coordinate system in which x is 0 at the left edge of the depletion region and increases as we move through the CCD. We first discuss Figure 5.5(b). For negative x, the material is assumed to be sufficiently conducting that $\mathcal{E}_x(x < 0) = 0$. The portion of the depletion region in the p-type silicon has a density of N_A negative ions. Using Poisson's equation in this region,

$$\frac{d\mathcal{E}_x}{dx} = -\frac{qN_A}{\kappa_0\varepsilon_0}, \quad 0 \leq x \leq \ell_1, \tag{5.14}$$

and applying the boundary conditions $\mathcal{E}_x(x = 0) = 0$ and $V(x = 0) = V_0$, and the definition $\mathcal{E}_x = -dV/dx$, we can show that

$$\mathcal{E}_x = -\frac{qN_A}{\kappa_0\varepsilon_0}x, \quad 0 \leq x \leq \ell_1, \tag{5.15}$$

and

$$V - V_0 = \frac{qN_A}{2\kappa_0\varepsilon_0}x^2, \quad 0 \leq x \leq \ell_1. \tag{5.16}$$

The portion of the depletion region in the n-type silicon has a density of N_D positive ions. Starting from Poisson's equation for this region,

$$\frac{d\mathcal{E}_x}{dx} = \frac{qN_D}{\kappa_0\varepsilon_0}, \quad \ell_1 \leq x \leq \ell_2, \tag{5.17}$$

and imposing the boundary conditions that both $\mathcal{E}_x$ and $V - V_0$ should be continuous across the boundary at $x = \ell_1$, we find that

$$\mathcal{E}_x = \frac{qN_D}{\kappa_0\varepsilon_0}x - \frac{q\ell_1}{\kappa_0\varepsilon_0}(N_D + N_A), \quad \ell_1 \leq x \leq \ell_2, \tag{5.18}$$

and

$$V - V_0 = -\frac{qN_D}{2\kappa_0\varepsilon_0}x^2 + \frac{q(N_D + N_A)\ell_1}{\kappa_0\varepsilon_0}\left(x - \frac{\ell_1}{2}\right), \tag{5.19}$$

$$\ell_1 \leq x \leq \ell_2.$$

We have now described the situation in the depletion region, but for completeness we continue the journey along increasing x into the remainder of the CCD. The magnitude of the discontinuity in $\mathcal{E}_x$ at ℓ_2 is equal to the inverse of the ratio of dielectric constants in the materials (assuming there is no surface charge trapped at the oxide/silicon interface). The field is constant within the insulator (again assuming there is no charge trapped within it). Within the metal electrode, the field will be zero. The potential varies linearly across the insulator and will be continuous at ℓ_2. The net bias across the detector is given by the difference between the voltage at the electrode and V_0 (V_0 would normally be at ground). These field and voltage behaviors are drawn in Figure 5.5 but will not be derived explicitly here.

The condition for formation of a buried channel is that the minimum in the electron potential energy should occur within the silicon. The location of this minimum is obtained by setting $dV/dx = 0$, where V is from equation 5.19. After a little manipulation, this condition leads to

$$\frac{\ell - \ell_1}{\ell_1} = \frac{N_A}{N_D},$$ (5.20)

where the value of x at the minimum is denoted by ℓ. A buried channel results if $\ell < \ell_2$.

To specify the operating parameters that allow a buried channel, we start with an expression for the voltage across the n-type layer. If the voltage at the p-type/n-type interface is V_{p1}, then by analogy with equation 5.16,

$$V_n - V_{p1} = \frac{qN_D}{2\kappa_0\varepsilon_0}(x - \ell_1)^2.$$ (5.21)

$\ell < \ell_2$ will be possible up to some maximum of $V_n = V_{max}$. The SiO$_2$ layer in the CCD is usually much thinner than indicated in the schematic diagram in Figure 5.5; as a result, we can take $V_{max} \approx V_{gmax}$, the maximum gate voltage that allows a buried channel. Setting $\ell = \ell_2$ in equation 5.20 and $x = \ell_2$ in equation 5.21, we solve this equation to find

$$V_{gmax} = \left(\frac{qN_D}{2\kappa_0\varepsilon_0}\right)(\ell_2 - \ell_1)^2\left(\frac{N_D}{N_A} + 1\right).$$ (5.22)

For example, if $N_D = 10^{15}$ cm^{-3}, $N_A = 10^{14}$ cm^{-3}, and $(\ell_2 - \ell_1) = 1$ μm, $V_{gmax} \approx 8.4$ V.

The construction of a buried channel CCD is governed by equation 5.22. Its operation also depends on maintaining the device at a sufficiently high temperature that the junction creating the buried channel does not freeze out. From a practical point of view, buried channels are only useful for CCDs that operate above 70–100 K.

Overfilling the well of a buried channel CCD will cause a portion of the collected charge to contact the oxide layer, causing a reversion to surface channel behavior. In most applications, the well depth should be defined as the maximum number of charge carriers that can be collected before charge contacts the oxide; the performance of the CCD is degraded beyond this point, for example by additional noise and by trailing of the image because of charges retained in surface state traps. A higher level of overfilling can cause the charge to spill out of the wells altogether and onto adjacent pixels. This "blooming" of the image occurs along columns, since the channel stops inhibit moderately saturated images from spreading to adjacent columns. Blooming can be suppressed by adding "anti-blooming gates" that collect excess charge around the electrodes and conduct it away without allowing it to fall into neighboring wells. These gates compete for real estate on the chip, so the effective fill factor is reduced about 30%, and they also result in a reduction in well depth. Extreme levels of over-illumination can fill large numbers of surface traps and lead to excessive "dark" current that can persist for long periods of time (days) so long as the CCD is held at low temperature.

5.2.1.4 Frontside Pinning

Another operating mode for CCDs can be explained by referring to Figure 5.5. Imagine that we maintain the voltage of the n-type layer but drive the gate voltage increasingly negative. We will eventually reach a condition where the voltage at ℓ_2, the Si–SiO$_2$ interface, is equal to that of the substrate, V_0 (= 0 V). Since the channel stops are at V_0, at this point holes flow from them into the Si–SiO$_2$ interface. As the gate potential is driven further negative, more holes flow to this interface and maintain it at V_0 so that the full gate potential falls across the SiO$_2$ layer and the potential well is maintained in the shape established by the potential on the n-type layer and the doping. This condition is described by saying the surface potential is pinned at the frontside of the CCD, and/or that the n-channel has inverted at the Si–SiO$_2$ interface.

Without frontside pinning, large dark currents can be produced when electrons are excited thermally into a trap at the Si–SiO$_2$ interface and then excited again into the conduction band. Under frontside pinned conditions, the large number of charge carriers at this interface fill the surface traps and suppress this dark current mechanism. However, as we have described the process so far, all the wells would be at the same depth and thus there would be no isolation of charge packets. To maintain packet isolation, at least one set of electrodes must have doping that is different from the rest in a way that maintains barriers against charge migration; the additional dopants are added by ion implantation. Their action to adjust the storage well depth is expressed in equation 5.2. Adjustment of the electrode design and use of implants to adjust the local surface potentials allows construction of multi-pinned-phase CCDs, which can collect charge under one set of electrodes while all phases are in an inverted condition. During readout, the voltages on the phases are manipulated to pass the charge to the output amplifier as in other CCDs. Such devices minimize requirements for cooling the detector to control dark current and can be operated in modes that eliminate blooming and provide a number of additional advantages (Janesick and Elliott 1992). Nothing comes for free, however. The well depths are reduced by a factor of two to three, to prevent charge from flowing over the implanted regions. Also, additional noise can result from "spurious charge" produced when the clocking to read out the CCD drives the holes away from the Si–SiO$_2$ interface with enough energy to release charge carriers by impact ionization, after which they collect in the CCD wells. This latter problem can be mitigated by limiting the voltage swing and speed of transition of the clock signals.

5.2.2 Charge Transfer Architectures

Several readout architectures for CCDs are illustrated in Figure 5.6. The simplest (shown in Figure 5.6(a)), called line address architecture, shifts the contents of the columns successively into an output register, or row, that transfers the charge packets to the amplifier. In this scheme, the array is illuminated while transfers occur. To avoid smearing the image, the CCD must either be read out quickly (in comparison with the time it takes to build up the picture) or be closed off with a shutter during readout. This requirement is usually no problem for very low light levels, for example, those encountered in astronomy. For applications where neither of these solutions is satisfactory, other architectures are used in which the charge is first shifted into a portion of the CCD that is protected from light and then read out. For example, in frame/field transfer architecture (Figure 5.6(c)), the charge is shifted quickly to an entire adjacent CCD section that is protected from light. Readout of this section can proceed at a slower rate as required (e.g., while another exposure is being obtained). However, in this architecture, smearing of images of bright sources occurs while the frame transfer is occurring. Interline transfer architecture shown in Figure 5.6(b) solves this problem by protecting every other column from light by a layer of evaporated metal. The charges are first shifted laterally into these columns and then shifted along these protected columns to the output register.

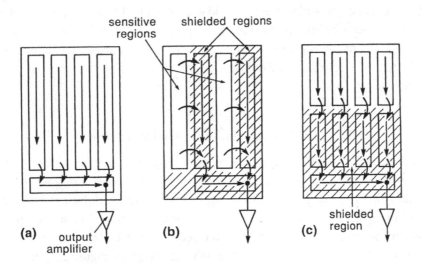

Figure 5.6 CCD readout architectures: (a) line address; (b) interline transfer; and (c) frame transfer. The hatched regions are shielded from light.

The interline register has dual contacts and control lines similar to those in an output register, to allow the two orthogonal transfer directions. Of course, the disadvantage of this approach is that half of the area of the array contributes no signal (one could make this criticism of the frame transfer devices also, but the shielded array portion will generally be placed to the side of the instrument field of view for them).

The readout scheme shown in Figure 5.6(a) allows the CCD to be operated in time-delay integration (TDI) mode. A fixed scene is swept over the array by moving the telescope or other optics in a manner that confines the motion of the image to occur along the columns of the array. The clocking along the columns is adjusted to move the charge at exactly the same rate as the scene, therefore building up the image while compensating for the motion of the object. Two advantages of this mode of operation are that the data are read out at a constant rate rather than in the bursts that come with reading the entire array, and corrections for fixed pattern noise (artifacts in the image due to imperfections in the uniformity of the detector array) are made easier by averaging the response over columns.

The CCD structure lends itself naturally to transferring charge in a linear fashion across the face of the array. For almost all applications, this constraint is acceptable. However, in tracking a scene moving rapidly in arbitrary directions, it would be helpful to move the electronic image of collected charge in any direction over the face of the detector to follow the scene motions and keep them from blurring the final result. This type of operation requires a CCD capable of TDI-like operation back and forth and in two directions. Burke et al. (1994) and Tonry and Burke (1998) demonstrate a 4-phase "orthogonal transfer" architecture with this attribute, as shown in Figure 5.7. To move charge to the right, gate 1 is held negative to act as a channel stop and gates 2, 3, and 4 are operated as a conventional 3-phase CCD. Charge can be moved to the left by reversing the clocking of these latter three gates. To move charge up, gate 2 is held negative as the channel stop and gates 1, 3, and 4 provide the 3-phase charge transfer. Charge is moved down by reversing the clocking.

5.2.3 Reading Out the Collected Charge

Most often, the collected charge packets are transferred sequentially to an output amplifier and read out as will be described in the next section. However, where the per-pixel collected charge is small, the net ratio of signal to noise can be improved by combining the signals from a number of adjacent pixels (at a loss of spatial resolution, of course). The CCD transfer process can be used for this purpose without introducing extra noise. For example, as shown in Figure 5.8, the columns can be advanced i pixels worth without clocking the output row, thus accumulating the contents of i pixels into the wells of the output row before this row is read out

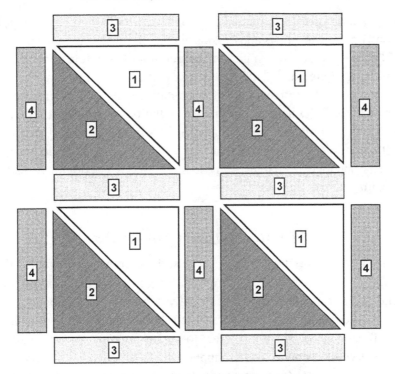

Figure 5.7 Orthogonal transfer CCD.

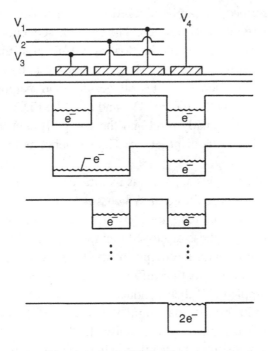

Figure 5.8 Charge combination in a CCD. The charge packets are combined under the electrode at V_4.

by the output amplifier. The combined pixel shape would then be rectangular with dimensions 1 pixel along rows and i pixels along columns. Similarly, the output row can be advanced j pixels worth into the output amplifier before a reading is taken, thus combining charge along rows. When combined with the combination along columns, the equivalent pixel size becomes i pixels along columns and j pixels along rows. If the CCD noise is dominated by the output amplifier read noise and these pixels were combined by reading them out *individually* and then averaging the results, the noise would be increased by a factor of $(ij)^{1/2}$ compared to the noise that is obtained if charge from the same pixels is combined on the CCD before reading out the signal.

5.2.4 *Measuring the Charge Packets*

At the end of the charge transfer process described above, the signal is delivered to an output amplifier, which can be grown on the same piece of silicon as the CCD. The input of the amplifier is electrically connected to a suitable structure to sense the charge from the CCD. Figure 5.9(a) shows a common approach, called a floating diffusion (FD). This term refers to a gate-free node where charge can be deposited but with no connection for application of an outside potential. The node is isolated from the rest of the output register by an output gate (OG). To the left of the OG, the figure shows a summing gate (SG) that can be used to combine charge packets as just described, or can be clocked in synchronism with the other gates just to pass the charge packets through. The readout sequence begins by setting the reset gate (R) positive to clamp the floating diffusion to the reset drain (RD) potential. R is then set to isolate the FD, and a reading is taken on the output of the source follower amplifier. The gate voltages are than manipulated to deposit a charge packet from the SG through the OG and onto the FD, and a second reading is taken. The amplifier is finally reset and the charge packet transferred through the R to the RD, where it is disposed of. The sequence is then repeated for the next packet.

An important feature of this approach is that the collected charge in each packet is measured *without* closing the reset switch so the kTC noise is frozen on C_S and does not degrade the net read noise. To illustrate, we consider an integrating amplifier circuit similar to those discussed previously in Chapter 4 and replace the detector and bias supply with the output of the CCD. Charge packets are transferred successively into the floating diffusion and onto the MOSFET gate. First, assume they are sampled and then conducted away by closing the reset switch, followed by a second sample to determine the zero level. In such a readout strategy, this circuit will be limited by kTC noise. If $T = 150$ K and $C = 0.1$ pF, $Q_N = 90$ electrons, many times higher than achieved with high-performance CCDs

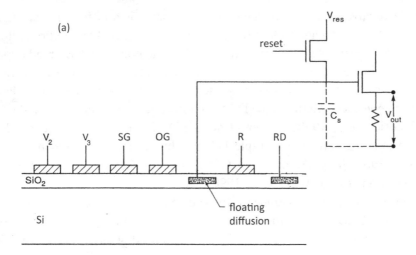

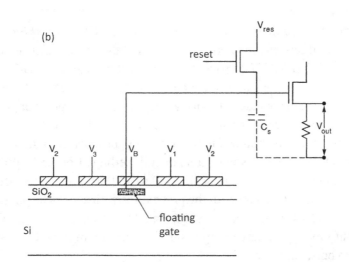

Figure 5.9 (a) Floating diffusion output amplifier; (b) floating gate output
amplifier.

using the alternative strategy that senses the collected charge without closing the
reset switch.

Where lower noise is desired, the floating gate (FG) arrangement in Figure 5.9(b)
can be used. In this arrangement, the transfer electrodes continue over and beyond
the diffusion that senses charge for the output amplifier. As the charge passes
over the FG electrode, it is sensed capacitively; that is, a complementary charge
is attracted to the floating gate. The floating gate is connected to the gate of the
amplifier FET that delivers the signal to the outside world. This process provides

for nondestructive readout of the charge: since charge is transferred over the floating gate and into a storage well in the CCD, it can be carried to additional floating gates and read again. Another implementation of the nondestructive read approach is a "skipper" readout. Rather than using the gates after the FG in Figure 5.9(b) to pass the packet to another output amplifier, their voltages can be manipulated to pass the charge back over the same FG to obtain another measurement, and so forth (e.g., Janesick et al. 1990; Moroni et al. 2011). The tradeoff is that the skipper approach is slower than the multiple FG one, but easier to implement and more compact. Using these techniques, CCD arrays can achieve read noises of an electron or even less (Janesick et al. 1990).

In many cases, one can continue to advance phantom charge packets to the output amplifier of a CCD after all the active pixels have been read out. This practice is called "overclocking"[2] and is useful for tracking any slow drifts in the behavior of the output electronics. In this regard, it is analogous to reference pixels used with many infrared detector arrays.

Where low noise is required with short readout times, CCDs can be made with multiple output amplifiers that read out subsections of the array. However, the electron multiplying CCD (EMCCD) provides a high-performance alternative. A gain section is added to the output register, where one of the three CCD phases can be run with an adjustable voltage. If this voltage is made large enough, charge is accelerated during the transfer sufficiently to create additional free charges by impact ionization, which are included in the transfer to produce gain. In practice the device is adjusted to have a very small gain per cell, perhaps 1 or 1.5%. With a gain register of about 600 cells, the net gains are 300 to 10,000. The gain process adds noise because of the statistical nature of the process (as with avalanche photodiodes, see Section 3.1.4.2). However, under very low light level conditions where the CCD must be read out rapidly, the large output signals overwhelm the CCD read noise and the signal to noise is improved.

5.3 Quantum Efficiency and Spectral Range

5.3.1 Front- and Back-Side Illumination

As Figure 5.1 is drawn, the electrode of the MOS capacitor would block photons coming from the right from illuminating the depletion region directly. If the electrode is transparent, which can be achieved by making it of very thin silicon and doping it heavily so it will be electrically conducting, then the depletion region can be illuminated through it (the oxide layer is very transparent through the visible

---

[2] Not to be confused with the practice in computer terminology of speeding up the clocking of a CPU or hard disk to obtain faster calculations.

and ultraviolet). This geometry is referred to as frontside illumination. Another important benefit of doped silicon electrodes is that electrical shorts between gates introduced by manufacturing flaws are usually not so highly conducting as to be fatal to the operation of the detector. Early CCDs that used metal electrodes suffered from very low yields because of such shorts.

Heavily doped thin silicon electrodes, however, are not transparent in the blue. As shown in Figure 2.4, the absorption coefficient goes up substantially at wavelengths shorter than 0.4 μm, as the photon energy approaches that for direct transitions at 4.1 eV. Even at wavelengths somewhat longer than this value, conventional silicon electrodes degrade the detector quantum efficiency. If silicon nitride is used as an insulator (in place of silicon oxide), it is found that significantly thinner silicon electrodes can be made and the transmittance improved. Electrodes can be made of materials that are more transparent than silicon nitride, such as doped indium tin oxide. The ultraviolet response can also be boosted by coating the CCD with a fluorescent dye such as Lumogen, which absorbs the ultraviolet photons and emits at longer wavelengths. However, none of these approaches can provide high quantum efficiencies, i.e., approaching 100%. In addition, a potential issue with frontside illumination is that the gate structures cause reflections and interference between layers, making the response nonuniform.

The problems associated with transparent electrodes can be circumvented with CCDs that are backside illuminated, that is, illuminated through the side opposite the electrodes, or from the left in Figure 5.1. Backside illumination leads to another set of issues that must be circumvented for good performance in the blue and ultraviolet. As viewed by the photons, the depletion region in a backside illuminated CCD lies behind a layer of silicon. Because of the very short absorption lengths for blue or ultraviolet photons, they generate charge carriers primarily near the back surface of the detector. The charge carriers must diffuse from these creation sites across any intervening silicon and into the depletion region. The physics of this process is very similar to that already discussed with reference to photodiodes, and the quantum efficiency will be as described in equation 3.23 (it can be worse than this equation predicts if significant numbers of charge carriers get caught in surface state traps at the backside of the silicon). A variety of techniques can be used to ensure that virtually all the charge carriers diffuse into the depletion region.

The most direct method to encourage full collection of charge carriers from blue and ultraviolet photons is to thin the detector. If much of the intervening silicon is removed, the photons will be absorbed close enough to the depletion region that the charge carriers will diffuse efficiently into it (however, if they have to diffuse very far, they may wander into a neighboring pixel, degrading the imaging performance). However, if the detector is thinned below an absorption length, its quantum efficiency will suffer because most photons will produce no charge carriers. The absorption lengths for silicon are a strong function of the wavelength (see Figure 2.4), so an optimum thickness will depend on the wavelength at which maximum efficiency is desired. For example, at $\lambda = 1$ μm, the absorption length is ~ 80 μm, while at $\lambda = 0.4$ μm, the absorption length is only ~ 0.3 μm. This large range of absorption lengths would pose an insoluble dilemma, except that the diffusion length for electrons in silicon is relatively long, up to 250 μm in intrinsic material at 150 K, and still 10–50 μm at the same temperature in the moderately doped material found in CCDs. It has been found that good blue quantum efficiency can be achieved by thinning to 15–20 μm. From comparison with the absorption length at 1 μm, however, it is clear that thinning to this degree will reduce the signals in the near infrared. In addition, where very thin detectors are needed, it is challenging to achieve uniform thickness and thus uniform response at the longer wavelengths.

A further important step in improving blue sensitivity is to treat the back surface of the CCD. Positive charge can accumulate at the backside due to broken bonds in the silicon itself and also due to impurities in the native oxide layer that grows on it. This charge must be counteracted or it will compete with the gates in collecting photoelectrons, which can recombine at the backside and are then lost to the detection process. One approach to solve this problem is called surface charging. For example, flooding the CCD with intense ultraviolet light is found to enhance the response. The ultraviolet photons cause photoemission of electrons in the bulk silicon, and some of these free electrons migrate to the backside surface, where they fall into traps in the oxide layer there. The trapped electrons persist after the ultraviolet radiation is removed and create an electric field that repels photoexcited electrons, preventing them from being trapped at the backside; instead, they diffuse toward the storage wells. Similar results can be achieved by growing a high-quality oxide layer with a low contamination of impurities on the backside surface and then adding a thin metal film that is a catalyst for disassociating oxygen, creating a layer of O_2^- ions bound to the film by chemisorption (Lesser and Iyer 1998). A variety of metals have been shown to be effective in this role: platinum, silver, and copper are examples (Lesser and Iyer 1998). A second approach is to provide for charging internal to the device. Molecular beam epitaxy can be used to deposit a layer of silicon including a monolayer of boron atoms, an approach called delta doping. Alternatively, boron or a similar dopant can be ion implanted into the backside of the CCD. In either case, the boron attracts electrons and the resulting field repels the photoelectrons from the back surface. Still another approach is to create a "biased flash gate" by adding an insulator layer and a transparent electrode, for example of indium tin oxide, and putting a potential on the electrode that repels the photoexcited electrons. When combined with appropriate thinning and an optimized antireflection coating, these techniques can produce good blue response with a high degree of uniformity and with quantum efficiency approaching 100%.

The performance of thinned CCDs suffers in the red beyond the effects of low quantum efficiency. Because the silicon is nearly transparent, most of the photons pass through it and some are reflected from the opposite side. These returning photons interfere with the incoming ones and produce strong interference patterns that appear as fringes in the image. The fringes can hamper the use of the CCD even where the wavelength-averaged quantum efficiency is acceptable.

5.3.2 Performance at Wavelengths Shorter than 0.3 or Longer than 0.7 μm

The absorption coefficient for silicon plunges for wavelengths $> 0.7\mu$m, as shown in Figure 2.4. Simply increasing the thickness of the CCD is an improvement for red performance, but if the absorbing layer is far thicker than the depletion region (typically about 5 μm thick) the charge collection may suffer and there will be a limit to the achievable quantum efficiency. In addition, lateral charge diffusion can result in signal wandering into adjacent depletion regions, leading to degradation of the quality of the image. Modern silicon processing can achieve extremely good control of impurities, leading to a better approach. From equation 5.2, if the minority impurity density (N_D), is made small, the width of the depletion region is large. CCDs can be manufactured using high-purity silicon for the absorbing layer; the field from the CCD electrodes then penetrates further into the material, allowing charge collection from depths of 40 to 50 μm. Further improvements are possible by adding an electrode to the backside of the device and applying a bias to increase the depletion depth, as in the fully depleted device in Figure 5.10. CCDs made in this manner are termed "deep depletion." In a deep depletion CCD, red photons can be absorbed efficiently over a depth where there is an electric field to aid in collecting the photoelectrons at the gate. With a device thickness of ~40 μm, quantum efficiencies as high as 80% can be obtained at 0.9 μm. Another strategy to achieve a deep depletion region is to implant a diode on the backside of the device similar to the diode used to establish the buried channel. The backside implant must be designed to be transparent to the incoming radiation. By placing a bias across the front- and backside diodes, it is possible to establish a field across the entire thickness of the device (Von Zanthier et al. 1998).

Silicon-based CCD quantum efficiencies drop at wavelengths shorter than 0.3 μm for a variety of reasons. The extremely short absorption lengths in the material exacerbate the charge collection issues in the blue; ultraviolet flooding, delta doping, or some equivalent technique is required for good charge collection. The refractive index changes substantially (by about a factor of 5) with wavelength

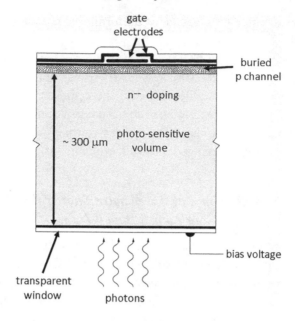

Figure 5.10 A deep depletion CCD pixel.

between 0.2 and 0.3 μm, and the real part of the silicon refractive index is less than 1 short of 0.2 μm, making classical broadband antireflection coatings very challenging. Excellent performance can nonetheless be achieved over selected moderate bandpasses (e.g., Hamden et al. 2016), and even over broad bands QEs of $\gtrsim$ 50% have been demonstrated (Nikzad et al. 2012).

However, the modest bandgap of silicon requires that the detectors be cooled to control dark current. Systems issues are another concern with silicon; the filters needed to block the visible light response of the detectors have significant absorption in the ultraviolet, and contamination collected by cooled silicon detectors absorbs in the ultraviolet. For these reasons, development of detectors using wider bandgap semiconductors such as GaN is of interest, as discussed in Section 3.1.2.2. In addition to relieving the problems in blocking visible light, such materials have the advantage of reduced dark current and generally do not require cooling below room temperature.

5.3.3 X-ray Detection

At low X-ray energies (up to ~10 keV), CCDs meet the goals for X-ray detectors well. Microchannel plates (next chapter) have also been used for X-ray astronomy in this energy range, but less widely than CCDs. Compared with CCDs, they have lower quantum efficiency (~30%) and energy resolution, but much faster time response. Many specifications of CCDs are similar whether they are used in the optical or the X-ray: similar pixel sizes ($\sim$10–20 μm) and array sizes, low read noise (2–4 electrons desired), and backside illumination (particularly for low energy X-rays).

Photons with wavelength shorter than $\sim$0.3 μm typically generate more than one electron–hole pair when they are absorbed in the CCD. The "quantum yield" is defined as the number of electron–hole pairs per interacting photon. For energetic photons (that is, those of wavelength shorter than 0.1 μm), it is found that 3.65 eV is required per electron–hole pair; the excess energy above the bandgap is dissipated as heat in the silicon lattice (Janesick and Elliott 1992). Therefore, X-ray detectors are generally operated in a single-photon-detection mode: each X-ray that strikes the CCD generates a number of electrons that are collected under the gates and soon after are transferred to the output amplifier. The number of electrons is proportional to the energy of the X-ray, so by reading out the CCD sufficiently rapidly that each X-ray is recorded separately, one also gets a low-resolution spectrum. Although operation in an integrating mode, as in the optical, would be feasible it loses information (and scrambles the spatial and spectroscopic properties of the source). At typical detection rates with existing X-ray observatories, it is necessary to read out every few seconds (e.g., 3.2 s for the AXAF CCD Imaging Spectrometer (ACIS)). These rates are obtained while preserving high *CTE* and low read noise by using frame transfer devices.

The number of free electrons yielded when an X-ray is absorbed is

$$N_e = \frac{E_X}{w}, \tag{5.23}$$

where E_X is the X-ray energy and w is the effective ionization energy as just discussed, $w \sim 3.65$ eV. The standard deviation of N_e is

$$\sigma_e = \sqrt{F N_e}. \tag{5.24}$$

The Fano factor, F, reflects the fact that the generation of the free electrons is not a Poisson-distributed process. In general, F is less than one when the electrons are produced in a correlated manner. In this case, the average size of the electron cloud is determined by the X-ray energy with relatively minor statistical fluctuations around this value, so $F \sim 0.12$. If the CCD read noise is σ_{read} and there are no other sources of extraneous noise, the spectral resolution is

$$R = \frac{E_X}{\Delta E_X} = \frac{N_e}{2.36 w \sqrt{\sigma_e^2 + \sigma_{read}^2}}, \tag{5.25}$$

where ΔE_X is the full width at half maximum of the distribution of output signals for an unresolved spectral line.

In practice, the response to an unresolved line is more complex than implied above because the absorption of the X-ray may occur in layers of the CCD other than the depleted region from which charge is collected, e.g., in the gate polysilicon or oxide layers. Consequently, the response described by equation 5.25 is accompanied by a low energy tail. There may also be a spectral feature 1.829 keV below the line energy (assuming it is larger than this value) due to escape of an X-ray emitted by an excited silicon atom. A second spurious feature called the fluorescence peak occurs at 1.739 keV when the silicon emits a fluorescent X-ray that is detected as a separate event. Fortunately, these issues account for only a modest part of the response of the CCD.

There are some additional cautions for interpreting X-ray CCD signals. If the X-ray flux onto the detector is too high (or the readout rate too low), more than one X-ray will frequently be absorbed by a pixel within a readout time, termed pileup. As a result, the signal will correspond to that for the sum of the energies of the two X-rays; that is, the spectrum of the source deduced from the CCD outputs will be distorted toward apparently enhanced high energies. This problem can be difficult to remove from the data; hence, observations should be designed carefully to avoid it if possible. A more benign problem occurs when an X-ray is detected while the CCD charge is being transferred. Because of the extended time for the free electrons to diffuse into the wells, the signal then appears as a streak across the reconstructed image. Because of its unique shape, such a signal can be readily identified and removed from the data. Yet another issue arises if the charge transfer efficiency (*CTE*) of the CCD is too low, resulting in increased noise. Unfortunately, the *CTE* can be degraded by cosmic-ray protons (and energetic electrons and photons). These particles can break bonds in the silicon crystal permanently, leading to crystal lattice damage that can trap free electrons as they are transferred over the damaged spot.

The efficiency of CCDs in the X-ray varies significantly with energy because of the absorption properties of X-rays in silicon, which reflects the general process in semiconductors. Photoelectric absorption is strong when the X-ray energy matches the binding energy of an inner shell electron, but below this energy the X-ray only undergoes Compton and Thompson scattering so the absorption coefficient falls dramatically just below the inner shell binding energy. It then rises as λ^3 toward lower energy (longer wavelength), as shown in Figure 2.5. Relatively thin CCDs – 10 to 20 μm thick – can have high absorption up to 3–5 keV. Normal front illuminated CCDs are suitable for operation from 1 to ~6 keV, and back illuminated devices are preferred below 1 keV because there is no absorption of the X-rays by the gate structure. Figure 5.11, curves (a) and (b) show typical quantum efficiencies for these types of device.

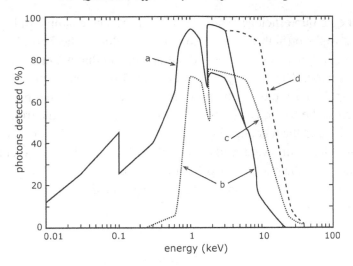

Figure 5.11 Quantum efficiencies of X-ray CCDs: (a) is back illuminated, (b) is front illuminated, (c) is front illuminated deep depletion, and (d) is back illuminated pnCCD.

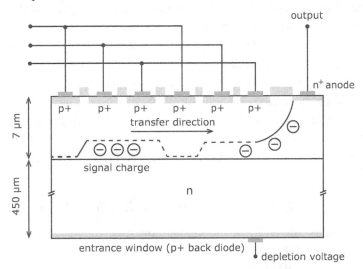

Figure 5.12 A pnCCD. After Hartmann et al. (2006).

However, depletion thicknesses of ~100 μm are desirable to cover the full range of operation of traditional Wolter telescopes such as Chandra and XMM, and even greater depths are necessary to extend the energy response toward 20 keV. Deep depletion CCDs, discussed in Section 5.3.2, can extend the response above 10 keV (Figure 5.11, curve (c)). An even deeper depletion region can be achieved with the pnCCD (Figure 5.12) (e.g., Strüder and Meidinger 2008). These devices can have useful quantum efficiency ($> 20\%$) to 20 keV. The example in Figure 5.12 is formed on a wafer of high-purity silicon, 450 μm thick. The 7 μm layer is grown on this wafer, and includes the n-doped output contact that maintains contact to the bulk of the silicon. The CCD gates are deposited on strongly p-doped structures and the backside has a monolithic p-doped surface, thus forming back-to-back pn junctions across the wafer. By a suitable choice of voltages on the front and back p-type electrodes and the n contact layer, it is possible to back-bias the two junctions both to deplete the entire wafer and to bring the minimum electron potential close to the CCD gates and the n+ layer. Conventional manipulation of the gate electrodes creates potential wells to collect charge and then transfers the collected charge to an output amplifier.

5.3.4 *Germanium CCDs*

CCDs depend on very special properties of silicon: (1) it can be grown in high-purity, virtually flaw-free crystals (critical to avoid effects like traps that interfere with charge transfer); and (2) it has a rugged and easily grown oxide layer that provides electrical insulation. Recent work has demonstrated high crystal quality for germanium, but its simplest oxide, GeO, is water-soluble and cannot withstand high temperatures (Leitz et al. 2017). The oxide GeO_2 is more stable, particularly if protected with films of more robust dielectrics. The necessary components for germanium CCDs have been demonstrated on small-scale arrays and individual devices (Leitz et al., 2017). If large format arrays can be developed, they would be useful both for the extension of response to ~1.7 μm (Figure 2.4) and for improved performance in X-ray detection at energies above 10 keV (Figure 2.5).

5.4 Optical/Infrared Detector Test Procedures

Test procedures for all types of the semiconductor detectors we have been discussing are similar. Of course, entire books are written to describe them (e.g., Vincent et al. 2015). We give a quick overview in part because it reveals some aspects of detector performance. Anyone actually carrying out testing of these devices should seek additional guidance.

5.4.1 *Measurement Approach*

Parameters of an amplifier/detector system that we may wish to measure include: (1) responsivity (as a function of frequency), (2) linearity, (3) dynamic range, (4) spectral response, and (5) noise (also as a function of frequency). Other performance parameters such as quantum efficiency and *NEP* can be derived from those listed. Central to such measurements is the availability of laboratory sources accurately calibrated in physical units, which we discuss first. We next discuss the approach for a detector with a non-integrating output, and then with an integrating amplifier.

5.4.1.1 Calibrated Sources

The most satisfactory way to calibrate a laboratory source is based as directly as possible on physical principles and thus depends largely upon measurements of physical constants and does not get undermined by multiple sub-calibrations. Infrared sources benefit from the simplicity of constructing a reasonable facsimile of a blackbody. Laboratory blackbody sources have a blackened chamber, usually conical, that is heated uniformly. Light escapes through a hole at the base of the cone that is much smaller than the diameter of the base. This arrangement is close to the textbook definition of a blackbody emitter, usually sketched as an integrating sphere with a small exit hole so that the escaping fraction is so small compared with the total energy density inside the sphere that the radiation remains in thermal equilibrium with the walls. Blackbody laboratory sources of this type are commercially available, with operating temperatures up to 1200 °C. The peak of the emission of a blackbody is given by the Wien displacement law:

$$\lambda_{max} = \frac{2900 \ \mu\text{m (K)}}{T},$$ (5.26)

that is, it is about 2 μm for the highest temperature standard blackbody source. Thus, these sources are not satisfactory for the visible range.

Instead, tungsten filament bulbs are used as visible light sources. However, in a conventional tungsten bulb, the filament erodes and tungsten is deposited on the envelope, both effects reducing the output. This degradation can be combated by filling the bulb with an inert gas with a small addition of a halogen and operating at very high temperature; the bulb envelope itself is made of quartz to survive the temperature. In this case, a halogen cycle chemical reaction re-deposits the evaporated tungsten back onto the filament, stabilizing its output. The U.S. National Institute of Standards and Technology (NIST) issues calibrated quartz tungsten halogen standard lamps that can operate at filament temperatures above 3000 K, calibrated in spectral radiance between 0.3 and 1.1 μm to 1 to 3%, depending on wavelength. However, these lamps still get out of calibration after some tens of hours of operation. Therefore, the fundamental calibration is based on silicon diodes. From equation 3.2, the responsivity of such a device depends only on its quantum efficiency and fundamental physical constants. This equation ignores reflection losses, and also is derived assuming there are no internal loss mechanisms to the diode that could absorb photoelectrons without producing a signal.

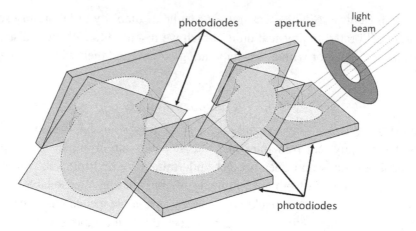

Figure 5.13 Operation of a six-element transmission tunnel-trap detector (detectors 3 and 6 are drawn as being transparent for clarity). After Brown et al. (2006).

Commercial photodiodes are now available in which the response agrees with theoretical estimates within 0.04% (between 0.46 and 0.63 μm), indicating that the internal losses are negligible (Saito 2012). Such diodes can be used in a tunnel-trap configuration as in Figure 5.13. They are arranged so all the light reflected off the nth photodiode strikes the $(n + 1)$th one, so that when the outputs are summed virtually all the reflected energy is accounted for. Because these devices operate at nearly 100% efficiency (> 99.5% between 0.5 and 0.9 μm, Woodward et al. 2018), there are no uncertain corrections that need to be applied to their signals, and their behavior is extremely stable. They can be used in arrangements with tunable lasers to calibrate accurately over a range of individual laser wavelengths to give an accurate spectral calibration, as well as working separately to calibrate at selected photometric bands. (Brown et al. 2006; Woodward et al. 2018).

5.4.1.2 Non-Integrating Detector System

The responsivity of a detector/amplifier system − the volts or amperes of output per watt of input signal − can be measured by having the system view one of these calibrated sources, calculating the input signal power, and measuring the output electrical signal. To be confident that the signal is derived only from the source, a second set of measurements must be taken with the output of the source blocked but no other measurement conditions changed, and the signal in this case must be subtracted from that when viewing the source. This requirement is most directly met by turning the source off; however, infrared blackbodies cool very slowly so instead a mechanical chopper is used to transmit and block the signal alternatively. In this case, the results must be corrected for any emission by the chopper itself, which can be significant at infrared wavelengths.

The most straightforward way to view the detector output is with an oscilloscope or equivalent computer interface. With corrections for the electronic gains of the detector support electronics and of any amplifiers in the test apparatus, it is then straightforward to compute the detector response. Because the detector views the source from a known distance, the power incident on it can be calculated as described in Chapter 1. In some circumstances, however, the signal modulated by a chopper wheel is measured with a meter without full time resolution of the waveform. In this case, we need to correct the calculation of the power received from the source for the fact that the signal contains components at overtone frequencies in addition to the chopped frequency, f. If we take the chopped signal to be a square wave of amplitude P_D calculated from equation 1.12, it can be represented by the series

$$P(t) = \frac{P_D}{2} + \frac{2P_D}{\pi}\left[cos(\omega t) - \frac{cos(3\omega t)}{3} + \cdots\right], \qquad (5.27)$$

where $\omega = 2\pi f$. The $cos(\omega t)$ term is the one at the fundamental frequency. For a cosine wave, the rms value is the amplitude divided by $\sqrt{2}$ or, from equation 5.27, the rms power received by the detector is $\sqrt{2}P_D/\pi = 0.4502\, P_D$. The responsivity of the detector is then

$$S(f) = \frac{V_{rms}(f)}{0.4502 P_D}. \qquad (5.28)$$

The factor we have derived for conversion of a square wave signal to an rms value at the fundamental frequency is called the waveform factor (sometimes it is also termed the modulation factor). Strictly speaking, our derivation for a square wave is only valid for the limiting case of an infinitesimal aperture. However, the error introduced by assuming square wave modulation is no more than 1% for aperture diameters up to 17% of the gap between teeth in the chopper wheel. Where larger apertures or other modulation schemes are used, consult Vincent et al. (2015) for a discussion of the appropriate corrections.

The response as a function of frequency can be determined either by analyzing the time-resolved detector output captured by a computer, or by repeating the above measurement at different rotation rates of the notched wheel, or by substituting wheels with different numbers of teeth. To determine the linearity of the detector, one must make accurate relative variations in the input signal power, either by varying the size of the aperture over the light source[3] and/or by changing the distance between the source and the detector. This same set of measurements can determine the dynamic range. The spectral response is measured by interposing

---

[3] Although caution must be used not to make this aperture so small that diffraction effects become important.

spectral filters between the source and detector. Greater resolution can be achieved by using a monochromator in place of filters.

One may also want to measure the noise in physical units. Fortunately, it is a lot easier to obtain a "standard noise source" than a calibrated signal source. A high-quality resistor that produces Johnson noise can be suitably amplified, and the result compared with the noise from the detector system. With measurement of the amplifier characteristics, the resistor input can be referred to physical principles directly. With this information and the responsivity, we can determine the noise equivalent power (*NEP*), the input power that can be detected at a ratio of signal to noise of one, with the noise determined in an equivalent 1 Hz bandwidth. The *NEP* is the noise (for example, in $\mathrm{VHz}^{-1/2}$) divided by the responsivity (in this case, in V W^{-1}). The detective quantum efficiency can be determined in a series of steps. First, the noise is measured while the detector views a constant level of signal power. Next, the response to this signal is measured and combined with the noise to yield the output signal-to-noise ratio. The level of signal power received by the detector can be calculated and used to derive the input signal-to-noise ratio as discussed in Chapter 1. The *DQE* is then derived as in equation 1.26. There are additional common performance metrics. Since *NEP* has the counter-intuitive behavior that smaller is better, sometimes the detectivity $= 1/NEP$ is quoted, abbreviated D. Particularly for photodiodes, the zero bias resistance, R_0, is another metric as discussed in Chapter 3, see equation 3.25. For many detector types, these metrics get better in inverse proportion to detector area. Metrics that compensate this dependence are D^*, the detectivity multiplied by the area, or the product of R_0 and area, the $R_0 A$.

5.4.1.3 Integrating Readout System

Properties of a detector/integrating amplifier combination, i.e., in a CMOS or infrared array or a CCD, can be measured using techniques similar to those just described. The output of such a detector is routinely digitized and input to a computer where the readout voltages are recorded as signals in digital numbers (DNs). A critical first step is to measure the gain of the detector/amplifier unit cell, sometimes called the pixel gain. The voltage gain of the FET (and following electronics) can be determined by injecting a calibrated voltage signal into the integrating node using the reset line and recording the response in DN. The voltage change produced by an electron deposited on the gate of the FET depends on the gate capacitance, or more properly the sum of the detector, gate, and any other capacitance terms in parallel. Once the capacitance has been measured or determined by analysis, the pixel gain is calculated as the electrons per DN.

There is a useful general technique to obtain the capacitance of the integrating node, C_S, as long as the readout does not add noise above the square root of the number of collected charges. In this technique, one compares the output noise, $\langle V_N^2 \rangle^{1/2}$, with the signal, V. The signal is proportional to the number of collected charges and the noise to the square root of this number, that is,

$$V = \frac{Nq}{C_S} \tag{5.29}$$

and

$$\langle V_N^2 \rangle^{1/2} = \frac{N^{1/2}q}{C_S}, \tag{5.30}$$

where N is the number of collected charges. Consequently,

$$C_S = \frac{qV}{\langle V_N^2 \rangle}. \tag{5.31}$$

In the more realistic case where there is a noise contribution from the amplifier, the same argument can be made by comparing signal and noise over a range of signal values so the amplifier noise can be separated from the statistical uncertainty in the number of charges. Caution is advised in this situation since some amplifiers have an amplitude-dependent noise component that could be confused with the statistical noise.

Although the above procedure is simple to implement, it can be seriously under-mined if there is a significant level of capacitive coupling between adjacent pixels, called interpixel capacitance (IPC), see Finger et al. (2005). By increasing the effective integrating node capacitance for a pixel, IPC artificially damps down the noise and leads to an underestimate of the pixel gain (i.e., fewer electrons per DN than the correct value). Typical arrays have sufficient IPC to cause factor of 2 errors for CMOS arrays and some tens of percent for infrared arrays. For further discussion and a description of means to measure and correct the effects of IPC, see Finger et al. (2005).[4]

The response is obtained as the difference of signals obtained with the cali-brated source alternately in view and hidden from view synchronously with the read cycle of the amplifier. Integrating systems often accumulate signals for a number of seconds between resets, in which case the measurements can be switched between source and nonsource manually; if the integration is much shorter, it may be necessary to obtain the data with a chopper synchronized with the integration cycle. Extending the measurements to determine the linearity and dynamic range is simplified with an integrating amplifier because one needs only measure the linear-ity and other properties of the integration ramp. These measurements are usually referred back to the equivalent detector outputs through the pixel gain.

---

[4] A brief overview is given in the next section also.

The metrics discussed above do not translate very transparently to high-performance detector arrays with integrating amplifiers (but see Problem 5.7), which are more typically described in terms of quantum efficiency, read noise, and dark current.

5.4.1.4 Imaging Performance

A basic requirement in imager design is to set the equivalent angular size of the array pixels to extract the maximum possible information. Pixels that project to angles that are too small will under-utilize the detector array and reduce the size of the field of view unnecessarily. Pixels that project to angles that are too large will lose information on small angular scales.

We can determine the required pixel size by determining the *MTF* of a detector array and seeing under what conditions the spatial frequency spectrum incident onto the array is preserved. We only have to compute the *MTF* of a pixel, since the full suite of pixels will have the same impact on the result. The function

$$rect(x) \begin{cases} = 1 & \text{for} \quad |x| \le 1/2, \\ = 0 & \text{otherwise} \end{cases} \tag{5.32}$$

is the response function for a pixel of width 1. We need to modify it for a pixel of width w:

$$rect(x/w) = 1 \quad \text{for} \quad |x/w| \le 1/2. \tag{5.33}$$

From Table 1.2, the corresponding Fourier transform is

$$F(w) = \frac{1}{\left|\frac{1}{w}\right|} \frac{sin\left(\frac{\pi f_s}{\pi\left(\frac{1}{1/w}\right)}\right)}{\pi\left(\frac{1}{1/w}\right)} = |w| \frac{sin(\pi w f_s)}{\pi w f_s}, \tag{5.34}$$

where f_s is the spatial frequency. The *MTF* is $\sqrt{FF^*}$ (equation 1.34), normalized to 1 at $f_s = 0$. Since

$$\frac{sin(x)}{x} \to 1 \quad \text{as} \quad x \to 0 \tag{5.35}$$

it follows that

$$MTF = \left|\frac{sin(\pi w f_s)}{\pi w f_s}\right|. \tag{5.36}$$

Some samples are shown in Figure 5.14. For illustration in the figure, we have defined units of frequency so $f_s = 1$ corresponds to the natural cutoff frequency of a diffraction-limited optical system, i.e., D/λ cycles/radian for a telescope of diameter D. Since $f_s = 1/P_S$ where P_S is the spatial period, the *MTF* at D/λ for sampling with two pixels gives a *MTF* at f_s of 0.64, i.e., with this pixel scale, the

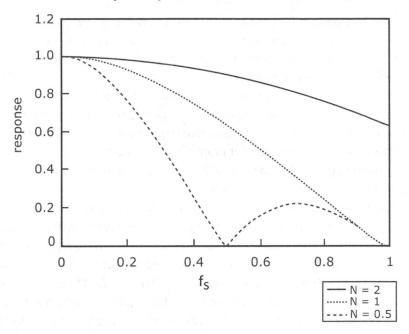

Figure 5.14 The *MTF* due to pixelization of a detector array, shown as a function of the number of pixels (N) across the FWHM of the diffraction-limited image of an optical system.

blurring through the pixelization of the detector array only modestly degrades the highest resolution details of an image. This case is shown in Figure 5.14 as N = 2. If we instead use a scale with one pixel matched to D/λ (N = 1), the attenuation of the fine details (high spatial frequencies) is severe, and at half a pixel per D/λ details that have one spatial period across a pixel cancel and give no signal, as indicated by the minimum at $f_s = 0.5$.

There are a number of other effects that can degrade the imaging performance of a detector array: (1) optical crosstalk, most important for arrays with thick pixels where the range of angles of the incoming photon stream results in it spreading over multiple pixels within the detector volume; (2) charge diffusion, where a photocharge created in one pixel diffuses into a neighboring one; and (3) interpixel capacitance (IPC), the capacitive coupling of a pixel to its neighbors. None of these effects lend themselves to straightforward calculation of *MTF*s (see Pinkie et al. (2013) for an example of the kind of modeling required). Nonetheless, we will describe briefly how they are determined. The first, optical crosstalk, is usually evaluated analytically, from a knowledge of the detector three-dimensional architecture and an optical raytrace (including effects of diffraction). The other two can be measured with a light source focused onto the array. Because it can be difficult to characterize a point image with the necessary accuracy, the measurements are sometimes done with a custom detector array with masking to isolate a single (or multiple) pixel(s), with the surrounding ones blocked from the light so any signals they record are from crosstalk. Alternatively, a bar occulter can be used; it is easier to characterize the edge performance and placement of a line occultation than a point source. Yet another approach is to use energetic particles – cosmic-ray muons or X-rays from a laboratory source – to create large numbers of free charge carriers. Many of these events will spread over multiple pixels and have to be identified from their asymmetric appearance and rejected. The single-pixel events that are left can be used to measure the reaction in the surrounding pixels. In the third case, IPC, there is another useful measurement technique. Because this behavior creates an electronic connection between neighboring pixels, their signals will be correlated – to an extent that is determined by the level of IPC. Therefore, an autocorrelation analysis of the pixel-to-pixel behavior can be used for an accurate determination of the IPC (see, e.g., Finger et al. (2005) for details). In a CCD the charge transfer efficiency is another measure of crosstalk; to be strictly correct, the measure is 1-*CTE*. A popular way to measure the *CTE* is to illuminate the detector with a ^{55}Fe X-ray source, with emission predominantly at 5.899 keV; absorption in silicon then produces 1620 free electrons. After rejecting "hits" on multiple pixels, the remaining events should have a constant amplitude independent of position on the CCD. If the *CTE* is poor, the pixels furthest from the output amplifier (and hence that need to go through the largest number of transfers to reach it) will have fewer remaining electrons.

5.5 Example

We consider a 500×500 pixel, 3-phase, buried channel CCD operated at 170 K with 25 μm electrodes. The output amplifier has an effective input capacitance of 2×10^{-13} F, and it is being read out as follows:

(1) The charge is transferred to the gate of the output amplifier and a reading is taken.

(2) The amplifier is reset.

(3) A second reading is taken.

(4) The signal is computed as the difference of these two readings.

To maximize the time spent integrating on signal, the array is read out in a total time of 0.2 s with 10^5 collected charges in each detector well. Assume the mobility scales as $T^{-3/2}$ between 300 K and 170 K, starting from the value in Table 2.3.

List the contributions to the noise that will be observed in the signal, and estimate each one (in electrons) for the pixel furthest (that is, most transfers) from the output amplifier.

(a) There will be a noise of $N^{1/2}$ in the number of collected charges, or 316 electrons.

(b) Because the signal is determined by resetting between samples, there will be kTC noise:

$$Q_N = (kTC)^{1/2} = 2.16 \times 10^{-17} \text{ C, or 135 electrons.} \qquad (5.37)$$

(c) There will be noise because of incomplete charge transfer. The transfer time is dominated by diffusion. To compute the time constant, we need the diffusion coefficient $D = \mu kT/q$. From Table 2.3, with mobility scaling as $T^{-3/2}$, we get

$$\mu_n = 1350 \text{ cm}^2 \text{ V}^{-1} \text{ s}^{-1} \left[\frac{300 \text{ K}}{170 \text{ K}}\right]^{3/2} = 3165 \text{ cm}^2 \text{ V}^{-1} \text{ s}^{-1}. \qquad (5.38)$$

We then obtain $D = 4.64 \times 10^{-3} \text{ m}^2 \text{ s}^{-1}$. From equation 5.9, with $L_e = 25 \times 10^{-6}$ m, we get

$$\tau_{th} = 5.46 \times 10^{-8} \text{s}. \qquad (5.39)$$

Estimating the capacitance from equation 5.5, with $X_0 = 0.1$ μm,

$$C = \frac{(25 \times 10^{-6} \text{ m})^2 (4.5)(8.85 \times 10^{-12} \text{ F m}^{-1})}{1.0 \times 10^{-7} \text{ m}} = 2.49 \times 10^{-13} \text{ F}, \qquad (5.40)$$

and, from equation 5.12,

$$N_0^{crit} = \frac{\pi (4.64 \times 10^{-3} \text{ m}^2 \text{ s}^{-1})(2.49 \times 10^{-13} \text{ F})}{2(0.3165 \text{ m}^2 \text{ V}^{-1} \text{ s}^{-1}))(1.60 \times 10^{-19} \text{ C})} = 35,800. \qquad (5.41)$$

Passing the charge from the furthest pixel through the output amplifier requires 500×3 transfers down the column and 500×3 transfers along the output row. The column transfers occur sufficiently slowly that the transfer efficiency should be high. The output row must make $500 \times 500 \times 3$ transfers in 0.2 s, or 1 in 2.67×10^{-7} s $= 4.89\tau_{th}$. We therefore lose a fraction $\varepsilon = e^{-4.89} = 7.52 \times 10^{-3}$ of N_0^{crit} for the thermally driven transfers (see Section 5.2.1.2), or a net loss of $7.52 \times 10^{-3} \times 35,800 = 270$ electrons per transfer. Since we started with 10^5 electrons, the net charge transfer efficiency is

$$CTE = 1 - \left(\frac{270}{10^5}\right) = 1 - 2.7 \times 10^{-3} = 0.9973. \qquad (5.42)$$

Substituting into equation 5.3,

$$N_{n,TL} = \left[(2)(2.7 \times 10^{-3})(1500)(10^5)\right]^{1/2} = 900 \text{ electrons.} \qquad (5.43)$$

The total read noise is then

$$N_{tot} = (316^2 + 135^2 + 900^2)^{1/2} = 963 \text{ electrons,} \qquad (5.44)$$

so the CCD is operating at about three times above the background-limited noise level of 316 electrons.

5.6 Problems

5.1 Comment on the feasibility of producing a buried channel CCD with doping to allow low noise operation as an extrinsic photoconductor in the mid-infrared spectral region (for example, near 10 μm).

5.2 Consider the CCD in the example, but illuminate it at a level that produces 1000 electrons per pixel in an integration. Describe (numerically where appropriate) the changes required in the operation of the chip to stay within 30% of background-limited operation.

5.3 Estimate the optical crosstalk for an array of silicon detectors 100 μm square and 0.9 mm thick, illuminated at f/2. Assume the absorption length is 0.9 mm and the refractive index of silicon is 3.4. Ignore reflection from the back face of the detector.

5.4 Consider a CCD that has a series of m pixels that have been illuminated uniformly. After n transfers with imperfect CTE, the output signal will show a deficit of ΔV_1 at the first pixel because of charge that has been lost to following pixels. The following pixels will show increasing signals, with the maximum, V_m, at pixel m. Show that the CTE is given approximately by

$$\left(1 - \frac{\Delta V_1}{V_m}\right)^{1/n}.$$

For n large and $CTE \sim 1$, derive the approximation

$$CTE \sim 1 - \left(\frac{1}{n}\frac{\Delta V_1}{V_m}\right).$$

5.5 Consider a CCD similar to the one in Figure 5.5. Let the thickness of the n-type region be 1 μm and the doping $N_D = 10^{16}$ cm^{-3}; and let the thickness of the p-type region be 40 μm and its doping 10^{14} cm^{-3}. Assume the diffusion length in the p-type region is 20 μm. Estimate the quantum efficiency in the blue spectral region (ignoring reflection losses and surface state charge trapping).

5.6 Suppose an array has detectors with high absorption efficiency of thickness 10 μm and with a diffusion length of 20 μm. What is the minimum pixel size so that the MTF of a Nyquist-sampled signal is ≥ 0.5?

5.7 Suppose a detector system is receiving a monochromatic signal at 0.6 μm of 5×10^{-18} W. The detector has a quantum efficiency of 50% and the integrating amplifier has a read noise of 50 electrons. The signal is integrated for

10 seconds and a second 10 second integration is taken off the source. What is the *NEP* of this system for this measurement?

5.7 Further Reading

Buil (1991) – good overview of use of CCDs in astronomy, including support electronics and software

Burke et al. (2005) – useful review of CCD and CMOS arrays

Durini (2015) – collection of good articles on CMOS and CCD sensors

Jerram et al. (2001), Hynecek (2001), and Strüder et al. (2008) – introductions to development of electron multiplying CCDs

Holland (2013) – general description of CCDs used for X-ray astronomy and other purposes

Holst (2008) – describes testing of infrared arrays

Howell (2006) – starting point for CCD imaging and data reduction in professional astronomy

Janesick (2001) – comprehensive discussion of construction, testing, and operation of CCDs for low light levels

Lesser (2015) – overview of state-of-the-art high-performance CCDs

Tabbert and Goushcha (2012) – broad technical overview of detection techniques in the optical

Theuwissien (1995) – very thorough discussion of CCD principles, operation, and construction

Tulloch and Dhillon (2011) – good overview of EMCCDs

6

Other Photodetectors

This chapter discusses an assortment of photodetectors that do not quite merit chapters of their own.

6.1 Photography

For more than a century, photography was the leading type of photoelectric detector. It provided many-pixel arrays with the ability (1) to integrate for hours if desired, allowing detection of unprecedentedly faint signals; (2) to store the results in a form that allowed archiving and analysis as needed; and (3) at least to allow limited manipulation of the results to bring out subtle features. However, it has been almost completely replaced by other detector types because of a number of inherent weaknesses: (1) detective quantum efficiencies are low, at best only a few percent; (2) it is not possible to "reset" the detector and take another exposure with exactly the same artifacts, an important capability with electronic detectors that allows them to reach detection levels far deeper that the few percent achievable with photography; and (3) accurate calibration is not possible both because of the inability to repeat measurements with the identical detector and because of nonlinearity and other image artifacts. Nonetheless, given its historic importance, we provide a short description of this process; those wanting more detail are referred to earlier editions of this book and to the references suggested at the end of this chapter.

Photography is based on chemical changes that are initiated by the creation of a conduction electron when a photon is absorbed in certain kinds of semiconductor. These changes are amplified by chemical processing until a visible image of the illumination pattern has been produced. A variety of photographic materials have been discovered, but those based on silver halide grains (tiny crystals of AgBr, AgCl, or AgBrI) have significant advantages in sensitivity and are used to the exclusion of alternatives. Although some details remain controversial, the silver

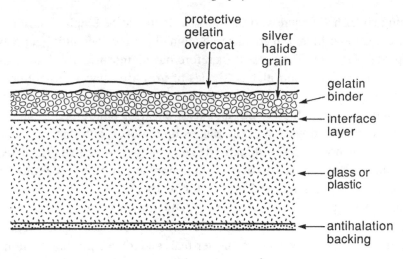

Figure 6.1 Cross-section of a typical photographic plate.

halide photographic process is now explained in terms common to other types of solid state detector.

A schematic cross-section of a photographic plate is shown in Figure 6.1. Silver halide grains are the active detectors; they are suspended in a gelatin binder to form the "emulsion," which is coated onto glass for mechanical stability. The bonding of the gelatin to the glass is improved by use of an interface layer, and the grains are often protected in red sensitive plates by adding an overcoat of gelatin. To avoid reflections from the back of the glass, it is coated with an "antihalation" backing that absorbs light that passes through the emulsion. Photographic film uses a flexible plastic support such as polyester or Mylar instead of glass to improve handling convenience at the expense of some dimensional stability. In the following discussion we will use "plates" as a short form of "plates or film." The gelatin suspension has many important attributes: (1) it is transparent to light (except ultraviolet); (2) it is compatible with the chemical processing that is required to produce images; (3) it swells when soaked in water, allowing the chemicals to contact the silver halide grains for development; (4) it is readily available; and (5) it can be made to have uniform properties.

The photographic process is centered on the silver halide grains. When photons strike a grain, they excite its atoms by raising electrons to the conduction band. A chain of events is triggered that results in the growth of a small silver speck, known as a development center, in the grain. During chemical development, these development centers act as catalysts that can blacken the entire grain through reduction of silver ions to silver atoms. Given time, the undeveloped grains would also blacken; this process is stopped by fixing the developed image by dissolving any remaining silver halide and washing it away. It should be emphasized that photographic detection is binary in character; a grain either receives sufficient photons to develop or it does not. The process therefore has an intrinsically limited dynamic range, and it provides controllable "shades of grey" in images largely because of the range of properties of the grains distributed over the plate. However, by intentionally introducing multiple types and sizes of silver halide grain, photographic materials are provided with dynamic ranges up to 400:1, and materials optimized for this performance aspect could have dynamic ranges some five times greater.

The inherent spectral sensitivity of photographic emulsions is limited to blue wavelengths. For example, the bandgap for AgBr ($E_g = 2.81$ eV) requires that a photon have a wavelength shorter than 0.44 μm to raise an electron into the conduction band and initiate the detection process. Two absorption regimes can be distinguished. For wavelengths shorter than about 0.32 μm, absorption occurs through direct electron transitions, and $a(\lambda) > 10^5$ cm^{-1}. For longer wavelengths, the photon energy can only be absorbed by indirect transitions, and $a(\lambda)$ is significantly reduced. From equation 1.18, we see that for grain diameters near 1 μm ($= 10^{-4}$ cm), $\eta \approx 0.9$ in the direct absorption region (where the losses are largely from reflection), but η will depend on wavelength in the indirect absorption region.

The Gurney–Mott hypothesis describes the processes subsequent to absorption of a photon by a silver halide grain. Photon absorption results in the elevation of a valence electron into the conduction band, producing an electron–hole conduction pair:

$$h\nu + Br^- \rightarrow Br + e^-. \tag{6.1}$$

In equation 6.1, the normal state of the bromine atom fully bonded into the crystal is shown as Br^-, indicating its attachment of an electron from a neighboring silver atom. The photo-produced hole at the neutral bromine atom is of little utility in the detection process. It can meet up with a photoelectron and recombine or it can ionize a neutral silver atom, in either case negating the detection. Fortunately, there are other possible fates: it can migrate to the grain surface and react with the gelatin, or it can unite with another hole, thereby forming a bromine molecule that can escape, eroding the crystal structure.

Usable detection occurs when the freeing of photoelectrons leads to the production of a stable silver molecule. After a photoelectron is created, it wanders through the grain by Brownian motion and may recombine with a halogen atom (the reverse of reaction 6.1); in this case, it is lost to the detection process. However, the electron may fall into a trap such as those that occur at flaws in the crystal structure. Once the electron is fixed at a trap, it attracts mobile silver ions toward this location; eventually it may combine with one of them to form a neutral silver atom,

$$e^- + Ag^+ \leftrightarrow Ag. \tag{6.2}$$

The single silver atom in reaction 6.2, however, is very unstable; it may instead decompose back into an electron–ion pair, reversing the direction in reaction 6.2. Alternatively, a wandering hole can oxidize the atom directly back to Ag^+. Moreover, there is a chance that the thermal excitation of the crystal will allow the electron to escape from the trap; in this case, the electron must again wander through the crystal, running the gauntlet of possible recombination and loss before it falls into another trap.

However, if we are lucky the single silver atom will act as a trap for a second electron,

$$Ag + e^- \rightarrow Ag^-. \tag{6.3}$$

The resulting negative silver ion can attract a positive ion from the crystal structure, combine with it, and produce a reasonably stable, though not yet developable, two-atom silver molecule,

$$Ag^- + Ag^+ \rightarrow Ag_2. \tag{6.4}$$

After its creation, the silver molecule can continue to act as a trap for photoelectrons or silver ions or both, and in this way it can grow as the exposure to light continues. The original molecule becomes the nucleus of a development center in the crystal. If this center reaches a critical size (three to four silver atoms), it renders the grain developable by catalyzing the reduction of silver ions to silver in the subsequent chemical processing.

The pattern of exposed grains on a plate is called the latent image. If it is to be useful, it must be amplified. In this case, amplification is accomplished by bathing the grains in a chemical reducing agent, or developer, that provides electrons to the silver ions in the grains, thus reducing them to metallic silver; other reaction products and any remaining silver ions are washed away. Developers are selected that utilize a strong catalytic action by the silver specks in the latent image so that the conversion proceeds much more quickly in the heavily exposed grains than in unexposed ones. The nature of this catalytic action is not entirely understood; one possibility is that the silver specks act as electrodes and conduct electrons from the developer into the grain, allowing them to combine with silver ions over the surface of the speck. After development, the amount of metallic silver in the latent image has been multiplied by as much as 10^8–10^9, an enormous gain.

Unfortunately, the detective quantum efficiencies of photographic materials are far lower than the absorptive quantum efficiency. The inefficiencies in going from photoelectrons to developable grains are such that about 10 to 20 photons must be absorbed by a grain in a reasonably short period of time for it to have a 50%

probability of developing. From this cause alone, the *DQE* is reduced by a factor of 10 to 20 from the absorptive *QE*. Two other effects can further reduce the *DQE*. First, the individual grains are distributed randomly over the plate. Two areas of equal size have randomly differing numbers of grains and hence have differing response even with identical exposure and development. Second, there are significant variations in grain properties. As a result, two plate regions with the same number of grains can be expected to exhibit nonuniformity even with identical exposure and development. These latter two effects are expressed quantitatively as the *granularity* of the plate, which refers to the variations in response measured through a fixed aperture for a plate with uniform exposure. As a result of the combination of the phenomena discussed here, the *DQE* as in equation 1.26 will be no more than 2–5%.

Despite the high grain quantum efficiency at short wavelengths, the overall efficiency of plates is reduced there because the gelatin absorbs photons in the ultraviolet, cutting off response for wavelengths short of 0.3 μm. Response can be extended further into the ultraviolet by using plates with extremely thin coatings of gelatin so that bare silver halide grains are exposed directly to the incoming photons. Such plates require special handling to prevent damage to the unprotected grains.

As normally practiced with silver bromide grains, the photographic process described above is only effective for blue and ultraviolet wavelengths. Extension to the red is achieved by adding a dye to the emulsion to produce "dye sensitization." Photons absorbed in the dye are capable of creating conduction electrons in the grain. One possibility, effective for relatively energetic photons such as visible light, is that the photon creates a conduction electron in the dye; the electron is then transferred directly into the conduction band of the halide crystal. Another possibility, which applies for red and infrared photons, is that the excited dye molecule transfers energy to the halide, which subsequently excites an electron into the conduction band through a multi-step process. In the spectral region of dye sensitization, only the grain surface, as opposed to the entire grain volume, takes part in the initial photon detection. In the case of *orthochromatic* emulsions, the response is extended only to green light using a green-absorbing dye. *Panchromatic* emulsions are more popular, with a red-absorbing dye adsorbed onto the silver halide grains. Infrared sensitive materials are prepared in the same manner, and can have response to wavelengths almost as long as 1.2 μm (Kodak 1-Z emulsions).[1]

A variety of methods have been used for color photography, but most modern approaches are based on a depth-wise superposition of emulsions such as in Figure 6.2. We will use this figure to illustrate the action of a positive color film.

---

[1] Kodak and other major manufacturers have discontinued making photographic plates, so this statement refers to history not to the present.

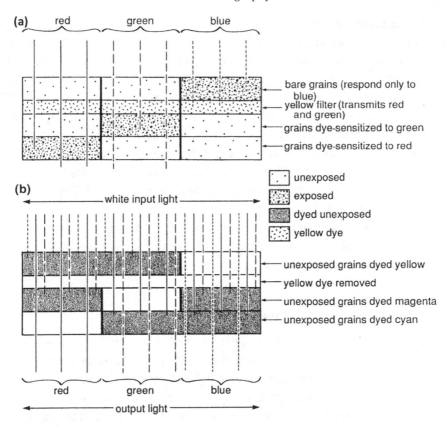

Figure 6.2 Illustration of operation of color photography: (a) shows the exposure of the film to the three primary colors; (b) shows the result after development.

The topmost layer of the emulsion in Figure 6.2 contains grains that are not dye-sensitized and respond only to blue light. A yellow filter removes the blue light to protect the underlying layers of grains (otherwise, the blue light would react with the silver halide in these grains to make them developable), but this filter transmits both green and red. One of the underlying layers is sensitized with a dye that responds to green but not red, and the other with a dye that responds to red but not green. Thus, when exposed to a colored scene, the three primary colors are recorded separately in the three layers of emulsion.

In the course of development, the yellow filter dye is removed and image dyes are produced in the emulsion layers, as shown in panel Figure 6.2(b). After initial development of the exposed grains, a second development forms the complementary dye in the unexposed grains in each layer. That is, yellow dye is produced in the unexposed grains in the blue-sensitive layer, magenta dye in the unexposed grains in the green-sensitive layer, and cyan dye in the unexposed ones in the red-sensitive layer. Magenta is a dye that absorbs only green but transmits red and blue; cyan removes red and transmits green and blue. At the end of the process, all silver has been removed and only the image dyes remain. As an example, where the film has been exposed to blue light, the top layer will be undyed and the lower ones will be dyed, respectively, magenta and cyan, which together transmit blue. Similarly, where the film has been exposed to green light, the top layer will be dyed yellow, the second layer will be undyed, and the bottom layer will be cyan; yellow and cyan together transmit green. Where the film has been exposed to red light, the bottom layer will be undyed and the top two layers will be dyed, respectively, yellow and magenta, which combine to transmit red. Thus, the colors of the illumination are reproduced directly.

There are many variations on the basic theme illustrated in this example. Negative films, for example, yield a picture that is converted to red, blue, and green only by transfer to positive photographic materials such as print paper.

6.2 Photoemissive Detectors

Photoemission refers to a physical process in which a photon, after being absorbed by a sample of material, ejects an electron from the material. If the electron can be captured, this process can be used to detect light. Photoemissive detectors use electric or magnetic fields or both to accelerate the ejected electron into an amplifier, enabling it to be detected as a current or even as an individual particle. These detectors are capable of very high time resolution (up to 10^{-9} s) even with sensitive areas several centimeters in diameter. They can also provide excellent spatial resolution either with electronic readouts or by displaying amplified versions of the input light pattern on their output screens. They have moderately good quantum efficiencies of 10–40% in the visible and near infrared; in some cases, higher values apply in the ultraviolet. They provide excellent performance at room temperature or with modest cooling, leading to a number of important applications. In addition, they remain among the highest-performance devices in the ultraviolet. They can be readily manufactured with 10^6 or more pixels. If the photon arrival rate is low enough that they can distinguish individual photons, the detectors are extremely linear.

6.2.1 Basic Operation

A simplest photoemissive detector is basically a vacuum tube analog of a photodiode, illustrated in Figure 6.3. The device is biased by placing a negative potential across its cathode (analogous to the p-type region of the diode) and anode (analogous to the n-type region). The vacuum maintained in the tube between cathode and anode is depleted of everything, not just free charge carriers; it corresponds

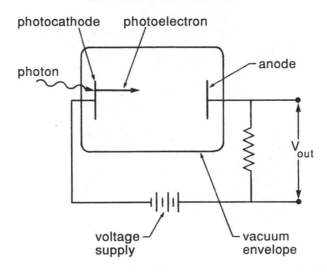

Figure 6.3 Operation of a vacuum photodiode.

to the depletion region of the diode. A conduction electron is produced in the cathode when it absorbs a photon with energy greater than the intrinsic bandgap of the cathode material. This electron diffuses through the material until it reaches the surface, where it may escape into the vacuum. The negative potential of the photocathode helps accelerate the emitted electrons into the vacuum, and the electric field between cathode and anode drives the electron across the vacuum to the anode, where it is collected and provides a signal. The resulting current is a measure of the level of illumination of the cathode. This kind of detector performs well even at room temperature because of the very high impedance of the physical vacuum that forms its depletion region.

   The key to achieving good performance with these detectors is to maximize the efficiency of the photocathode, which is determined by three mechanisms: (1) the absorption efficiency for photons; (2) the transport losses of electrons as they migrate from the absorption sites to the photocathode surface; and (3) surface barrier losses resulting from the inhibition of electron emission from the photocathode. To elaborate on its operation and the selection of suitable materials for its construction, we first describe the simplest form of photoemission as originally described by Einstein – that from a metal. Consider the metal–vacuum interface; the band diagram is shown in Figure 6.4. Because the conduction and valence states are adjacent, the available electron states will be filled to the Fermi level. An electron on the surface of the metal will require some energy to escape into the vacuum; we define the energy level in the vacuum to be $E = 0$. The work function, W, is defined as the energy difference between the Fermi level and the minimum escape energy. If a photon with $E > W$ is absorbed, it can raise an electron high

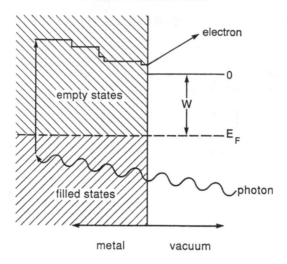

Figure 6.4 Energy band diagram of the photoelectric effect in a metal.

into the conduction band. As this electron diffuses through the material and begins to thermalize, it will lose energy to the crystal lattice. As long as it is not in thermal equilibrium, it is referred to as a "hot electron." If it reaches the metal surface with energy $E \geqslant 0$, it has a reasonably high probability of escaping into the vacuum, as described by the theory of the photoelectric effect. As discussed here, however, the process is not very useful for a photon detector because the high reflectivity of metals would lead to very low quantum efficiency.

Semiconductors are used as photoemitters in detectors because they have much lower reflectivity than metals. The more complex properties of these materials require expanded notation to describe photoemission. Refer to the energy level diagram in Figure 6.5, which applies to a p-type material. As before, W is the energy difference between the Fermi level and the minimum energy the electron needs to escape the surface. The electron affinity, χ, is the difference between E_c (the minimum energy level of the conduction band) and the minimum escape energy. Unless the material has a very small value of $E_F - E_v$ or is at a high temperature, the Fermi level contains few electrons, so a photon of energy W can rarely produce photoemission. Usually an energy $\geqslant (E_g + \chi)$ is required. Finally, even for electrons that reach the surface with sufficient energy, the escape probability is always less than unity. Escape probabilities are difficult to predict theoretically, and are normally determined through experiment. The simplest photoemissive materials useful for photon detectors are semiconductors that have $(E_g + \chi) \leqslant 2\,\text{eV}$ and electron escape probabilities of about 0.3. Unlike metals, these materials have reflectivities of only 15–50%. Thus, in the most favorable cases, the quantum efficiencies are 50–85% times 0.3, or 15–25%.

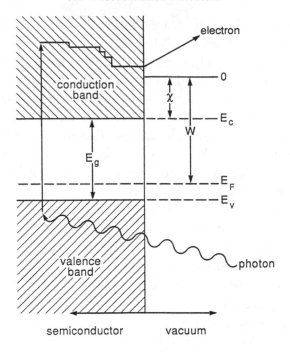

Figure 6.5 Energy band diagram of the photoelectric effect in a semiconductor.

Photon absorption occurs at some depth in the material, as has been emphasized in Figures 6.4 and 6.5. To escape from the material in the reflection geometry illustrated in these figures, the free electron must diffuse back through this depth to the surface without losing too much energy to the crystal lattice. Transparent photoemitters are often used rather than reflective ones. For them, the thickness of the photoemitter must be sufficient for good absorption of the incident photons but not too thick to inhibit the diffusion of electrons out the other side. In either case, to make an effective detector, the absorption length for photons needs to be of the same order or less than the diffusion length for electrons. Practical photocathodes can be produced in which virtually all the photoelectrons contribute to the signal (within the limits set by the reflectivity and the electron escape probability discussed above).

As we have seen, the absorption coefficient, $a(\lambda)$, in semiconductors is typically between 10^4 and $\geqslant 10^5$ cm^{-1} for direct transitions and between about 10^2 and $\geqslant 10^3$ cm^{-1}, or about a factor of 100 smaller, for indirect transitions. For hot electrons, the diffusion length, L_H, is difficult to predict theoretically. Measurements indicate that a typical value is 10^{-2} μm for direct transitions and ~1μm, or a factor of 100 greater, for indirect transitions. Thus, to first order, the absorption lengths and diffusion lengths compensate, and the probabilities that photoexcitation will produce electrons that can escape from the surfaces of direct and indirect semiconductors are comparable.

Hot electrons will eventually thermalize, falling down to energy levels near the bottom of the conduction band. In this case, the diffusion length is given by equation 3.17 and is of the order of 1–10 μm for direct semiconductors that are suitable for high-quality detectors. That is, the diffusion length for thermalized electrons is two orders of magnitude larger than for hot ones. The thermalized electrons can therefore reach the surface from far deeper in the photocathode than can the hot ones discussed so far; unfortunately, for most materials the thermalized electrons do not have sufficient energy to escape from the surface once they get there. This problem is particularly severe for detectors operating in the red and near infrared because the photon energy is already low. The emission of low-energy electrons can be enhanced substantially, however, by using multi-layer photocathodes; the most extreme forms have negative electron affinities (NEA), enabling thermalized conduction electrons to escape (Spicer 1977; Zwicker 1977; Escher 1981; Zhang and Jiao 2019).

As an example, consider what happens if we use GaAs as the photoemitter and coat it with a thin layer of Cs_2O. For the sake of illustration, assume that we begin with two distinct materials (GaAs and Cs_2O) and bring them into physical and electrical contact as shown in Figure 6.6; in reality the Cs_2O is evaporated onto the GaAs. Assume that the GaAs is doped p-type and the Cs_2O n-type (neutral or n-type GaAs will not produce satisfactory results – see Problem 6.1). The energy levels of the two materials prior to contact are shown in Figure 6.6(a). When the two materials come into electrical contact with each other, the charge carriers will flow until the Fermi levels are equal, producing the situation illustrated in Figure 6.6(b). In this energy band structure, *any* electron excited into the conduction band of the GaAs has sufficient energy to escape from the material if it can diffuse through to the thin layer of Cs_2O before it recombines in the GaAs. Therefore, a photon needs to have only $E \geqslant E_g$, not $E \geqslant (E_g + \chi)$, to produce a free electron, thus providing good response in the red and near infrared. Similar benefits can be obtained with a thin coating of Cs rather than Cs_2O. An additional advantage of this configuration is that the escape probability is raised to about 0.45, providing peak photocathode quantum efficiencies of 35–40%. Although originally NEA photocathodes could only be manufactured to operate in the reflective mode, transmission mode architectures have now been developed (e.g., Zhang et al. 2019).

Another approach to good infrared performance is to grow a pn-junction on the photocathode along with a transparent electrode that is used to bias it so the bands are bent to encourage electron escape (Niigaki et al. 1997).

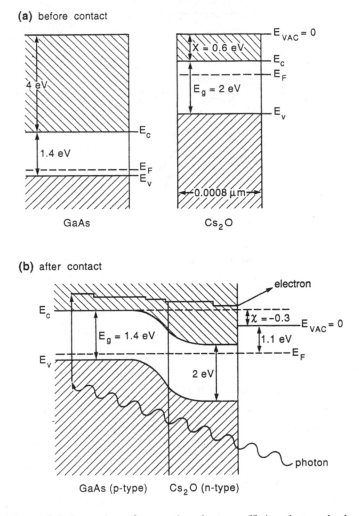

Figure 6.6 Operation of a negative electron affinity photocathode.

Figure 6.7 shows the spectral response of a variety of photocathodes, including some that have negative electron affinity. As with other semiconductor-based detectors, the spectral response near the long wavelength cutoff is significantly affected by the temperature of operation (Budde 1983). High-performance photomultipliers for the ultraviolet are available with CsI and CsTe photocathodes (utilized in the X-ray and from ~0.1 to 0.2 or 0.1 to 0.3 μm respectively in the ultraviolet). As with some other ultraviolet detector types, their long wavelength cutoffs are at short enough wavelengths to significantly ease out-of-band rejection in spectral filters. They are sometimes called "solar blind" detectors, since they

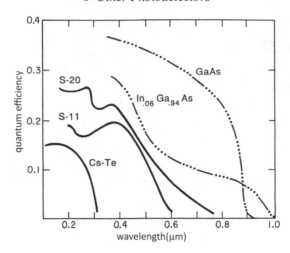

Figure 6.7 Spectral response of some representative photocathodes. The solid lines are for conventional materials, and the dot-dash lines are for negative electron affinity ones. After Zwicker (1977).

respond only at wavelengths where the atmosphere is reasonably opaque to sunlight. Photocathodes have also been developed for the near infrared, using the pn-junction architecture in InP/InGaAsP (responding to 1.4 μm) and InP/InGaAs (to 1.7 μm), although the quantum efficiencies are low ($\lesssim 3\%$).

Mechanisms other than photoexcitation can also release electrons from the surfaces of photoemissive materials, producing dark current that interferes with the observation of low levels of light. At room temperature, dark current is usually dominated by the emission of thermally excited electrons. For the sake of simplicity, we assume that all electrons reaching the surface with sufficient energy to escape are emitted and that emitted electrons are immediately replaced by other conduction electrons (assumptions appropriate for a metal). A calculation similar to that described in Section 2.2.6 predicts a dark current of

$$I_d = q \left(\frac{4\pi m_e}{h^3}\right) (kT)^2 A e^{-\frac{W}{kT}}, \tag{6.5}$$

where A is the area of the photoemissive surface and W is the energy difference between the Fermi level and the minimum energy the electron needs to escape the surface. This expression is called the Richardson–Dushman equation. It normally overestimates I_d for semiconductors, but it remains a useful approximation.

Equation 6.5 suggests two means of controlling dark current. The first is to cool the detector. Temperatures of 193–253 K (−80 to −20 °C) usually reduce the thermal component of the dark current below other components; the required temperature depends on the photoemissive material. The second means, applicable only to non-imaging devices, is to restrict the detector area, which can be done by making it small in the first place or, for some detectors, by using external magnetic fields to deflect photoelectrons from all but a small area of the photoemissive surface away from the collecting and amplifying apparatus.

In addition to the thermal component of dark current, there are nonthermal components that are not removed by cooling. For example, electrons are released when high energy particles (cosmic rays or local radioactivity) pass through the photoemissive material; these electrons produce very large output pulses. When residual gas molecules in the detector vacuum strike the photocathode surface, they also produce large pulses called ion events. The device may also have a detectable level of electrical leakage, which allows currents to pass directly from the voltage input to the signal output terminal. The latter two items must be controlled during the manufacture of the detector by: (1) establishing and maintaining a high-quality vacuum within the device to eliminate residual gas; (2) designing the device with extremely high electrical resistance between the input and output; and (3) maintaining a high level of cleanliness so the high resistance between input and output established in manufacture is not compromised by contamination. For any given device, if they are not controlled, these nonthermal dark current mechanisms can significantly decrease the performance from the theoretical value, which is based primarily on the photocathode quantum efficiency.

When low light levels are observed, the number of photoelectrons escaping from the photocathode needs to be amplified. Electron optics can be used to guide the electrons precisely while they are in the vacuum, bringing them to the inputs of high-performance amplifiers. These high-performance amplifiers and optics come in a variety of forms and will be described in Section 6.2.3. It is the accuracy and relative ease with which electrons can be guided and amplified in the vacuum that accounts for much of the power of photoemissive detectors.

6.2.2 Quantitative Results

We can make quantitative performance predictions for photocathodes, particularly for those with negative electron affinity, based on the properties of photodiodes discussed in Chapter 3, although we must keep in mind that for the photodiode we assumed that the probability that the electrons would escape into the depletion region was unity. In contrast, we have seen that this probability will be only about 25–50% for the photoemissive detector.

For example, suppose φ photons s^{-1} fall on the photocathode, producing $n_e = \eta \varphi$ electrons s^{-1}, where η is the quantum efficiency. The signal current is then

$$I_s = q\eta\varphi. \tag{6.6}$$

Representing the power in the incident flux as $\varphi hc/\lambda$, the responsivity (sometimes termed the "radiant sensitivity") is

$$S = \frac{\eta \lambda q}{hc}, \tag{6.7}$$

which is identical to equation 3.2. Similarly, the noise in the photocurrent can be obtained from equation 3.3 to be

$$\langle I_N^2 \rangle = 2q^2 \, \varphi \eta \, df. \tag{6.8}$$

So, as discussed in Section 5.4.1.2, dividing the responsivity by the noise, we obtain the *NEP*:

$$NEP = \frac{hc}{\lambda} \left(\frac{2\varphi}{\eta} \right)^{1/2}. \tag{6.9}$$

The performance may instead be limited by the dark current. If the dark current is Poisson in behavior, that is, caused by single-electron events similar to those responsible for the photocurrent, we have (instead of equation 6.8)

$$\langle I_N^2 \rangle = 2q I_d \, df, \tag{6.10}$$

where I_d is the dark current. and the *NEP* becomes

$$NEP = \frac{hc}{\eta \lambda} \left(\frac{2 I_d}{q} \right)^{1/2}. \tag{6.11}$$

If there is a substantial dark current component from ion events, cosmic-ray events, or other causes that do not obey the same Poisson statistics as the photon events, the noise will be degraded and the *NEP*s will be greater than just derived.

The analogy with photodiodes can be extended to details of photoemissive detector operation. In particular, the quantum efficiency of a photocathode can be determined from the material properties and the thickness of the photoemissive layer using a calculation exactly parallel to the one for the quantum efficiency of the diode, see Section 3.1.3.2 and the example later in this chapter.

6.2.3 Practical Detectors

6.2.3.1 Photomultiplier Tube (PMT)

A commonly used photoemissive detector is the photomultiplier, illustrated schematically in Figure 6.8. The voltages required for operation of the photomultiplier are provided by a single supply, the output of which is divided by a chain of resistors as shown in Figure 6.9. Photons eject electrons from a photocathode that is held at a large negative voltage (of order 1500 V relative to the anode). The electrons are accelerated and focused by electric fields from additional electrodes ("electron optics") until they strike dynode 1, the surface of which is made of a material that emits a number of electrons when struck by one

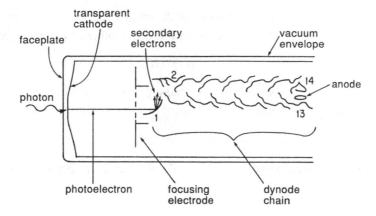

Figure 6.8 Design of a photomultiplier. After Csorba (1985).

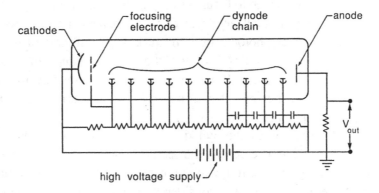

Figure 6.9 Voltage divider for a photomultiplier to give fast pulse response under situations where there is a high output current.

high energy electron. These secondary electrons are accelerated into the surface of dynode 2 by the potential maintained by the resistor divider chain. At this second dynode they produce more electrons. Dynode-to-dynode voltages of 100–200 V give good yields of secondary electrons without unnecessarily increasing the unwanted side effects of high voltages (for example, corona and breakdown). This process of acceleration and electron emission is repeated down the chain of dynodes, and the amplified pulse or current is collected at the anode. The electron pulse initiated by the photoelectron travels together down the dynode chain; by designing the tube geometry carefully, the output pulse width can be made as short as a few nanoseconds, allowing single photons to be counted with extremely high time resolution. Specialized designs are capable of even faster time response (Donati 2000).

To be specific about the action of the dynode amplifier, assume that φ photons s^{-1} strike the photocathode in some time interval t; the number of electrons emitted is

$$n_1 = \varphi \eta t = \eta n_0, \tag{6.12}$$

where n_0 is the total number of photons striking the cathode in time t. Assuming that dynode 1 has gain d_1, the signal emerging from this dynode will be

$$n_2 = d_1 \eta n_0 = d_1 n_1, \tag{6.13}$$

and the net signal after amplification by m identical dynodes each with gain d is

$$n_{out} = d^m n_1. \tag{6.14}$$

This kind of amplifier is called an electron multiplier.

The image dissector scanner is a variation on the photomultiplier. In this device, the electron optics are designed to focus the output electrons from the photocathode onto a plate with only a small aperture for them to pass through. As a result, the dynode chain receives photoelectrons from only a small area of the cathode. Deflection coils in the electron optics allow the sensitive region to be shifted in position across the photocathode. If the sensitive spot is scanned over the cathode, the image dissector scanner operates as an areal detector array, but one that is sensitive to only one pixel at a time.

The process of emission by the dynodes is very similar to the photoemission that occurs at the cathode, except that the escaping electrons are produced by the interaction of an energetic electron with the material rather than by the interaction of a photon. Nonetheless, the same considerations apply regarding the escape of excited electrons from the material; as a result, similar materials (including some with negative electron affinity) are used for both the cathode and the dynodes.

Although in the simplistic discussion above, the photocathode properties determine the sensitivity of a photomultiplier, a variety of other effects also play a role. First, electrons released by the cathode must be focused on the first dynode; losses in this process (perhaps just a few percent if adequate photocathode to dynode voltage is applied and there is no deflection by magnetic fields) appear as a reduction in the effective quantum efficiency. In the following discussion, we combine this loss with the photocathode quantum efficiency into an effective quantum efficiency, η'. From equation 6.12, the intrinsic signal to noise at the entrance to the first dynode is then

$$\left(\frac{S}{N} \right)_0 = (\eta' n_0)^{1/2}. \tag{6.15}$$

The losses in the electron optics are increased when electrons are deflected by magnetic fields; for example, unshielded photomultipliers can show variations in response for different orientations relative to the Earth's magnetic field. Artificial sources of magnetic fields, such as electric motors, can also modulate the collection efficiency of the electron optics. Fortunately, thin sheets of high permeability magnetic material ("mu metal") can provide adequate magnetic shielding.

In addition, the signal-to-noise ratio is reduced if the dynodes have inadequate gain. Assuming that the electron multiplication in the dynodes follows Poisson statistics, the noise in the signal current emerging from the first dynode (equation 6.13) will have two components: $d_1(n_1)^{1/2}$ (the amplified noise in n_1) and $(n_2)^{1/2}$. Combining these two noise sources quadratically, the total noise is

$$(d_1^2 n_1 + n_2)^{1/2} = (d_1^2 \eta' n_0 + d_1 \eta' n_0)^{1/2}. \tag{6.16}$$

The signal to noise emerging from dynode 1 is then (from equations 6.13, 6.15, and 6.16)

$$\left(\frac{S}{N}\right)_1 = \frac{d_1 \eta' n_0}{(d_1^2 \eta' n_0 + d_1 \eta' n_0)^{1/2}} = \frac{(\eta' n_0)^{1/2}}{\left(1 + \frac{1}{d_1}\right)^{1/2}}$$

$$= \left[\eta' n_0 \left(\frac{d_1}{d_1 + 1}\right)\right]^{1/2}. \tag{6.17}$$

Taking the expression in equation 6.17 as the input to a similar calculation, it can be shown that the signal to noise emerging from a chain of dynodes is

$$\frac{S}{N} = \frac{(\eta' n_0)^{1/2}}{\left[1 + \frac{1}{d_1}\left(1 + \frac{1}{d} + \frac{1}{d^2} + \cdots\right)\right]^{1/2}}$$

$$\approx \left[\frac{\eta' n_0}{1 + \frac{1}{d_1}\left(\frac{1}{1 - 1/d}\right)}\right]^{1/2} = \left[\eta' n_0 \left(\frac{d_1'}{d_1' + 1}\right)\right]^{1/2}, \tag{6.18}$$

where we have taken the gains of the subsequent dynodes to be equal to each other and to d. Here

$$d_1' = d_1 \left(\frac{d - 1}{d}\right). \tag{6.19}$$

From equation 6.18, it is clear that good performance depends on the first dynode having a large gain, d_1; large gain for the remaining dynodes is desirable but less critical. Comparing equation 6.18 with equation 6.15, it can be seen that the effective quantum efficiency, η', delivered at the output current is degraded compared to the value delivered to the input of the dynode chain. The factor by which it is reduced is

$$f_i = \frac{d_1'}{d_1' + 1}. \tag{6.20}$$

For example, if $d_1' = 3$, the detective quantum efficiency is reduced by about 25%.

Fortunately, dynodes made of materials with negative electron affinity, such as cesiated gallium phosphide, can yield $d_1 = 10 - 20$. Because they must carry larger currents than the initial dynode, the later dynodes are often made of a material with more robust electrical properties but lower gain. Equations 6.19 and 6.20 show that such designs can largely avoid degradation in signal to noise from the action of the dynode chain. To realize the potential of these designs, it is important to maintain a large accelerating voltage in the first few stages of the photomultiplier so that these stages operate at maximum gain. At large signal levels, space charge effects inside the tube can affect the linearity. For this reason, as well as to avoid damage from excess currents, it is often desirable to remove the signal from an intermediate stage (and not supply voltage to the following ones) if the signal is large.

Our mental image of photomultipliers is that they are large fragile glass devices. However, they now are also available (e.g., from Hamamatsu) as rugged modules with integral resistor divider chains that can be as light as 10 grams and as small as 7 mm thick and 2 cm on a side. By supplying an array of anodes and a suitable design for the electron multiplier unit, photomultipliers can also be manufactured that operate as small (2×2 to 8×8 pixel or 1×32) arrays (Hamamatsu 2007). Each pixel in these arrays has the very fast response typical of the single-output tubes, with pulse widths as short as 2–3 ns.

Although the analysis above is based on sensing the current in the photomultiplier, these detectors are frequently operated in a pulse counting mode. In this case, fast electronic circuits at the photomultiplier output identify all output pulses above some chosen threshold value. The threshold is selected to weed out small pulses produced by noise in the electronics or other events not associated with photon detection. With high-quality photomultipliers, a range of threshold settings exists in which virtually all photon events are accepted but with rejection of the great majority of noise events from the dynode chain.

A disadvantage of this approach appears at relatively high illumination, when the pulse rate can become sufficiently large that pulses begin to overlap. A variety of problems then occur; for instance, pulses above the threshold are missed because they arrive too close in time to other pulses, and pulses below the threshold can be counted if they arrive in coincidence. The result is that the photomultiplier is no longer linear; the current mode of operation is therefore generally preferred for high levels of illumination.

At low illumination, however, pulse counting is usually preferred, primarily because the mechanisms that produce large pulses (for example, ion events or cosmic rays striking the photocathode) do not contribute disproportionately to the measurement. This high degree of noise immunity is not shared by current mode operation, in which the entire charge in a pulse contributes current. Even under ideal conditions, pulse counting has an advantage in terms of achievable signal to noise because to first order it is immune to the variations in pulse shape and size that occur due to statistics in the multiplication process. In other words, each event above the threshold (that is, each detected photon) is counted equally.

Nonetheless, the action of the dynode chain also degrades the signal to noise in pulse counting, in this case when the statistics of the multiplication process drop the pulse height below the detection threshold. Again, it is important for the first dynode to be operated with high gain; assuming the dynode multiplication follows Poisson statistics, the probability of no response whatever is

$$P(0) = e^{-d_1}. \tag{6.21}$$

For example, $d_1 = 3$ would yield $P(0) = 0.05$. For any realistic setting of the pulse threshold, there would be significant additional losses beyond $P(0)$.

So far, we have concentrated on quantum efficiency and related parameters that influence signal to noise in photomultipliers, but stability and linearity are also important in most applications. The voltages required by the photomultiplier are typically provided by a circuit similar to the one shown in Figure 6.9. The dynode multiplication d varies roughly linearly with applied voltage; hence, with the total tube gain of d^m, we derive a condition on the power supply stability necessary to obtain a given stability in the gain of the dynode chain:

$$\frac{G + \Delta G}{G} = \left(\frac{d + \Delta d}{d}\right)^m = \left(\frac{V + \Delta V}{V}\right)^m, \tag{6.22}$$

where G is the total gain, V is the power supply voltage, and for simplicity we have taken all the dynode stages to have equal gain. Simplifying through the approximation that $\ln(1 + x) \approx x$ for small x, we find that

$$\frac{\Delta V}{V} = \frac{1}{m}\left(\frac{\Delta G}{G}\right), \tag{6.23}$$

that is, to maintain gain stability of 1% requires a power supply with stability better than 0.1% for a typical tube having $m \approx 10$. Where the photomultiplier is used in current mode, the gain stability affects the accuracy of measurements directly. In pulse counting mode, it is possible through judicial choice of the accepted pulse threshold value to provide significantly reduced sensitivity to gain variations.

If it is also important that the output current of the tube be linear with input signal, care must be taken in designing the resistor chain that supplies voltages and currents to the dynodes so the signal current flowing in the tube does not significantly alter the gains of the amplifier stages. A common rule of thumb is that the current through the resistor divider chain should be at least ten times the maximum output current of the tube. A more quantitative approach is to replace the dynode-to-dynode signal currents with equivalent resistances (determined from the currents and dynode-to-dynode voltages) and to analyze the resulting circuit to be sure that the voltages at the dynodes are not affected beyond the tolerance set by the desired degree of linearity in gain.

Particularly at low light levels (for which the signal tends to be resolved into a series of pulses), the analysis suggested above needs to be based on the peak current in a pulse, which may be far larger than the time-averaged current. Because it is impractical to supply the divider chain with the large currents required for the pulses, it is common to install capacitors in parallel with the last few resistors in the divider chain. The capacitors are selected to store sufficient charge to maintain gain stability for the duration of a pulse.

From the above discussion, even at low light levels pulse counting has only a slight advantage over current mode operation in ideally behaved cases. However, it is often preferred because of the noise immunity, stability, and high degree of linearity it affords. A much more detailed analysis of these issues is provided by Young (1974).

Photomultipliers have often been used to obtain results requiring very high accuracy and repeatability. In these cases, some additional issues arise besides linearity and the tradeoffs between current mode and pulse counting, many of which are discussed by Young (1974) and Budde (1983). Not only is the response nonuniform over the face of the photocathode, but there may be subtle changes in spectral response. Thus, the precise spectral response of the tube depends on the operating temperature, the size of the illuminated area, the effects of magnetic fields, and other factors. In addition, the properties of the device generally degrade slowly with time, even in applications where it is not abused.

In summary, these detectors have a number of important advantages. They can provide extremely precise and repeatable measurements. They can have large sensitive areas and still provide excellent performance at room temperature or with modest cooling. Photomultipliers are available with photocathodes up to 200 mm in diameter. Small arrays are also available, although they are subject to strong competition from Geiger mode avalanche photodiodes.

6.2.3.2 *Microchannels*

A different form of electron multiplier from a dynode chain can be made from thin tubes of lead-oxide glass, typically of 5–45 μm inner diameter and with a length-to-diameter ratio of about 40–60. The inner surfaces of the tubes are fired in a hydrogen atmosphere, leading to a breakdown of the glass surface and formation of a layer of PbO. A large voltage is maintained across the ends of the tubes; the tubes must be manufactured in such a way that a suitable voltage gradient is set up along their lengths. The tubes are fused together and mounted in close-packed

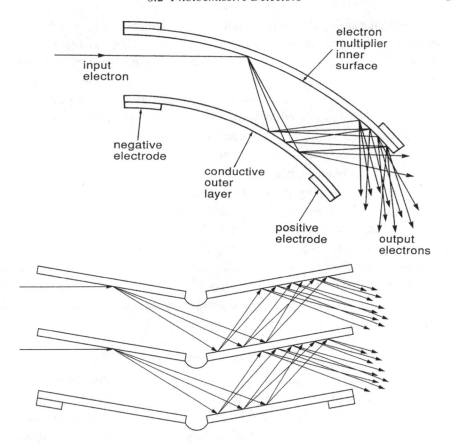

Figure 6.10 Operation of a single curved-channel or C-plate microchannel (upper) and a Z-plate microchannel (lower).

arrays called microchannel plates (MCPs) that hold them at an angle to incoming electrons, see Figure 6.10. The PbO acts as an electron multiplier; collisions with it by the incoming electrons create secondary electrons. The microchannel plates are designed so these electrons are accelerated by the voltage along the device until they strike the walls again and produce more electrons; Figure 6.10 shows two possible layouts. The Z-plate version has proven relatively easy to manufacture and dominates the commercial offerings.

In a photon detector, electrons from a photocathode are accelerated and directed into microchannel plates by electron optics. The net result is similar to the multiple amplifications in a dynode chain and can produce gains nearly as large. Largely because of their compact dimensions, pulses amplified by microchannels tend to be very fast. As in the preceding discussion on dynode resistor chains for photomultipliers, one requirement on the interior surface of the glass tubes is that it be able to conduct sufficient current to replace the charge lost in the electron cascade without varying the voltage significantly. This goal can only be achieved for relatively small total currents (< 1 μA). Using currently available materials, the gain for one amplification in a microchannel cannot be made much larger than two or three; from equation 6.20, the intrinsic signal to noise of the photocathode is thus degraded significantly. Microchannels also tend to have relatively high dark currents, $\approx 10^{-11}$ A (compared to $\approx 10^{-16}$ A for a dynode chain). As a result of these limitations, microchannels are usually not competitive with dynode chains for single-channel low-background detectors such as photomultipliers.

However, microchannels can be operated with gains of 10^4 to 10^6 whilst retaining spatial or imaging information. Since MCPs can have active areas up to 10 cm or more in diameter, with spatial resolution set by channels of diameter 5–25 μm and inter-channel spacings of 8–40 μm, they are an essential element in important types of image intensifier as discussed in the following section. In these applications, they impose a significant cost in quantum efficiency because photoelectrons that strike between the tubes – typically $\sim 40\%$ – are absorbed without being detected. This penalty is overruled by their unique ability to amplify imaging information.

6.2.3.3 Image Intensifiers

A photocathode, electron focusing optics, and a phosphor output screen or a suitable recording medium can be combined in a number of ways to intensify light whilst retaining imaging information. Various implementations of image intensifier will be discussed below in the order of increasing complexity. The first four are arranged so that photoelectrons produced at a photocathode are accelerated through a vacuum into a phosphor, where they are absorbed and produce a number of output photons, providing gain over the single photon that was initially absorbed by the cathode. That is, they literally amplify the image. The most sophisticated of these designs produce gains of 100 or more. As with other photoelectric detectors, very high performance can be achieved at room temperature because of the high resistance of the vacuum in the device. Nonetheless, to achieve extremely low noise, these devices are cooled to reduce the thermal emission of electrons from their photocathodes. For demanding low-light-level work, this type of device has additional shortcomings that have led to development of image intensifiers where a purely electronic output device replaces the phosphor, to be discussed at the end of this section.

The first two ingredients in the image tube recipe should be familiar. The third – a phosphor – is a material that de-excites by releasing energy in the form of light. In the present case, the excitation is provided by energetic electrons. A common family of phosphors is based on ZnS or ZnSCdS, which is activated with silver and chlorine. The silver acts as a p-type impurity and the chlorine as an n-type

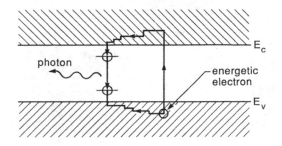

Figure 6.11 Process of luminescence of a phosphor.

one; in equilibrium, the chlorine donates its electron to the silver so both impurities are charged. The luminescence process is illustrated in Figure 6.11. The energetic electron creates free electrons and holes in the sulfide. These charge carriers wander until they are captured by a charged impurity atom. The de-excitation (with emission of light) occurs when the electron transfers from the chlorine to the silver to re-establish the equilibrium condition. Typically, a well-designed phosphor has an efficiency of about 25%; that is, about 25% of the energy lost by the electrons appears as light, with the remainder being deposited as heat.

To prevent the emitted light from escaping back into the image tube where it would be detected by the photocathode, the phosphor is coated on the cathode-facing side with aluminum. This coating can be thin enough that it does not impede the passage of the energetic electrons. The aluminum also acts as an electrode, maintaining the voltage potential of the phosphor.

In the simplest form of image intensifier, the phosphor/anode (positive electrode) is placed close to the photocathode (see Figure 6.12(a)). The electrons are accelerated over this short gap, which must be kept small enough to prevent them from spreading appreciably in the lateral direction. This device is called a proximity focused image intensifier. Its performance is limited by the size of the electric potential that can be held without breakdown across the small gap between cathode and anode. In addition, the small gap between photocathode and phosphor makes it particularly difficult to avoid contaminating the input signal with amplified light from the output.

Another type of device uses a microchannel plate as an electron multiplier between the cathode and anode (phosphor) (see Figure 6.12(b)). The microchannels overcome the breakdown problem of the proximity focused device and provide much higher gain through their extra stage of multiplication. A disadvantage is that the effective quantum efficiency is reduced through a number of effects: (1) 25–45% of the photoelectrons are absorbed in the walls of the glass tubes between the microchannels and are lost to the detection process; (2) the

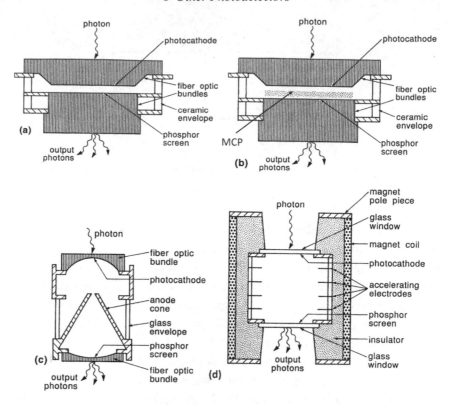

Figure 6.12 Cross-sectional diagrams of a variety of image intensifier types: (a) proximity focused; (b) Gen II or III; (c) electrostatically focused; and (d) magnetically focused. After Csorba (1985).

multiplication per stage is sufficiently low that there is a significant probability that the photoelectron will produce no secondary electron when it strikes the wall of the microchannel; and (3) the low gain increases the noise as described by equation 6.20. The microchannel plate may also produce gain nonuniformities. Nonetheless, because they provide large light amplification in a small volume, these devices are popular and are called second generation, or "Gen II," image intensifiers. A similar design with an advanced photocathode is sometimes termed a third generation intensifier.

A third class of image intensifier uses the field between the cathode and anode to accelerate the electrons as in the proximity focused tube. However, the separation between cathode and anode is increased compared to that in the proximity focused device, and a suitable set of electrodes is placed surrounding the space between the cathode and anode (see Figure 6.12(c)). These electrodes act as an electron lens by creating an electrostatic field to deflect the electrons and to keep them from spreading excessively in the lateral direction, refocusing them onto the phosphor at the end of their flight. The velocity dispersion of the photoemitted electrons perpendicular to the tube axis is not corrected by this type of electron optics, so it limits the sharpness of the image formed on the anode. In addition, the electrostatic electron optics suffer from significant pin-cushion distortion. Because the optics in these tubes suffer substantial field curvature, the photocathode and the phosphor must lie on curved surfaces to obtain sharp images. Flat output faces are obtained by matching the curved face plates to dense bundles of fiber optics. Despite these limitations, these devices are used widely because they are simple and efficient. They are called first generation, or "Gen I," image intensifiers.

A fourth system is similar to the Gen I intensifier except it uses magnetic instead of electrostatic focusing (see Figure 6.12(d)). As before, the electrons are accelerated from cathode to anode by the electrostatic potential maintained between these electrodes. The space between the electrodes, however, is surrounded by either an array of permanent magnets or an electromagnet that produces an axial magnetic field. The defocusing due to the dispersion in electron emission velocity is reduced by an order of magnitude over that achieved by electrostatic focusing; image diameters can be less than 10 μm. Moreover, pin-cushion distortion is reduced. The chief remaining distortion is caused by nonuniformity in the magnetic field, which can result in the outer parts of the image being rotated with respect to the inner parts. This "S" distortion can be held to $\leqslant 0.1$ mm over a 40 mm diameter tube with carefully designed systems. The major disadvantage of this kind of intensifier is that its magnet makes it heavy and bulky. Assuming that the tube is to be cooled to reduce thermal emission, an electromagnetic coil requires substantial cooling power.

To illustrate the operation of electron optics used in many varieties of photoemissive detector, we consider a magnetically focused image intensifier (see Figure 6.12(d)). If the coils around the evacuated region are large enough (and conform to other requirements), they will produce a highly uniform magnetic field, $\mathcal{B}$, running down the axis of the tube from cathode to anode (or vice versa). This field will have no effect on the electron velocity component that runs along this axis, $v_{\parallel}$, but any velocity components perpendicular to the tube axis, $v_{\perp}$, will be subject to a force perpendicular to $v_{\perp}$ and of magnitude $qv_{\perp}|\mathcal{B}|$. Similarly, if the electrostatic accelerating field is purely along the tube axis, it will influence $v_{\parallel}$ only and leave $v_{\perp}$ unchanged. The electron will therefore undergo a spiral motion, with the radius of the spiral determined by the balance between magnetic and centripetal forces:

$$r = \frac{m_e\, v_{\perp}}{q\,|\mathcal{B}|}. \qquad (6.24)$$

In one complete cycle of this spiral, the electron will return to a position exactly equivalent to the one where it started, but translated along the tube axis by

$$d_c = \int_0^{t_c} v_{\parallel}(t)\, dt, \tag{6.25}$$

where t_c is the time required for one complete cycle. This time is

$$t_c = \frac{2\pi r}{v_{\perp}} = \frac{2\pi m_e}{q|\mathcal{B}|}. \tag{6.26}$$

Note that t_c is independent of $v_{\perp}$. Thus, d_c is the same for all electrons, regardless of their initial values of $v_{\perp}$. With suitable choices of magnetic and electrostatic field strengths and tube dimensions, it is possible to relay the image as it emerges from the photocathode accurately back to the anode.

If we assume that the emission velocity of the electron from the cathode is negligible then

$$v_{\parallel} = -\frac{q}{m_e}\mathcal{E}t. \tag{6.27}$$

To determine the design of a tube where the electrons conduct one spiral in the magnetic lens, we can substitute $\mathcal{E} = -V/d_c$ in equation 6.27 (where V is the voltage between the cathode and anode), integrate the result as in equation 6.25, and substitute for t_c from equation 6.26 to obtain

$$d_c = \left(\frac{2m_e}{q}V\right)^{1/2}\frac{\pi}{\mathcal{B}}. \tag{6.28}$$

A standard application of the Biot–Savart law yields the magnetic field in an ideal solenoid (for example, far from edge effects) in the form

$$\mathcal{B} = \frac{\mu_0 I_S N}{L}, \tag{6.29}$$

where I_S is the solenoid current, N is its number of turns of wire, and L is its length. Equations 6.28 and 6.29, together with the characteristics of the photocathode and phosphor, determine the first-order design and performance of the image intensifier, as illustrated in the example after this section.

Within the limitations discussed in the preceding section, electrostatic electron optics can produce results similar to those illustrated here with magnetic optics. Because the mathematics for this case is relatively complicated, we will not treat it in detail; see Csorba (1985) for details.

Because the types of image tube discussed above all have a phosphor as their output, they have a number of common characteristics. Imaging areas are typically a few to a hundred square centimeters, with resolution elements about 50 μm on a side. The time response is determined by the decay of the emission from the phosphors, which can have exponential time constants of several milliseconds.

Phosphors emit over a very broad solid angle. It is difficult to design a conventional optical system that captures a large fraction of this output. This problem can be counteracted by using fiber optic exit bundles to conduct the light without allowing it to spread. A single fiber optic is a thin rod made of a low-index glass core clad with a high-index glass. A photon introduced into this rod at a small angle relative to the rod axis is relayed down the rod through total internal reflection off the high-index cladding. Fiber optic plates are manufactured by bonding fiber optics into close-packed bundles. Although the net transmission efficiency of these plates is only 50–60%, they are used as faceplates for many image intensifiers. For example, we have already mentioned that they accommodate the field curvature of Gen I tubes to provide flat entrance and exit windows. For all tube types, the output of the fiber optic bundle can be received by a photographic plate or electronic detector in contact with the flat surface of the fiber bundle, providing efficient transfer of the image to the recording medium.

Since image intensifiers are often stacked to increase the net system gain, it has been found convenient to contact the exit faceplate of one tube to the entrance of another through fiber optic bundles. When tubes are cascaded in this manner, there is an inevitable degradation of resolution due to the convolution of the finite resolutions of each of the components, including misalignment of fibers from one plate to its mate. Output spot sizes from such chains of intensifiers are typically 80–100 μm. When coupling by conventional optics is unavoidable, many intensifier stages are cascaded to overwhelm the optical inefficiencies with gain. Such systems have degraded resolution and increased distortion compared with systems having fewer stages.

As with other detector arrays, image intensifiers have fixed pattern noise. The primary causes are (1) the granularity in their phosphors, (2) the large-scale nonuniformities in both the photocathode and the phosphor, and (3) the pattern in which the optical fibers are arranged on the input and output. A further complication in flat fielding arises because of source-induced background signal. We have mentioned with regard to proximity focused tubes the possibility that the output light will reach the photocathode and produce a signal. Although the problem is most severe for this tube type because of the small gap between photocathode and phosphor, it can also occur for the other image intensifier types that have phosphor output screens. Since source-induced backgrounds need not be uniform over the photocathode and can vary in a complex relationship to the pattern of input illumination, they can violate the assumptions that allow simple and accurate correction of fixed pattern noise. In situations requiring accurate flat fielding, signal-induced background signals can seriously limit the performance of image intensifiers.

In addition to random photon noise and noise from thermally generated electrons, noise pulses occur in all types of image intensifier when residual gas atoms in the tube acquire a charge and are accelerated onto the phosphor or other output medium to produce ion events. For best performance at very low light levels, it is usually necessary to test and select commercial image tubes that have particularly low ion event rates. Tubes with high ion event rates can be used as second and third stages in image tube stacks; in this application, the ion events become unimportant because of the lower gain for them compared with that for photon events in the first stage. An unavoidable source of large events is the output from cosmic rays and other energetic radiations that strike the photocathode or phosphor.

Historically, a broad variety of techniques have been developed to record and store the outputs of image intensifiers: photographic plates, image dissector scanners, photographic emulsions inside the vacuum of the unit, diode arrays. To a large extent, this variety has now narrowed to CCDs and CMOS arrays. Where long integrations (and low time resolution) are acceptable, and cooling can be provided to suppress dark current, modern CCDs and CMOS arrays produce better performance across the visible and near infrared when operated bare. However, an intensified CCD/CMOS can operate at the photon noise limit even at very low light levels with only moderate cooling (albeit with reduced quantum efficiency compared with a bare CCD/CMOS). In addition, the increase in CCD read noise with rapid readout rates can make an intensifier stage necessary to reach the photon-noise limit when high time resolution is needed, either for detection of single photon events or because of the variability of the object under observation.

6.2.3.4 Direct Electronic Readouts

When an electrical rather than a visual output is desired, an image intensifier can dispense with the phosphor screen and register the electron pulses on some suitable electronic device. With this approach the output of an image intensifier can be read out quickly enough that the events due to individual photons can be distinguished (assuming sufficiently low photon rates). The relatively high noise that may result from reading out the electronic detector quickly is overcome by the high gain of the image intensifier, so each photon event results in many electrons being received by the readout device. In addition, centroiding algorithms can be used to locate each photon event to within a fraction of its diameter; typically, an input photon can be located to about 0.1 of the full width at half maximum (FWHM) of the output spot if the event rate is low enough that multiple events do not confuse the centroiding algorithm. When used in this way, the effective resolution of the tube is significantly increased over that obtained by integrating a long exposure on the output device before reading it out.

One approach places a CCD inside the vacuum of the tube. Electrons released by the photons striking the photocathode are accelerated and focused by electrodes at high voltage, and bombard the CCD in an electron image of the pattern of illumination of the cathode. For high time resolution, frame transfer CCDs are used and are typically read out at about once per 0.1 s. These "electron bombardment CCDs" (EBCCDs) circumvent the shortcomings of phosphor outputs, and can provide detective quantum efficiencies of $\sim 40\%$ in the ultraviolet (Joseph 1995 and references therein).

In another example, if an individual anode is provided for each microchannel in an intensifier with a microchannel plate, the full imaging information is retained. The implementation of this approach, however, would be undesirably complex for reading out a large number of microchannels. A variety of encoding schemes can be employed (see Lampton 1981; Timothy 2013) that use a more modest number of anodes to provide positional information. This type of image tube is most suitable for recording single photon events, and is used widely in the ultraviolet and X-ray spectral regions.

A high-performance implementation of this concept is to use overlying grids of vertical and horizontal wires at the output of the microchannel plate combined with output circuitry that determines the address in each grid of an electron pulse that deposits charge on the wires. This device is termed a multi-anode microchannel array (MAMA) detector and is described by Cullum (1988), Timothy and Morgan (1988), Timothy (1988), Joseph et al. (1995), Joseph (1995), and Timothy (2013). They have been made in formats up to 1024×1024 pixels with *DQE*s greater than 10% from 0.19 to 0.27 μm and greater than 20% near 0.12 μm. Dark counts can be $<10^{-5}$ s^{-1} and read noises are not of concern because of the high gain for each detected photoevent provided by the microchannel plate.

An alternative to the MAMA device uses delay lines to encode the pulse position (Lampton et al. 1987; Doliber et al. 2000). In one possible arrangement, a single distributed RC delay line zig-zags over the entire sensitive area of the anode, interleaved with wedge electrodes to maintain the electric field. The position of an event can be determined by comparing its times of arrival at the two ends of the line. Another approach uses a double set of delay lines, at the sides of the sensitive area. Wedge electrodes are placed to cover the sensitive area and are connected to a series of positions on these delay lines. The charge from a detected event divides between the nearest wedge electrodes and is carried to the delay lines; the wedge involved can be identified by the delay and the relative pulse sizes can be used to refine the position estimate for the event. Compared with MAMA imagers, delay line units have relaxed construction tolerances but also, because relatively large output pulses are required to drive the delay lines, have larger problems with signal saturation.

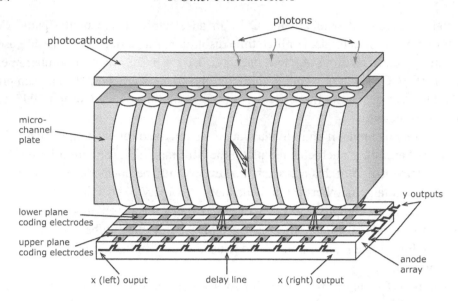

Figure 6.13 The GALEX image intensifier with microchannel and delay line readout.

Figure 6.13 shows this approach, as applied for the detectors for the GALEX space mission. This intensifier operates in the near-ultraviolet (177–283 nm) with a crossed delay line readout (the full array has 4k × 4k pixels). When the photocathode releases an electron, it enters one of the microchannels. The gap between the photocathode and the entrance to the microchannels is only 300 μm, so the positional information is retained. Orthogonal electronic charge collectors are placed at the output of the MCP and convey the signal to external delay lines interconnecting all of the collectors for each coordinate. By measuring the time interval between the emergence of the two signals at the opposite ends of a delay line, it is possible to locate where the signal originated along its coordinate and thus where the original photon hit the photocathode.

The far-ultraviolet (134–179 nm) detector in GALEX is similar in concept, except that the photocathode is deposited directly on the inside of the microchannel tubes. A similar approach can be used in the X-ray (e.g., the High Resolution Camera (HRC) on Chandra). Both of these examples use CsI for the photocathode, but other materials can also be useful such as KBr and CsTe. However, none of these materials yields quantum efficiencies in the ultraviolet above the ~10% range. Photocathodes with higher quantum efficiency are under development using GaN and its alloys, especially $Al_xGa_{1-x}N$. All of these materials also have reduced response in the visible and hence ease substantially the problem of blocking the visible part of the spectrum; for example, the bandgap for GaN is 3.4 eV while that of $Al_xGa_{1-x}N$ varies from 3.5 to 6 eV as x goes from 0 to 1.

6.3 Example

We will consider a magnetically focused image intensifier as an illustration. We will use a GaAs-based photocathode, and we will design the tube to have a gain of a factor of 100. We assume a photocathode diameter of 40 mm and a distance from cathode to anode/phosphor of 60 mm.

First, consider the design of the photocathode. Typical properties of GaAs are as follows:

reflection coefficient: 0.16
optical absorption coefficient: 2×10^4 cm^{-1}
carrier recombination lifetime: 10^{-9} s
diffusion coefficient: 150 cm^2 s^{-1}
electron escape probability: 0.4

We note that the absorption length $= 1/(2 \times 10^4$ cm$^{-1}) = 0.5$ μm. The diffusion length is

$$L = (D\tau)^{1/2} = \left(150\,\text{cm}^2\,\text{s}^{-1}10^{-9}\,\text{s}\right)^{1/2} = 3.87 \ \mu\text{m}. \qquad (6.30)$$

Comparing, we see that the absorption length is much shorter than the diffusion length, as required to make a good photocathode; the high absorption allows us to treat the quantum efficiency approximately as given in the extreme represented by equation 3.23. If we take the absorption to go exponentially, we can show that the photocathode with the highest product of absorption factor and diffusive-limited quantum efficiency has a thickness of 1.5 μm, where $\eta = \eta_{ab} \times 0.929b = 0.883b$. We estimate b as the product of the quantity (one minus the reflection) of the photocathode with the electron escape probability, or $b = 0.336$. Therefore, $\eta = 0.297$.

To obtain a gain of 100, every electron emitted from the photocathode must produce on average $100/\eta = 337$ output photons. If we simplify the calculation by assuming that the phosphor emits light only at 0.55 μm, then each photon requires an energy of 2.25 eV. If the phosphor efficiency is 25%, the electrons must be accelerated to an energy of 337×2.25 eV$/0.25 = 3039$ eV, that is, the potential between cathode and anode should be 3039 V. There is no allowance in this estimate for possible dead spots in the phosphor or other losses. To be prudent, we will design the tube for a potential of 5000 V.

To obtain a uniform magnetic field in the focusing region, we will design the coil to be twice the diameter of the photocathode, that is, 80 mm, and twice the length between cathode and anode, that is, 120 mm. Equation 6.29 allows us to estimate the required field strength:

$$\mathcal{B} = \left(\frac{2m_e}{q}V\right)^{1/2}\frac{\pi}{d_c} = 1.602 \times 10^{-2} \ \text{weber}\,\text{m}^{-2}, \qquad (6.31)$$

where $d_c = 60$ mm and $V = 5000$ V. A reasonable number of turns for the coil is 1000, so we can compute the required current from equation 6.29,

$$I_S = \frac{LB}{\mu_0 N} = 1.53 \text{ A}. \tag{6.32}$$

A reasonable wire diameter in the coil is 1 mm (giving a coil eight windings deep with no allowance for insulation). The total length of wire is $2\pi r N = 2\pi(0.044 \text{ m}) \times (1000) = 276.5$ m. The conductivity of copper at room temperature is 6.5×10^7 $(\Omega \text{ m})^{-1}$. From equation 3.1, we can estimate that the resistance of the coil winding is 5.42 Ω, and the power dissipated in the coil is $I_S^2 R = 12.7$ W. This power must be removed from the tube; if it is to be operated at a reduced temperature to reduce dark current from the photocathode, the cooling system must also contend with the heat conducted into the tube from the ambient temperature surroundings. Therefore, a substantial cooling system is required.

6.4 Quantum Well Detectors

The photodiodes we discussed in Chapter 3 are made using the same semiconductor material throughout but with doping to modify its properties in different ways on opposite sides of a junction. Another type of junction can be based on joining different semiconductor materials; in this case, not only can we make use of doping of the materials to adjust the position of the bands on either side of the junction, but the bandgap itself can change across the junction. Such a junction is called a heterojunction.

To make a successful heterojunction, it is required that the crystal properties of the two materials match closely. Although there are other material pairs that can also be used, a particularly attractive set is GaAs and $Al_xGa_{1-x}As$. The aluminum atoms can replace gallium atoms in GaAs with virtually no effect on the crystal lattice structure, but the bandgap increases with increasing amounts of aluminum. At a GaAs to $Al_xGa_{1-x}As$ junction, there is a discontinuity of the conduction band edge by approximately x eV and of the valence band edge by approximately $0.15x$ eV. Thin layers of these materials can be deposited in a highly controlled fashion using molecular beam epitaxy (see Section 3.1.2.3).

Consider Figure 6.14. It shows the edge of the conduction band for a structure with a thin layer of GaAs grown between two pieces of GaAlAs (we do not need to be specific about the relative amount of Al and Ga because the discussion will be qualitative). If conduction electrons are created in the GaAlAs, some of them will wander into the GaAs, will lose energy there, and will no longer have sufficient energy to re-enter the GaAlAs. The electrons are therefore trapped in a potential well. The situation can be described through the standard square potential well

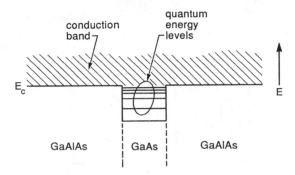

Figure 6.14 The formation of a quantum well in a layer of GaAs between layers of GaAlAs.

problem in quantum mechanics. A similar calculation to the one below for tunneling through the potential barrier shows that, if the walls of the well are infinitely high, the captured electrons must occupy energy levels given by

$$E_n = \left(\frac{\hbar^2 \pi^2}{2m_e{}^* W^2} \right) n^2, \tag{6.33}$$

where n is an integer, m_e^* is the effective mass of the electron, and W is the width of the well. For GaAs, $m_e^* = 0.067 m_e$. Quantum well behavior can also be obtained with holes in the valence band, with obvious modifications in the formalism. Energy levels are sketched in the quantum well in Figure 6.14. If we have a multilayered structure with thin layers of GaAlAs alternated with thin ones of GaAs (see Figure 6.15(a)), we obtain multiple quantum wells such as in Figure 6.15(b).

Multiple quantum well devices can be used in a variety of ways for infrared detectors, generally described together as "quantum well infrared photodetectors" (QWIPs). If the walls of the wells (that is, the layers of GaAlAs) are made sufficiently thin, then electrons can pass from one well to another by quantum mechanical tunneling through the wall between them. We can derive simple constraints on this process by considering tunneling through a potential barrier. We refer to Figure 6.16 and consider an electron incident on the potential barrier from zone 1 to the left. We want to know the probability that the electron will tunnel through the potential barrier and appear in zone 3 on the right.

The behavior of the electron is described in quantum mechanics by the time-independent Schrödinger equation:

$$\left(-\frac{\hbar^2}{2m_e} \frac{\partial}{\partial x} + U(x) \right) \psi(x) = E\psi(x), \tag{6.34}$$

where $\hbar$ is Planck's constant divided by 2π, m_e is the electron mass, U is the potential energy, E is the total particle energy, and ψ is the wave function. We will

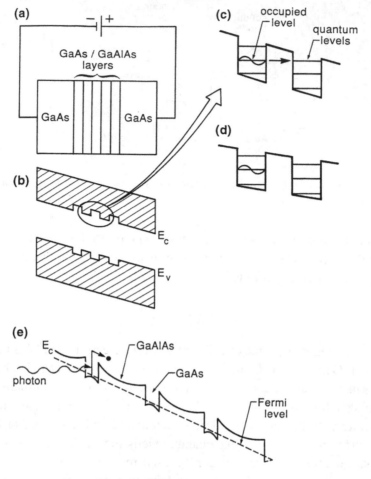

Figure 6.15 Operation of a quantum well detector. The structure of the detector and the bandgap diagram are shown schematically in (a) and (b). The transition of electrons between adjacent wells is shown in (c) and (d). Panel (e) shows the use of doping to lift the Fermi level into the bottoms of the wells to provide a high density of charge carriers for photon absorption.

assume that $E < U_2$; otherwise the electron would be able to cross the barrier without tunneling. The solution to equation 6.34 in zone 1 is

$$\psi_1(x) = Ae^{jk_1x} + Be^{-jk_1x}, \tag{6.35}$$

where

$$k_1 = \frac{(2m_eE)^{1/2}}{\hbar}. \tag{6.36}$$

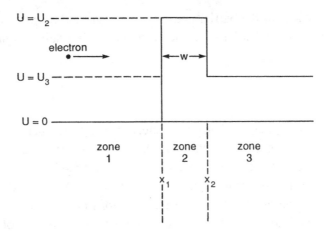

Figure 6.16 Quantum mechanical tunneling through a barrier.

Here, the term with coefficient A represents the electron incident onto the barrier from the left and that with coefficient B represents the electron reflected off the barrier. Similarly, the solution in zone 2 is

$$\psi_2(x) = Ce^{\kappa x} + De^{-\kappa x}, \tag{6.37}$$

where

$$\kappa = \frac{[2m_e(U_2 - E)]^{1/2}}{\hbar}, \tag{6.38}$$

and in zone 3 the solution is

$$\psi_3(x) = Fe^{jk_3x}, \tag{6.39}$$

where

$$k_3 = \frac{[2m_e(E - U_3)]^{1/2}}{\hbar}. \tag{6.40}$$

Equation 6.39 is the wave function of an electron traveling to the right after tunneling through the barrier. $A, B, C, D,$ and F are constants that can be related through the boundary conditions – namely, that both $\psi(x)$ and $\partial\psi/\partial x$ should be continuous from one zone to another. We wish to determine the relationship between wave function amplitudes for the input in zone 1 (amplitude $= A$) and the output in zone 3 (amplitude $= F$). This relationship can be shown to be

$$\left|\frac{F}{A}\right| = \frac{4k_1\kappa}{(k_1^2 + \kappa^2)^{1/2}(k_3^2 + \kappa^2)^{1/2}} e^{-\kappa w}, \tag{6.41}$$

where w is the width of the barrier and we have neglected terms of order $e^{-\kappa w}$ in comparison with those of order $e^{\kappa w}$. The probability of tunneling is just

$$P_T = \frac{k_3}{k_1} \left| \frac{F}{A} \right|^2. \tag{6.42}$$

The height of the potential barrier is determined by the quantum well formulation, but we can take a typical value to be 0.1 eV; substituting for $U_2 - E$ in equation 6.38, it can be seen from equations 6.41 and 6.42 that reasonably large tunneling probabilities occur for barrier widths of $w \sim 10^{-8}$ m. Also, with walls of $\sim 10^{-8}$ m, the overlap of electron wave functions causes the energy levels in the wells to spread into bands. The optical and electrical properties of the resulting "superlattice" depend strongly on the thickness and spacing of the layers, and they can be radically different from those of bulk materials. There is active research into how to make use of superlattices as photon detectors (Razeghi and Nguyen 2014).

As an example, if a bias is placed across the device as in Figure 6.15(b), tunneling then becomes highly favored if the energy state of the electron in its well is aligned with an open energy state in the adjacent well. An absorbed photon can excite an electron to an upper energy level within a well, a bound-to-bound transition, and can cause it to tunnel to a matching energy level in the next well. This situation is illustrated in Figure 6.15(c) and contrasted with Figure 6.15(d) where the energy levels are not aligned. The electrons freed by tunneling then cascade through the series of quantum wells and conduct an electrical current through the device.

In the above operating mode, the photon absorption occurs through a transition between two moderate-width quantum bands; as a result, the spectral response of the detector is relatively narrow, a behavior that is generally undesirable. Worse, the conditions that allow efficient tunneling of photo-excited electrons also allow for tunneling in general, including of ground state electrons (from the unexcited levels). Thus, the dark currents of such detectors are high and cannot be reduced sufficiently by cooling to achieve good performance. Therefore, work on quantum well detectors has moved to the bound-to-quasibound operating mode in which an absorbed photon lifts an electron out of the well into the continuum states above it and the detector bias carries the electron over the potential wall. In this particular type of detector, the structure is adjusted to raise the tops of the wells to be as close as possible to the lowest continuum states. These devices have only a modestly wider spectral spectral response, not much wider than for the bound-to-bound devices. To further broaden the spectral response, multiple wells tuned to different wavelengths can be used (Guériaux et al. 2011). Broader response could be obtained with bound-to-continuum devices, but they tend to have higher dark current than the bound-to-quasibound ones.

As shown in Figure 6.15(e), the material for these detectors is heavily doped with a donor-type impurity to lift the Fermi level into the bottom of the quantum well so that the GaAs regions have a relatively high concentration of conduction electrons, providing free electrons for photon interactions. Thermally excited conductivity is inhibited by the potential barrier formed by the GaAlAs. Detectors operating in this fashion have response over wavelength ranges that are 15 to 35% of the center wavelength of the spectral response. They can operate at relatively small bias fields, decreasing the tunneling dark current (see equation 6.38). Also, since well-to-well tunneling is no longer part of the detection process, the separation between wells can be made an order of magnitude larger, $\sim 10^{-7}$ m, greatly suppressing tunneling (equations 6.41 and 6.42) and the accompanying dark current.

The height of the potential barrier and hence the wavelengths of response can be adjusted readily in the quantum well devices by controlling the amount of Al in the AlGaAs layers. Current development emphasizes the 10 μm region, but response to ~ 28 μm has been demonstrated (Perera et al. 1998). A short wavelength limit of ~ 3 μm is imposed by the maximum barrier heights that can be achieved with heterojunctions. Quantum mechanical selection rules for the electron transition require that the photons be introduced into the detector in a direction along the quantum well layers to be absorbed efficiently. Conventional front illumination of the detector will produce little response, to first order; useful quantum efficiencies of ~ 10–20% require side illumination. Lithography is used to pattern the detectors with structures to achieve this illumination geometry; Figure 6.17 shows a relatively simple arrangement just for illustration. However, current approaches center on phenomena such as Bragg reflection or two-dimensional lamellar gratings (e.g., Andersson and Lundqvist 1992; Chen et al. 1996 and references therein).

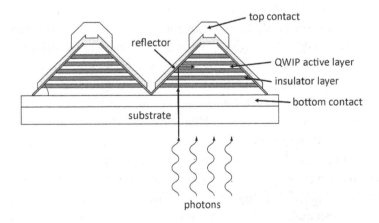

Figure 6.17 Illumination of QWIP detectors.

The responsivity of a multiple quantum well detector has an interesting behavior. As with the conventional photoconductor, we take the photoconductive gain to be the ratio of the number of charge carriers generated by photon absorption to the number that are collected at an electrode of the device. We define P_e to be the probability for an electron excited by photon absorption to escape from a well, and P_c to be the probability that the electron is captured in one of the wells it must pass over as it flows toward the positive electrode. Then the photoconductive gain for a single well is

$$G_{ph} = \frac{P_e}{N\,P_c},$$
(6.43)

where N is the number of wells the electron passes over. The photocurrent per well is obtained from equation 2.14 as

$$i_{ph} = \varphi q \eta G_{ph} = \varphi q \eta_i \frac{P_e}{N\,P_c},$$
(6.44)

where η_i is the quantum efficiency of the well. The photocurrent from a detector with N wells is approximately

$$I_{ph} \sim N i_{ph} = \varphi q \eta_i \frac{P_e}{P_c}.$$
(6.45)

That is, the photocurrent is approximately independent of the number of layers in the detector, N. The tunneling dark current noise grows as $\sqrt{N}$ (Liu 2000). Thus, the dark-current-limited signal-to-noise ratio of detectors with equal quantum efficiency will go inversely as $\sqrt{N}$. Usually this signal-to-noise ratio is the fundamental limit for quantum well detectors that are not background limited. If the detector is background limited, then we get the usual relation that the signal-to-noise ratio improves with $\sqrt{\eta}$, where η is now the quantum efficiency of the full detector, i.e., all the wells together. At low absorption, η is proportional to the absorption per well times N. The best overall performance is therefore obtained with the largest possible absorption and a minimum number of wells.

Quantum well detectors do not offer the same performance at the level of individual devices as can be obtained with direct bandgap intrinsic semiconductors (for example, HgCdTe). A more detailed performance comparison can be found in Kinch and Yariv (1989), and Levine (1990). However, there are many parameters that can be adjusted in quantum well devices to optimize various aspects of their performance. They are manufactured for certain applications taking advantage of their good uniformity, low cost, and high operable pixel yields – but where the ultimate in performance, e.g., high quantum efficiency and low dark current, is not an overriding consideration. Their characteristics are particularly appropriate in the

10 μm region where high thermal backgrounds overwhelm their dark current and a modest cost, reliable detector array has a number of practical applications.

Circuits built around GaAs MESFETs can serve as the readout integrated circuits (ROICs) for QWIP detector arrays in megapixel formats. Unlike silicon, GaAs does not have a native oxide. The lack of an easily grown insulator layer somewhat reduces the design flexibility of GaAs integrated circuits compared with silicon. Rather than the MOSFET, the unit cell is a metal-semiconductor field effect transistor (MESFET; see Chapter 9) that is similar to a JFET in that the transistor operations are built in pn-junctions. However, the ability to fabricate the QWIP detector and its readout electronics on the same chip is a significant advantage in producing low-cost arrays.

6.5 Energy-Resolving Detectors

A number of approaches, mostly centered on superconductivity, can provide detectors with outputs that reveal the energy of a detected photon. These devices will be discussed in Chapter 7.

6.6 Problems

6.1 Using bandgap diagrams, discuss why n-type GaAs cannot be used effectively for a negative electron affinity photocathode.

6.2 Suppose you are making a measurement using a photomultiplier with very large dynode gain (so differences due to dynode gain can be ignored). The signal you are measuring corresponds to an average of 100 events per second. In addition, there is an excess noise process that produces an average of 1 pulse per second that is 100 times as large as a photon pulse. All the events arrive with a Poisson distribution in time (that is, the uncertainty in the number of pulses received in a time interval is the square root of the expected number). Compare the signal to noise that will be achieved in 100 seconds by current measurement and by pulse counting.

6.3 A photomultiplier with cutoff wavelength of 1 μm has dark current of 10^{-6} A when operated with a net gain of 10^6 at a temperature of 300 K. Assume that the effective quantum efficiency is 0.1 and that you want to operate the tube in pulse counting mode and to be degraded from background-limited sensitivity by no more than 10% with a background of 100 photons s^{-1} onto the photocathode. Assuming the dark current is described accurately by the Richardson–Dushman equation (equation 6.5), determine the temperature at which you need to operate the tube.

6.4 Assume the output of the phosphor of an image intensifier is Lambertian. Suppose that the gain of the intensifier is a factor of 50 and that its output is coupled to another intensifier by an $f/2$ relay lens. What is the effective gain of the intensifier/lens combination as seen by the second intensifier? Assuming a photocathode quantum efficiency of 10%, comment on the usefulness of this combination.

6.7 Further Reading

Choi (1997) – comprehensive discussion of quantum well detectors

Csorba (1985) – comprehensive description of photomultipliers, image intensifiers, and television tubes

Donati (2000) – relatively modern description of photomultipliers

Eastman Kodak Co. (1987) – guide to the selection and use of photographic materials for scientific applications, full of useful information

Eccles, et al. (1993) and Smith and Hoag (1979) – thorough discussion of use of photographic materials at very low light levels

Escher (1981) – review of properties of NEA photoemitters

Gunapala and Bandara (2000) and Liu (2000) – two articles in the same volume that give an approachable overview of quantum well detectors

Hamamatsu (2007) – excellent discussion of photomultipliers, theory and practice

James and Higgins (1960) – advanced introduction to the fundamentals and practice of photography

James (1977) – advanced treatment of the photographic process, not particularly suitable as an introduction

Krause (1989) – short introduction to principles behind color photography

Levine (1993) – definitive review of quantum well infrared photodetectors, but not very suitable as an introduction

Newhall (1983) – a classic history of photography

Polyakov (2013) – overview of use of photomultipliers

Razeghi and Nguyen (2014) – review of superlattice/quantum well detection state of the art

Spicer (1977) – another good review of NEA photocathodes

Sturmer and Marchetti (1989) – introduction to the principles behind photography

Timothy (2013) – thorough article on microchannel plates

Zwicker (1977) – another in-depth discussion of photocathodes

7

Superconducting Detectors

Superconductive materials have a number of attributes that provide unique opportunities for advanced detectors. First, changes in the internal state of the detectors can be made by applying very small levels of energy, about three orders of magnitude less than needed to free charge carriers in semiconductors. Second, there are a number of quantum mechanical effects that have prominent influence on the behavior of superconductors, such as profuse tunneling through thin insulating barriers, and these can be exploited for high sensitivity detectors. They are particularly important at long wavelengths, i.e., the submm- and mm-wave, where semiconductor photodetectors are not available. Although photodetectors (and bolometers) have wide application at shorter wavelengths, superconducting detectors also have applications in the X-ray, visible, and near infrared based on their very fast response and moderate spectral resolution within an imaging array (e.g., Szypryt et al. 2017).

7.1 Superconductivity

7.1.1 General Description

Superconductivity occurs in certain materials below a critical temperature, T_c. A detailed theory for superconductivity was provided by Bardeen, Cooper, and Schrieffer (1957) and is known as the BCS theory. Fortunately, more simplistic approaches are sufficient for our discussion (see Tinkham (1996) for more information).

The basis for the BCS theory is that, at very low temperatures, even a weak attractive potential can hold electrons in pairs. As an electron moves through the superconductor crystal lattice, it deforms the lattice by its attraction of positive charge. This process creates an energy minimum for a second electron of opposite spin, resulting in an attractive force between the two with a binding energy per electron of $\Delta(T)$, which is at its maximum at $T = 0$:

$$2\Delta(0) \approx 3.5kT_c, \tag{7.1}$$

where k is Boltzmann's constant. The resulting bound electrons are referred to as a Cooper pair. Although the individual electrons are fermions, which obey the Pauli exclusion principle, the bound pairs have an integer net spin and can behave as identical particles, i.e., bosons, without having to obey this principle. Cooper pairs can form over relatively large distances. The value of this binding distance (also known as the coherence length) depends on the composition of the material, but it can be of the order of 0.1 to 1 μm. As a result of their large binding distances, all Cooper pairs in a superconductor interact with each other to form a collective condensate. In a quantum mechanical sense, they all share the same state and can be described in total by a single wave function. As a result, the energy to break a single pair is influenced by all of the electrons, resulting in the pairing not being destroyed by the thermal oscillations of the lattice if the temperature is sufficiently low. The crystal distortions that lead to the binding energy of the Cooper pairs are not fixed to the crystal lattice but can move. Because they share the same state, all the Cooper pairs tend to move together. If they are given momentum by an applied electric field, there is no energy loss mechanism except for that resulting from the breaking up of the pairs, which itself is energetically unfavorable. Therefore, they all participate in any imposed momentum and can conduct electric currents with no dissipative losses, i.e., no resistance. Quantum effects can also substantially influence the behavior of *unbound* electrons in a superconductor, leading to their description as "quasiparticles."

As more and more Cooper pairs form, the number of available states for their interactions decreases so the energy reduction obtained by creating a pair gets smaller, until the process is no longer favored. Thus, the maximum binding energy for a pair is limited, equal to $2\Delta(0)$ (equation 7.1). This leads to behavior analogous to that of semiconductors as shown in Figure 7.1, i.e., only energy excitations above the "bandgap" can interact with the material. As the temperature rises, an increasing number of Cooper Pairs will be broken by thermal excitation; also the bandgap decreases, going to zero at T_c, see Figure 7.2. We consider the case where the temperature is sufficiently low that the bandgap is nearly temperature-independent. We can then speak of a "valence" band occupied by Cooper pairs, and a "conduction" band occupied by the quasiparticles. The bandgap is 2Δ and is only a few milli-electronvolts wide, making possible a variety of very sensitive electronic devices. Breaking of a Cooper pair creates two quasiparticles in the "conduction" band; we therefore envision this process occurring on Cooper pairs at the Fermi level, just

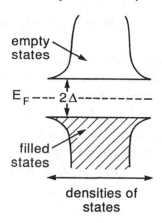

Figure 7.1 Pseudo-energy-level diagram for a superconductor. The energy 2Δ is required to break a Cooper pair. The density of available states is extremely large at the top of the "valence" and bottom of the "conduction" bands because of the boson nature of Cooper pairs and indicates a potentially high rate of transfers from one state to the other, i.e., from Cooper pairs to quasiparticles.

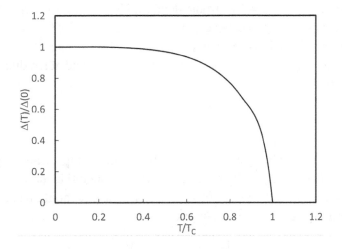

Figure 7.2 Behavior of the Cooper pair binding energy, Δ, with temperature.

Δ below this band. The concentration of thermally generated quasiparticles can be written similarly to equations 2.35 or 2.37:

$$N_T \sim 2N(0)V(2\pi \Delta kT)^{1/2} e^{-\Delta/kT}, \tag{7.2}$$

where $N(0)$ is the density of states at the Fermi energy, and V is the material volume. From equation 7.1, $\Delta \sim 1.75kT_c$. Therefore, the concentration of thermally generated quasiparticles goes as $e^{-1.75T_c/T}$, i.e., rapidly decreases as the temperature drops below T_c, making for optimum semiconductor-like behavior, i.e., little "dark current."

The analogy with the semiconductor energy level diagram should not be taken too far, however. There is no analogy to the dopants that play such an important role in applications for semiconductors. In addition, the quantum effects that dominate superconductivity are insignificant in semiconductors. Besides the small gap width, the superconductivity gap is distinguished by the very large number of permitted energy states that lie just at the top of the "valence" and bottom of the "conduction" bands. This high density of states is possible because Cooper pairs are bosons.

7.1.2 Superconductivity Electrical Resistance

The simplest view of superconductivity would predict an extremely abrupt transition in resistance at T_c. In that case, the local quantum mechanical fluctuations in the superconducting state would be electrically noisy and would limit some uses of superconductors as sensitive electronic devices. However, the observed behavior is more complex; as an example, we consider a cylindrical wire held near T_c. Assume the wire is above T_c, so it has normal conduction, which we take to be uniform over the cross-section of the wire. To compute the magnetic field at a radius r within the wire, we apply Ampère's law:

$$\frac{1}{\mu_0} \oint \mathcal{H}dl = I, \tag{7.3}$$

where μ_0 is the permeability constant, $4\pi \times 10^{-7}$ weber/amp-m, and I is the current enclosed within the integral. By symmetry, the field should be tangent to a circular path at a radius r from the center of the wire, and also it should be constant along this path. This equation yields in our case

$$\frac{1}{\mu_0}\mathcal{H}\,(2\pi r) = I_0\frac{\pi r^2}{\pi a^2}, \tag{7.4}$$

where I_0 is the total current carried by the wire and a is the radius of the wire. We obtain for the strength, $\mathcal{H}$, of the magnetic field

$$\mathcal{H} = \frac{\mu_0 I_0 r}{2\pi a^2}. \tag{7.5}$$

The behavior described by equation 7.5 is modified when the wire temperature is below T_c. Superconductors have the property that they expel magnetic fields up to a critical value $\mathcal{H}_c$, and conversely if they are exposed to a field greater than $\mathcal{H}_c$, their superconductivity ceases (the Meissner effect). A magnetic field is also reduced exponentially as it penetrates into a superconductor; the London penetration depth is the e-folding distance (London and London 1935),

$$\lambda_L(T) = \sqrt{\frac{m_e}{\mu_0 n_{cp} q^2}} \frac{1}{\left[1 - (T/T_c)^4\right]^{1/2}}, \tag{7.6}$$

where m_e is the mass of the electron, q is the electronic charge, and n_{cp} is the density of electrons bound in Cooper pairs, i.e., twice the density of Cooper pairs. For illustrative purposes, we can take equation 7.5 to act just over a depth of λ_L; at T_c (and above), this would be over the full radius of the wire, but the depth decreases as the temperature is reduced below T_c. We focus on this surface layer. The London penetration depth at $T = 0$ is a characteristic of an individual superconductor, e.g., $\sim$200 nm for NbN. Given the value for a given material, equation 7.6 can be used to determine n_{cp}, which is an important parameter in some of the following discussion.

$\mathcal{H}_c$ depends on temperature:

$$\mathcal{H}_c(T) = \mathcal{H}_c(0)\left[1 - (T/T_c)^2\right]. \tag{7.7}$$

Let the temperature of our wire decrease through T_c. Because of the small value of $\mathcal{H}_c$ just below T_c, the magnetic field at the surface of the wire will exceed $\mathcal{H}_c$ and the wire will continue to conduct normally there. However, from equation 7.5, $\mathcal{H}$ will decrease linearly with radius within the surface layer of the wire and at some radius will no longer exceed $\mathcal{H}_c$, so within this radius the wire can become superconducting. In this region, the wire assumes a structure in which normal and superconducting zones alternate. The size of this partially superconducting region increases as the temperature is reduced below T_c, as indicated by the increase in $\mathcal{H}_c$ (equation 7.7), so a greater proportion of the current can be carried by superconductivity, resulting in a reduction of resistance with decreasing temperature. Eventually λ_L decreases sufficiently that the superconductivity in the core of the wire dominates. The resistance in the surface layer is

$$\frac{R}{R_n} = \frac{1}{2}\left\{1 + \left[1 - \left(\frac{T_c - T}{\delta T_c}\right)^2\right]^{1/2}\right\}, \quad T_c - \delta T_c \leq T \leq T_c, \tag{7.8}$$

where R_n is the normal resistance and $\delta T_c = I(dI_c/dT)^{-1}$; $I_c = 2\pi a\mathcal{H}_c/\mu_0$ is the maximum current at which the entire wire can be superconducting (compare equation 7.5 with $r = a$). For additional details, see Tinkham (1996). Equation 7.8 shows that the wire resistance falls rapidly, but not infinitely rapidly, over a narrow range of temperature just below T_c. At $T_c - \delta T_c$, the resistance is $1/2\,R_n$, and it abruptly falls to zero below this temperature as the entire volume becomes

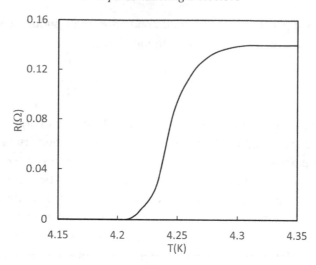

Figure 7.3 Resistance of a superconductor (in this case, mercury), as a function of temperature.

superconducting. Figure 7.3 is an illustration of the change of resistance of a superconducting film with temperature.

Superconductivity can also be lost if a device carries too high a current density, i.e. if the current exceeds the critical current:

$$I_c = 2\pi a \mathcal{H}_c, \tag{7.9}$$

where a is the radius of the (assumed) cylindrical conductor. I_c is controlled by the size of the conductor, a in equation 7.9, as well as the material properties through their influence on $\mathcal{H}_c$.

7.1.3 Kinetic Inductance

The density of thermally generated quasiparticles decreases exponentially with Δ/kT, where Δ is the Cooper pair binding energy per electron. Thus, for a superconductor well below the critical temperature, the great majority of the electrons are bound in Cooper pairs. When photons are absorbed with energy greater than the Cooper pair binding energy, 2Δ (0.4–4 meV depending on the material), they break the pairs and release excess quasiparticles. For incident power, P, the number of excess quasiparticles, η_{qp}, in a device is

$$\eta_{qp} \propto \frac{\eta P \tau_{qp}}{\Delta}, \tag{7.10}$$

where η is the energy transfer efficiency from the incident radiation to quasiparticle production ($\sim$0.57) and τ_{qp} is the lifetime of the quasiparticles. If the thermally generated quasiparticle density is low (aided if the superconductor is in the form of a thin film), the excess quasiparticles can dominate the behavior of the material.

Free charged particles are accelerated by any high-frequency electric field; because of conservation of momentum, it takes a finite time for them to react. As a result, they impose a phase lag similar to that created in a conventional electrical circuit by an inductor, an effect described as kinetic inductance. Although kinetic inductance occurs in any conductor, it is normally overwhelmed by the resistance. In a superconductor, this goes to zero and the inductance becomes a useful property. A class of detector is based on this process (discussed in Section 7.3). The breaking of the Cooper pairs by incident photons creates excess quasiparticles and affects the reactance of the detector to increase the inductance (and the resistance), because the quasiparticles block the Cooper pairs from occupying some of the electron states.

To demonstrate this process, we set the kinetic energy of the Cooper pairs, KE_C, equal to an equivalent energy due to the kinetic inductance, IE_C, or

$$KE_C = \frac{1}{2}\left(2m_e \langle v_{cp}\rangle^2\right)(n_{cp}lwd) = IE_C = \frac{1}{2}L_K I^2, \tag{7.11}$$

where m_e is the mass of an electron, $\langle v_{cp}\rangle$ is the average velocity of the Cooper pairs, n_{cp} is the number density of Cooper pairs, l is the length, w the width, and d the depth of the superconducting film, so its volume is lwd, L_K is the effective kinetic inductance, and $I = 2\langle v\rangle n_{cp}wde$ is the supercurrent through the film. Thus,

$$L_K = \frac{m_e}{2n_{cp}q^2}\frac{l}{wd}, \tag{7.12}$$

showing how the breaking of the Cooper pairs increases the kinetic inductance.

According to the Ginzburg–Landau (GL) theory of superconductivity, the density of Cooper pairs is

$$n_{cp}(T) \approx n_{cp}(0)\left(1 - \frac{T}{T_c}\right), \tag{7.13}$$

so we find (Annunziata et al. 2010)

$$L_K(T) = \frac{m_e}{2q^2}\left(\frac{l}{wd}\right)\left(\frac{1}{n_{cp}(T)}\right) \approx \frac{m_e}{2q^2}\left(\frac{l}{wd}\right)\left(\frac{1}{n_{cp}(0)\left(1 - \frac{T}{T_c}\right)}\right)$$

$$= L_K(0)\left(\frac{1}{1 - \frac{T}{T_c}}\right) \tag{7.14}$$

that is, the kinetic inductance becomes smaller as the physical temperature is reduced below the transition temperature; for a given physical temperature, the kinetic inductance is larger for lower T_c.

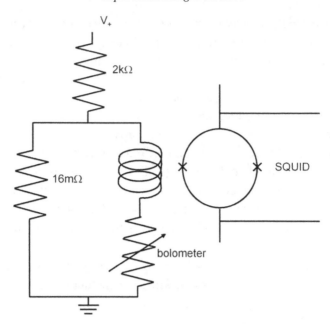

Figure 7.4 SQUID readout (to right). To illustrate a typical circuit, we show the SQUID as it might be used for a superconducting bolometer. The bolometer signal is coupled into the SQUID by the inductor in the circuit to the left.

7.2 Superconducting Electronics

Conventional silicon-based FET transistors are well matched to semiconductor detectors, particularly when the detector operates at a temperature above $\sim$20 K where there is sufficient thermal excitation within the FET to support stable operation. However, the low resistance of superconducting detectors matches poorly to the high impedances of silicon FETs, and the low temperatures of operation compromise their performance. A more suitable readout is a superconducting quantum interference device (SQUID), as discussed by Richards (1994). Figure 7.4 shows the conventional symbol for the SQUID: a loop, broken by two x's for Josephson junctions. As originally hypothesized by Josephson, a junction of two superconductors with an intervening insulator can carry a current of Cooper pairs with no resistance up to a critical value, I_0, above which it experiences resistive losses. The Cooper pair current across a Josephson junction is a sinusoidal function of the superconducting phase difference between the two sides of the junction. This behavior arises because of the long-range quantum interaction of Cooper pairs. To maintain phase coherence around the loop of a SQUID requires that the phase change of the wave function be $2n\pi$, where n is an integer. The phenomenon is analogous to the behavior of a two-slit optical interferometer.

That is, the output voltage of a SQUID biased to carry a current I is dominated by interference effects, and is given by (Tinkham 1996)

$$V = \frac{R}{2} \left[I^2 - (2I_0 \cos (\pi \Phi / \Phi_0))^2 \right]^{1/2}, \tag{7.15}$$

where R is the resistance of the junctions (including any shunt resistance added to avoid excess noise), Φ is the magnetic flux through the SQUID, and Φ_0 is the quantized unit of magnetic flux, $\Phi_0 = h/2q \sim 4 \times 10^{-15}$ weber. Thus, the behavior of the loop current depends sensitively on the correspondence of the magnetic flux through the loop to quantized values, providing a large change in I–V characteristics near $2I_0$. The quantization of magnetic flux in units of Φ_0 is a consequence of this quantized behavior of the phase of the electron pseudo-wavefunction. Interference of the superconducting wavefunction around the loop results in a voltage response on the output of the SQUID that is a very sensitive function of the current applied to the input coil, since the magnetic flux through the loop is regulated by this current. When its output is made linear by using feedback, the device works as a useful amplifier.

Because there is no superconducting equivalent to the transistor, the SQUID takes its place as the basic building block of superconducting electronics, where it is the foundation of a variety of complex circuits and applications (see, e.g., Van Duzer and Turner 1998; Clarke and Braginski 2006; or Ruggiero and Rudman 2013).

7.3 Microwave Kinetic Inductance Detectors (MKIDs)

A rapidly developing form of detector utilizes the kinetic inductance of super-conductors – microwave kinetic inductance detectors (MKIDs). The leading "microwave" in MKID indicates the operating frequency of the readout electronics; in principle, these devices can be used in the microwave through the X-ray. However, the greatest interest has been in their use at the low-frequency end of this range.

7.3.1 MKID General Concept

MKIDs are based on a resonant circuit, an example of which is shown in Figure 7.5. Both the capacitor and the inductor store energy, which can pass back and forth between them. For the moment, set $R = 0$. The complex impedance of the circuit is then

$$\frac{1}{Z} = j \left(\omega C - \frac{1}{\omega L} \right), \tag{7.16}$$

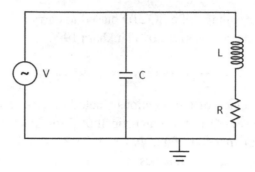

Figure 7.5 Example of a resonant circuit.

where ω is the angular frequency. The real part is then

$$|Z| = \sqrt{(ZZ^*)} = \frac{\omega L}{\omega^2 LC - 1}. \tag{7.17}$$

The resonant frequency is

$$\omega_0 = \frac{1}{\sqrt{(LC)}}, \tag{7.18}$$

where $|Z|$ nominally goes to infinity. At this frequency, the input impedance of the circuit is maximized and the oscillation can continue over many cycles without additional input. In fact, the timescale for decay of the oscillation is set by the value of the resistance, which dissipates the resonating current.

For a MKID detector, the inductor is a superconductor, so absorption of a photon by it splits Cooper pairs to create quasiparticles. From equation 7.12, the result – a decrease in n_{cp} – is an increase in the kinetic inductance. This process reduces the resonant frequency (equation 7.18). If the circuit is connected as a load onto a transmission line carrying a microwave signal close in frequency to the resonance, this change can be sensed by its effect on the signal to complete the detection process.

Both the inductive and capacitive components of the resonant circuit of an MKID need to be fabricated using thin film lithography in superconducting materials. Coplanar waveguides (CPWs) and microstrips are usually preferred because of their ease of fabrication, see Figure 7.6. One approach for a detector is shown in Figure 7.7, in which the inductance and capacitance are laid out as lumped elements. The closely spaced interleaved fingers at the bottom of the figure provide capacitance, while the long meandering section in the middle provides inductance. There are two inductive terms, one just due to the geometry of the meander and the other the kinetic term. When photons are absorbed in the inductive section that have energy greater than 2Δ, they break Cooper pairs into quasiparticles.

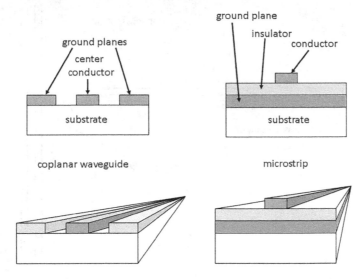

Figure 7.6 Coplanar waveguide (CPW) (left), microstrip (right).

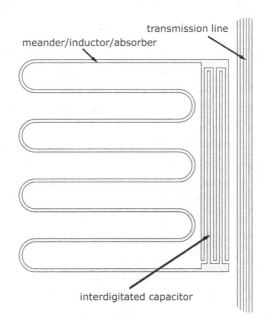

Figure 7.7 A lumped element pixel in a MKID. The shaded regions show super-conductor layers; the white background is an insulating substrate. The rightmost parallel traces are the CPW feedline that carries the microwave signal that is used to sense the illumination level onto the pixel. To its left are interleaved fingers in the pixel capacitor, and the meandering section at the left provides the pixel inductance. This figure is illustrative and is not to scale, nor are the numbers of traces in a given element necessarily realistic.

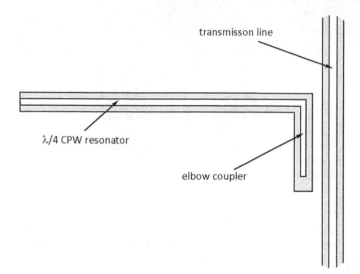

Figure 7.8 A $\lambda/4$ coplanar waveguide (CPW) shunt-coupled transmission line resonator, with an elbow coupler to the transmission line. The shaded regions show superconductor layers; the white background is an insulating substrate.

The result is an increase in both the resistance and the kinetic inductance. The inductive response can be made significantly larger than the resistive one (e.g., Schlaerth et al. 2009); to make it also large enough to stand out over the fixed magnetic/geometric inductance, the traces in the meander need to be narrow and thin. In this case, the result is a significant reduction in the resonant frequency and a change in other resonant properties. This type of resonator can be illuminated directly with the design of the meandering section optimized for absorbing photons, providing a large fill factor for the absorbing component. That is, it becomes a pixel in a detector array.

An alternative approach is to fabricate the pixel as a waveguide of length $\lambda/4$ and shorted at its end (Figure 7.8). It can be shown that the resonant properties of such a device are similar to those of the lumped element pixel; in this case, the length of the line determines the resonant frequency. The elbow trace at the right is used to establish the capacitive coupling from the pixel to the transmission line. Since the components are more explicit in the lumped element device, our discussion will focus on it.

7.3.2 MKID Circuit Behavior

We now return to Figure 7.5 and describe its performance as a MKID pixel in further detail. The resonating performance of the circuit is described by the quality factor, Q, a dimensionless parameter that characterizes how long the circuit continues to resonate once it has been stimulated. We define n as the number of oscillations the circuit makes before its energy content is reduced by $1/e$; that is, if the exponential decay time constant is τ and the oscillation frequency is f, $n = f\tau$. By definition,

$$Q = 2\pi n = \omega_0 \tau, \qquad (7.19)$$

where ω_0 is the angular frequency at resonance. A high Q corresponds to a low rate of energy loss compared with the total stored energy in the circuit; that is, the resonance persists with the oscillations decreasing only slowly. We can also describe τ as the rate at which the energy stored in the circuit is lost, or the ratio of the amount of energy stored in capacitor and inductor divided by the dissipative power in the resistor. As the circuit in Figure 7.5 oscillates, power is passed back and forth between the capacitor and inductor with the rms current being equal in each side of the circuit. The internal quality factor, Q_i, is then

$$Q_i = \omega_0 \tau = \omega_0 \frac{E_{stored}}{P_{lost}} = \omega_0 \frac{I_{rms}^2 L}{I_{rms}^2 R} = \omega_0 \frac{L}{R}. \qquad (7.20)$$

Because slow decay corresponds to a high level of "purity" in the oscillating signal, Q also describes the range of oscillating frequencies exhibited by the circuit; a high level of Q indicates that the bandwidth is narrow, i.e., the oscillations are confined closely to the resonant frequency.

Figure 7.5 describes a single MKID pixel, but to read out its signal it needs to be integrated into a circuit like Figure 7.9. In the vernacular of high-frequency

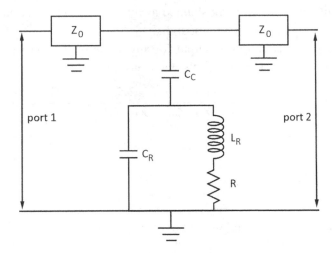

Figure 7.9 MKID unit cell and surrounding circuitry. The 2-port has an input transmission line with impedance Z_0. The unit cell is coupled to it through the capacitor C_C and is shown as lumped capacitive, inductive, and resistive elements.

electrical engineering, this is a two-port device. That is, it has one pair of input lines
(= the input port) and one pair of output lines (= the output port). In this case, a
varying signal near the resonant frequency is applied to the input port, conveyed
past the unit cell by a transmission line fabricated, for example, as a coplanar
waveguide, and at the output port can show modifications to the input signal by
the unit cell circuit. The connection through which any modification is made is the
capacitor, C_C.

The circuit in Figure 7.9 has a frequency-dependent impedance, $Z_R(\omega)$, that
shunts energy introduced into port 1 to ground before it reaches port 2. To quantify
this behavior, we need to determine this impedance:

$$Z_R(\omega) = \frac{1}{j\omega C_C} + \left[1 \Big/ \left(\frac{1}{j\omega L_R + R} + j\omega C_R \right) \right]$$

$$= \frac{1}{j\omega C_C} + \frac{j\omega L_R + R}{1 - \omega^2 C_R L_R + j\omega C_R R}. \tag{7.21}$$

The load impedance, Z_L, looking into the 2-port can be expressed as

$$\frac{1}{Z_L} = \frac{1}{Z_R} + \frac{1}{Z_0}. \tag{7.22}$$

The behavior of a two-port circuit is often described by a 2×2 "scattering
matrix." The terms S_{11} and S_{22} represent signals reflected off the left (input) and
right (output) sides of the circuit (as drawn) with a matched load on the other side
(as shown for the output in Figure 7.9). S_{21} and S_{12} are transmission in the forward
and backward directions, respectively (and again into a matched load). In the case
of the MKID, we are primarily interested in S_{21}, that is, the signal that passes by
the MKID unit cell without being shunted away and is received at the output, as a
function of the frequency of the input signal.

To determine S_{21} we ratio the output for any value of Z_R to that with Z_R very
large so that no signal gets lost over the resonator shunt:

$$S_{21} = 1 - \frac{1}{1 + 2Z_R/Z_0} = \frac{2Z_L}{Z_L + Z_0} = \frac{2}{2 + \frac{Z_0}{Z_R}}. \tag{7.23}$$

We start with equation 7.23 and derive S_{21} for small deviations of the frequency
from ω_0. Since

$$\frac{1}{Z_R} = j\omega C + \frac{1}{j\omega L}, \tag{7.24}$$

we can then substitute in equation 7.23 to obtain, after some manipulation,

$$S_{21} = \frac{1}{1 + jQ\left(1 - \frac{\omega_0^2}{\omega^2}\right)} \tag{7.25}$$

since, allowing for the matched loads of Z_0 on the input and output, for this whole circuit $Q = \omega_0 Z_0 C / 2$. Substituting $\omega_0 + \delta\omega$ for ω where $\delta\omega$ is small compared with ω_0, and ignoring small terms,

$$1 - \left(\frac{\omega_0^2}{\omega^2}\right) = 1 - \left(\frac{\omega_0^2}{\omega_0^2 + 2\delta\omega\omega_0 + \delta\omega^2}\right) \approx 1 - \left(\frac{1}{1 + 2\frac{\delta\omega}{\omega_0}}\right) \approx 2\frac{\delta\omega}{\omega_0}, \quad (7.26)$$

thus making equation 7.25 take the form

$$S_{21} = \frac{1}{1 + 2jQ\frac{\delta\omega}{\omega_0}}, \quad (7.27)$$

making the amplitude of S_{21}

$$|S_{21}|^2 = \frac{1}{1 + Q^2 \left(\frac{\delta\omega}{\omega_0}\right)^2} \quad (7.28)$$

and the phase angle

$$\theta = tan^{-1}\left(2Q\frac{\delta\omega}{\omega_0}\right), \quad (7.29)$$

while the resonant frequency is from equation 7.18

$$\omega_0 = \frac{1}{\sqrt{(L(C_C + C_R))}}. \quad (7.30)$$

Through the effect on Q, changes in the inductance affect the resonant properties as shown in Figure 7.10. In our case, as the kinetic inductance is increased by the absorption of photons and breaking of the Cooper pairs, the profile of the resonance becomes broader and lower in peak amplitude, and shifts toward lower frequencies. Equally importantly, the phase of the oscillation changes as indicated by equation 7.29.

These effects are more conveniently represented by putting S_{21} on the complex plane (right side of Figure 7.10). Absorption of photons changes the phase by $\delta\theta$ and the radius by δA. The absorption of the photon can be sensed by measuring the change in phase or in amplitude (radius), sometimes called a dissipative readout.

Multiple MKID pixels are fabricated along a single transmission line as shown in Figure 7.11. Assuming lumped element pixels, each is designed for a specific (and different from its siblings) frequency by adjusting the dimensions of the interleaved finger capacitors. Resonant frequencies of $\lesssim 10$ GHz are used, and the resonance of a pixel can be made sufficiently sharp (high Q) that it is less than 1 MHz wide. Therefore, up to 1000–2000 pixels can be multiplexed onto a single transmission line without significant overlap of response (assuming they all have high Q and operate at accurately determined frequencies). To guard further against overlap,

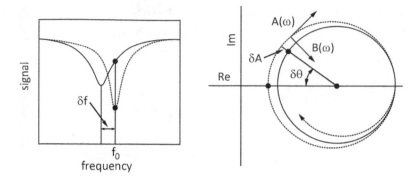

Figure 7.10 Effect of a photon flux on the resonant properties of the MKID resonant circuit. It is assumed that the flux has increased from the dashed to the solid curves. To the left is shown the shift toward lower resonant frequency and the reduction in Q (i.e., the broadening of the range of resonant frequencies). The changes can be shown more diagnostically in the complex plane, where the resonance traces a circle. This is shown to the right where the same process is described by plotting S_{21} in the complex plane. After Baselmans (2012).

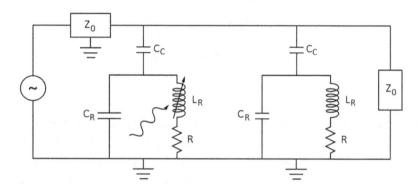

Figure 7.11 Lumped element representation of a 2-pixel MKID detector array. The cell to the right has a smaller capacitance than the one to the left to adjust it to a different resonant frequency. A photon absorbed by the cell to the left has caused its inductance to change.

i.e., crosstalk between pixels, the pixel resonant frequencies are not incremented monotonically along the transmission line; instead, they are set so there is a large difference in resonant frequency between any pixel and its immediate neighbors. A comb of frequencies is passed down the transmission line, each matched to one of the pixels. The output is read out by a high-frequency HEMT amplifier (Section 9.2.3.1). The outputs of the detectors are then multiplexed in the frequency domain, and the signals from each one can be extracted by demodulation of the HEMT output, either by sensing the amplitude of the load on the transmission line, or the phase. In practice, phase measurement is generally used because it provides better noise performance. Although the output electronics can be complex, detector arrays made in this way have the major advantage that they do not require the complex cryogenic circuitry described in Section 7.2.

7.3.3 Quantum Efficiency

A mm-wave MKID can be made highly absorbing to incoming photons by fabricating it to have the ohms/square that matches the impedance of free space, which is $\zeta_0 = (\mu_0/\varepsilon_0)^{1/2} = 377\ \Omega$, where μ_0 is the absolute magnetic permeability and ε_0 is the permittivity of free space. An efficient absorber that has low volume and hence little heat capacity is a thin layer of metal adjusted in thickness to have this resistance. (A given film thickness will have the same resistance for any square surface area, see Problem 7.2.) The theoretical foundation of this approach can be found in Stratton (1941) or Born et al. (1999), and it is given in practical form for this application by Clarke et al. (1977). Given that the meandering inductive portion of an MKID has structure significantly smaller than the wavelength to be detected, it can be fabricated to meet this goal (as illustrated in Figure 7.12).

The latter reference considers the case of a metal film deposited on the back of a dielectric substrate such as sapphire or diamond, a geometry appropriate for a specific type of bolometer but the physics applies generally. Assuming normal incidence, the transmittivity and reflectivity at the first (unmetallized) surface are

$$t_1 = \frac{4n}{(n+1)^2} \tag{7.31}$$

and

$$r_1 = \frac{(n-1)^2}{(n+1)^2}, \tag{7.32}$$

respectively, where n is the index of refraction of the dielectric. The transmittivity, reflectivity, and absorptivity at the second (metallized) surface are

$$t_2 = \frac{4n}{(n+1+x)^2}, \tag{7.33}$$

$$r_2 = \frac{(1-n+x)^2}{(1+n+x)^2}, \tag{7.34}$$

and

$$a_2 = \frac{4nx}{(1+n+x)^2}, \tag{7.35}$$

respectively, where $x = \zeta_0/R_0$, and R_0 is the resistance per square area of the metal film. Although in general there can be complex interference phenomena

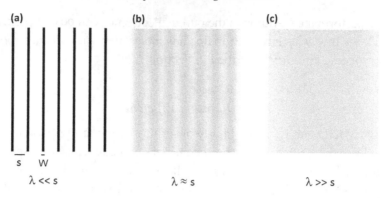

Figure 7.12 Appearance of a parallel grid of conductors to photons of different wavelengths relative to the grid spacing. When the wavelength is much larger than the spacing, the grid appears as a uniform surface impedance of value dependent on the grid spacing, the width of the grid lines, and the grid material.

in the dielectric plate, a simple and efficient case occurs when $r_2 = 0$ (that is, $x = n - 1$): (1) there are no multiple reflections, and (2) the absorptivity is independent of wavelength. For the case of a sapphire dielectric, $n \sim 3$, this case gives $t_1 = 0.75$, $a_2 = 0.667$, and a net absorption efficiency of $t_1 a_2 = 0.50$. Thus, the quantum efficiency using such an absorber would be 50%. However, MKIDs also must share the detector surface area between the absorber (i.e., the serpentine inductor) and the readout components (i.e., the capacitor and transmission line), so the effective quantum efficiency will generally be lower than this value.

To implement this approach, the meander of superconducting traces in a lumped element MKID needs to be made to appear as a uniform sheet resistance. Our usual view of a conductive grid makes this requirement seem impossible; see Figure 7.12(a) where the wavelength is much shorter than the grid spacing. However, as the grid becomes closer and closer spaced compared with the wavelength (Figure 7.12(b, c)), the effective sheet resistance, R_{eff}, becomes uniform, given by an averaging over the effects of the grid lines and the spaces between them:

$$R_{eff} = R_S \frac{w}{a}, \tag{7.36}$$

where R_S is the sheet resistance (ohms/square) of the grid line material, a is the width of a grid line, and w is the line-to-line spacing (i.e., the pitch). The sheet resistance can be manipulated by changes in the thickness and width of the meander lines, making optimization of the meander a multi-parameter but still manageable task.

The absorption of the lumped MKID element can be further increased by placing a "backshort" behind it. The detector will typically be constructed on one side of a

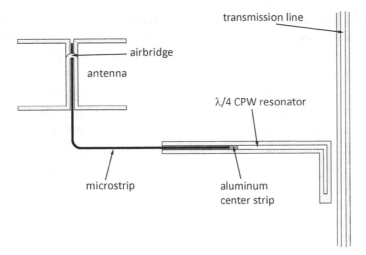

Figure 7.13 A $\lambda/4$ coplanar waveguide (CPW) shunt-coupled transmission line resonator, with an elbow coupler to the feedline (as in Figure 7.8). The twin slot antenna (to the left) receives the photons and their energy is conveyed to a microstrip through the airbridge, and then to the shorted end of the MKID resonator.

silicon wafer. If it is on the "frontside," so the incoming radiation strikes it first, a backshort can be created by making the other side reflective and adjusting the thickness of the wafer so maximum positive interference occurs at the meander/absorber, e.g., to 1/4 wave optical thickness. If the MKID is on the "backside" (i.e., the incoming radiation passes through the wafer before encountering the MKID unit cell) then an additional reflector with a suitable (e.g., $1/4\ \lambda$) spacing can be placed behind the unit cell for a similar function.

These lumped element MKIDs can provide filled detector arrays, i.e., where the pixels are fabricated adjacent to each other and fill a focal plane so they can be flood illuminated with no significant loss of efficiency (other than that due to the area of the interleaved capacitor, which does not participate in the absorption). Where a filled array is not necessary, there is another approach to feeding the photon energy into the MKID. One can use a microstrip antenna to receive the photon and couple its energy to the MKID inductor, as sketched in Figure 7.13. The antenna signal is coupled to a microstrip by the airbridge. For efficient conveyance of the signal from the antenna to the resonator, the microstrip is made of a superconductor with a moderately high T_c, e.g., niobium allows efficient operation up to about 750 GHz. The energy can be coupled efficiently into the resonator using a low T_c superconductor, e.g., aluminum, for the shorted end of the resonator; from equation 7.1, if the Cooper pair binding energy is low and they are easy to break, this approach amplifies the signal leading to detection down to relatively low energy photons.

For example, for aluminum this occurs for frequencies > 90 GHz. Because the antenna occupies real estate in the detector array, this approach requires that the MKID pixels be fed by, for example, individual lenses to fill in the inter-pixel spaces and provide uniform response.

Quantum efficiencies for optical/near-infrared MKIDs are somewhat lower. The most favorable materials for the inductor have absorption levels in layers of the appropriate thickness of no more than 40% at 0.4 μm, 30% at 0.7 μm, and 15% at 1.6 μm, setting an upper limit to the achievable quantum efficiency (Szypryt et al. 2016). They also have the competition for real estate between the inductor/absorber and the capacitor and readout components.

7.3.4 Performance Measurement

Some attributes of MKIDs can be measured in ways analogous to those already discussed for semiconductor detectors in Section 5.4, e.g., quantum efficiency, noise, signal to noise. However, the properties of these devices do require different types of measurement for pixel yield and, when used in the ultraviolet, optical, or near-infrared, spectral resolution.

Multiplexing many MKIDs onto a single transmission line requires that the resonant frequencies of the pixels be well-controlled so there are no overlaps. The success in achieving this goal can be determined by conducting a frequency scan of the multiplexing transmission line and locating all the resonances of the pixels it serves. If two pixels lie too close to each other in frequency, there can be too much crosstalk, making it impossible to distinguish their signals. Such cases become dead pixels in a MKID array. In the case of a lumped element pixel (Figure 7.7), the value of the capacitance is determined by the photolithography and can be controlled well. However, the inductance depends critically on the quality and uniformity of the superconducting films applied in the meander, and on their dimensions, i.e., thickness. From equation 7.14, the inductance depends on T_c of this film. A popular material for the superconductor is TiN, deposited by sputtering off a Ti target within a nitrogen-rich atmosphere. In these films, T_c depends critically on such processing details as the flow rate used to maintain the N_2 environment (Szypryt et al. 2016). Variations in T_c in TiN MKID arrays are a significant current limitation on the control of the pixel operating frequencies. Two approaches might solve this problem: (1) improving the processing of the TiN films; or (2) finding another material that provides a more uniform T_c. The problems 7.3–7.5 allow evaluating the dependence on film thickness.

Because MKIDs are read out very fast, they respond to single photons. In this case, the effect on the pixel resonance depends on the energy of the photon (i.e., not on the number of photons that have recently been absorbed as in conventional operation of a CCD in the optical). For the mm-wave, the spectral resolution can be determined by modeling and electronic measurements. In the optical range, the spectral resolution of a MKID can be determined by measuring its response to photons from a laser that yields a monochromatic output at a precisely known wavelength (or to a number of lasers simultaneously with well-separated output wavelengths). It depends primarily on the internal Q_i of the pixels; typical resolutions in the ultraviolet-optical-infrared are $R = \lambda/\Delta\lambda \sim 8$. However, values of Q_i above $\sim 100{,}000$ do not seem to yield improvements in resolution. The issue is thought to be noise caused by fluctuations in the field strength at the resonator due to the existence of two energy states at the interface between the superconductor and substrate of a pixel. Amplifier noise in the readout electronics also limits the achievable spectral resolution.

7.3.5 Applications of MKIDs

MKIDs have had a major impact on mm- and submm-wave astronomy because they can provide many-pixel high-performance detector arrays, circumventing the complexities of cryogenic multiplexing electronics as required by the rival technology, transition edge sensor bolometers (next chapter). An example of current applications is the NIKA2 instrument on the IRAM telescope, which includes three 1000-pixel MKID arrays (Calvo et al. 2016). Plans are under way to increase the pixel count in this instrument substantially through development of smaller-pixel MKIDs (Shu et al. 2018).

In the optical and infrared, MKIDs face strong competition in many appications from semiconductor detectors, i.e., CCDs and photodiode arrays. Demonstration instruments have been developed with 20,000 pixels and spectral resolution of $\lambda/\Delta\lambda \sim 10$ (Szypryt et al. 2017). However, such an instrument must compete with integral field spectrometers with mega-pixel CCD arrays. Another application that exploits the intrinsic advantages of the MKID is to use the simultaneous wavelength resolution and imaging at very high readout speeds to control seeing speckles and enhance post processing to achieve high image contrast for, e.g., exoplanet searches (Meeker et al. 2018).

7.4 Other Superconducting Detectors

7.4.1 Superconducting Tunnel Junctions (STJ)

We have emphasized the suitability of semiconductors for visible light detection because their bandgap energies match that of a visible photon. The result is that an absorbed photon usually produces just one charge carrier; no information is yielded beyond the detection event itself. Superconductors have bandgaps about a thousand

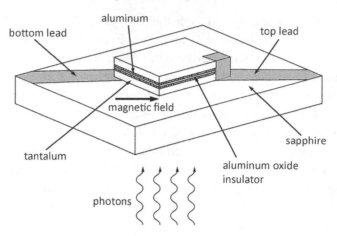

Figure 7.14 Simple superconducting tunnel junction detector.

times smaller than semiconductors. In a superconductor-based detector, a visible photon would produce 1000 or more charge carriers, in proportion to its energy. Measuring the amplitude of the event would therefore determine the wavelength of the photon. A superconducting tunnel junction detector takes advantage of this potential gain in information. Fundamentally, it consists of a sandwich of two layers of superconductor, separated by a sufficiently thin insulator to allow tunneling ($\sim 10^{-9}$ m thick, from the discussion in Section 6.4 and in view that an insulator bandgap is a few eV) and is termed a superconductor–insulator–superconductor (SIS) junction. A detector based on this arrangement is illustrated schematically in Figure 7.14.

A photon is absorbed by breaking Cooper pairs in a superconductor layer of this device, transferring its energy to the resulting quasiparticles in the "conduction" band. Assuming the photon is in the visible or X-ray, the quasiparticles are elevated high into the "conduction" band, where they free additional electrons through ionization and through plasmon emission and decay (plasmons arise from the quantized oscillation states of the free electron gas). As a result, the energy is converted into a cloud of electrons that decay down to the energy level of the bottom of the "conduction" band; since they are all at this level, their number is proportional to the energy of the absorbed photon. Only at this point does tunneling become significant, given the high (formally infinite) density of states. The tunneling process itself is illustrated in Figure 7.15 where, for simplicity of illustration, we have ignored the decay of the initial free quasiparticle and just show the excitation directly to the bottom of the conduction band, and again for illustration assume only a modest energy offset between the two sides of the insulator. Of course, quantum effects have a wonderful ability to make things more complicated, and thus there

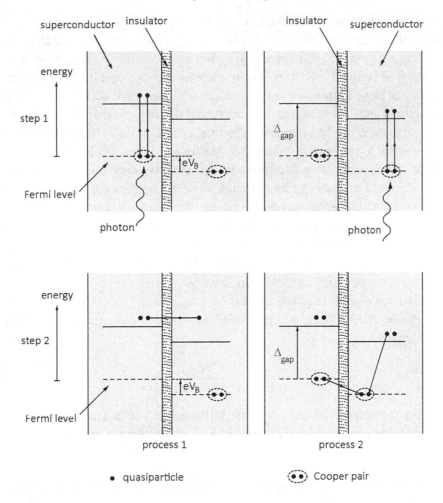

Figure 7.15 Two ways for tunneling to occur in a STJ. In process 1, a photon breaks a Cooper pair and two quasiparticles are elevated into the "conduction" band on the left side in step 1, and in step 2 one of them tunnels through the insulator to the right side. In process 2, a photon breaks a Cooper pair on the right side of the insulator in step 1. In step 2, a Cooper pair on the left side breaks and one of the resulting quasiparticles tunnels to the right side to join with a quasiparticle de-excited from the "conduction" band on that side. The end results are identical: a quasiparticle on each side of the device, with a Cooper pair on the right side.

is more than one way for tunneling to occur. Figure 7.15 shows two possibilities. Two more are the reverse of those shown, but are energetically unlikely if the bias voltage (V_B) is significant.

With no voltage applied across the junction, the Cooper pairs can flow from one superconductor to the other to carry small currents (the Josephson effect) because the two superconductors share the same energy states. A magnetic field is applied to suppress the Josephson effect and a small voltage applied across the SIS junction, $V < 2\Delta/q$. If the STJ is held at a very low temperature, so long as it is in the dark and well below T_c virtually no current will flow through it. If it now absorbs a photon, a large number of quasiparticles will assemble at the bottom of the "conduction band," from where they can tunnel through the insulator to the empty excited state band on the other side. The total charge in the resulting current pulse is proportional to the photon energy (see Martin and Verhoeve 2013).

To be more quantitative, we first consider the operating temperature required for this device. Thermally generated quasiparticles can penetrate the insulator and appear as a dark current, contributing noise. Thus, there is a premium in running the SIS junction at temperatures well below T_c to suppress quasiparticles (see equation 7.2). The best results are obtained when the thermally generated current is well below the photon-generated one, which for low light level detection requires an operating temperature no more than about 10% of T_c.

The average energy required to create a single quasiparticle is $\sim 1.7\Delta$ (Martin and Verhoeve 2013). Thus, the number of quasiparticles, N_0, created by a photon of wavelength λ is

$$N_0(\lambda) = \frac{hc}{\lambda \, 1.7\Delta} \sim \frac{700}{\lambda(\mu m) \, \Delta(meV)}. \tag{7.37}$$

We define the spectral resolution as the full width at half maximum of the distribution of pulse amplitudes for photons of a fixed wavelength, which is $2.36 \sqrt{N_0}$. Thus,

$$\frac{d\lambda}{\lambda} = 0.089 \, (\lambda(\mu m) \, F \, \Delta(meV))^{1/2}, \tag{7.38}$$

where F is the "Fano factor." The Fano factor adjusts for the loss of quasiparticles that do not tunnel through the insulator, for example by recombining as Cooper pairs. It has a typical value in this circumstance of ~ 0.2. For example, a niobium-based junction (with $\Delta = 1.55$ meV) has a potential resolution at 0.5 μm of $d\lambda/\lambda = 0.035$, or 3.5%.

We now use the simple STJ detector in Figure 7.14 for illustration. Assume it is based on superconducting tantalum, with $\Delta = 0.664$ meV. A base film of tantalum is deposited on a substrate of sapphire. A thin aluminum layer is deposited on the tantalum, which is oxidized to create a ~ 1 nm thick aluminum oxide insulator followed by another aluminum-based layer and a top film of tantalum. Figure 7.14 greatly simplifies the leads to the device; in practice the SIS sandwich is covered with an insulating SiO_2 layer and vias etched through the layer allow contact to be made by a final metalized layer. The device is operated at ~ 0.3 K and is illuminated through the sapphire. A magnetic field is applied to the STJ to suppress the Josephson current.[1]

In addition to the basics already described, the aluminum layers provide gain in this detector. Because the bandgap in aluminum ($\Delta = 0.172$ meV) is much smaller than for tantalum, the quasiparticles tend to get concentrated in the aluminum layer right next to the insulator, and to multiply through the cascade-type process already discussed. Further multiplication occurs because, when a quasiparticle tunnels through the insulator, it stays close to the insulator and has a high probability of breaking up a Cooper pair on the side it just came from and bonding with one of the resulting quasiparticles that also tunnels through the insulator. This process leaves a third quasiparticle that can also tunnel through the insulator, break a Cooper pair on the side it left, and so forth. Thus, the creation of a quasiparticle by photon absorption can lead to a net charge of many electrons crossing the insulator. Equation 7.38 becomes

$$\frac{d\lambda}{\lambda} = 0.089 \left[\lambda(\mu m) \left(F + \frac{n+1}{n} \right) \Delta(\text{meV}) \right]^{1/2} \tag{7.39}$$

if n ($n \geq 2$) charge carriers tunnel through the insulator on average for each quasiparticle created by the absorbed photon. Because of the statistical nature of the process, it carries an increase in noise. However, if the resolution of the device without gain is limited by amplifier noise, the operation with gain can be important in achieving the best possible system level resolution.

Detectors built as in Figure 7.14 and with $n \sim 30$ produce the expected spectral resolution, e.g., $\Delta\lambda/\lambda = 0.064$ at 0.98 μm and $\Delta\lambda/\lambda = 0.042$ at 0.50 μm. Quantum efficiencies of 70–80% are achieved in the ultraviolet, blue, and visible, with a drop toward the red because of reflection by the tantalum film (Martin and Verhoeve 2013). However, the uniformity of the pixel responses in arrays is substantially poorer than with semiconductor arrays such as CCDs.

STJs must be read out fast enough to detect single photons. JFET-based readout amplifiers are used to provide both the required low noise and the speed. Since JFETs cannot operate below ~ 60 K, the signals from the detector are brought out of the cryostat where a room temperature amplifier circuit is placed; typical power requirements are $\gtrsim 100$ mW per channel. Arrays of STJs have been built in which a number of detectors are connected in series to a single output amplifier, with the disadvantage that the noise performance is degraded compared with single-pixel readout. Nonetheless, excellent performance has been achieved with 10×12 pixels (Martin et al. 2006).

---

[1] Suppressing all the junctions simultaneously in a large array is difficult since the junctions may require different field levels for adequate suppression.

However, true multiplexing of STJs has not yet been demonstrated; it would require development of a readout circuit that can operate at very low temperatures and with far lower power dissipation than currently. The most commonly discussed multiplexing schemes use radio-frequency single-electron transistors (RF-SETs) as pre-amplifiers. A RF-SET (Schoelkopf et al. 1998) has an architecture somewhat like that of a FET, with a double junction structure defining a source and drain and the current between them regulated by charge deposited through a capacitor on a metallic island between them. Because of cryogenic quantum effects, they can utilize quantization of charge on a small conducting "island" to operate as high-performance electrometers. They are the electrostatic siblings of the better known SQUIDS. They have sub-femtofarad input capacitance, making them suitable as readouts for the low capacitance of STJs. They also have very low power dissipation, highly advantageous for multiplexing detectors that must operate at very low temperature. However, RF-SETs are complex devices and not yet technically mature enough to support large format arrays.

The use of STJs for energy-resolved imaging is therefore feasible for single sources and small surrounding fields, but not yet technically feasible for large imaging fields. As a result, their development and deployment suffers from the competition of the nearly perfect photoconductive detectors available in the ultraviolet, visible, and infrared, as discussed in the preceding chapters. With the use of integral field units (IFUs) to feed conventional spectrometers, energy-resolved imaging is widely practiced already, and in a format that can provide more flexibility in spectral resolution – and higher resolution if desired – than afforded by STJs (Eisenhauer and Raab 2015). Nonetheless, STJs provide a capability that could have interesting applications, for example in taking advantage of their potentially very high time resolution.

7.4.2 Superconducting Nanowire Single-Photon Detectors

Superconducting nanowire single-photon detectors (SNSPDs) can provide very fast single-photon detection over the visible and near infrared. They are most commonly fabricated in niobium nitride (NbN), which has a relatively high superconducting critical temperature ($\sim$10 K, allowing cooling with atmospheric pressure liquid helium) and a very fast detection recovery time (i.e., the time to cool sufficiently after a detection to be sensitive again: < 100 picoseconds). The quantum efficiency of NbN devices drops precipitously beyond $\sim$2 μm (Korneev et al. 2005); extending the response to longer wavelengths requires using other superconducting materials.

Typically, a SNSPD consists of a superconducting nanowire of order 5 nm thick, 100 nm wide and multiple μm long. Because it is virtually impossible to focus photons on the 100 nm width, the wire is arranged as a meander that covers the area of a pixel, 10 to 20 μm on a side. It is maintained well below its superconducting critical temperature and is biased to maintain a DC current just below its superconducting critical current (equation 7.9). When a visible or near-infrared photon is absorbed by the nanowire, it breaks multiple Cooper pairs to create a local "hot spot" that is no longer superconducting. The higher resistance of the hot spot forces the bias current to flow around it, but that causes the current density in the surrounding regions to exceed the critical current, creating a resistive barrier across the entire nanowire. This diverts most of the bias current to an output amplifier, connected across the length of the nanowire, to produce a voltage pulse registering the detection. The resistive barrier also blocks the bias current from continuing to flow through the nanowire, expediting the region around the hotspot recovering to the superconducting state. The time for this to happen is typically set by the inductive time constant of the nanowire, that is by the kinetic inductance of the nanowire divided by the impedance of the readout circuit.

Devices using NbN have been constructed with detection efficiencies of ~67% at 1.06 μm when antireflection coated and placed in optical cavities to enhance the absorption (Rosfjord et al. 2006).[2] Output pulse widths are ~150 ps (FWHM; readout electronics limited), photon detections are well separated for times of arrival differing by about 600 ps, and jitter – the uncertainty in the photon arrival time – is less than 50 picoseconds (Zhang et al. 2003).

7.4.3 Quantum Capacitance Detectors (QCD)

The quantum capacitance detector (QCD) is based on the single Cooper-pair box (SCB), a superconducting quantum circuit.The basic operation of a QCD, illustrated in Figure 7.16, begins with submillimeter radiation absorbed by an antenna being coupled onto a superconducting absorber. Its energy then breaks a number of Cooper pairs and the resulting density of quasiparticles within the absorber is coupled to a SCB with two Josephson junctions arranged as in a SQUID. The SCB includes a tiny superconducting island with a capacitance typically of a few femtofarads (fF = 10^{-15} F). The charge state of this island is set by the macroscopic number of conduction electrons on it. Therefore, the quantum superposition of charge states can be controlled by adjusting its potential. This is achieved with the gate capacitor, C_G, to produce a sharp capacitance change when a quasiparticle tunnels from the absorber onto the island. The island is capacitively coupled to a superconducting resonator circuit; when the capacitance of the island changes,

---

[2] The cavities increased the absorption over that of bare devices by a factor of 2.44 and the antireflection coating by a factor of 1.05, so the bare device efficiency is ~26%.

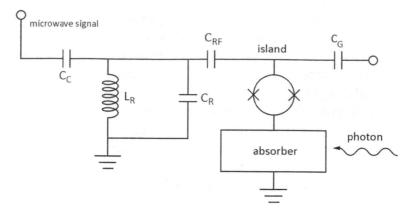

Figure 7.16 Quantum capacitance detector schematic. After Shaw et al. (2009).

it causes a shift in the resonant properties of this circuit. Many of these readout circuits, tuned to different frequencies, can be linked through a transmission line, allowing an array of QCDs to be multiplexed in the frequency domain. Like the MKID, this allows large-scale multiplexing with the frequency or phase shifts of the individual pixels measured by room temperature electronics.

In comparison with the mature and very high performance devices available in the optical and near-infrared wavelength range, there has been more room for improvement and hence a broader variety of different approaches under development for the submillimeter, such as the MKIDs discussed above and the quantum capacitance detector (QCD) described here, as well as superconducting hot electron bolometers to be discussed in the next chapter. Nonetheless, the performance levels and available array architectures with MKIDs and bolometers (next chapter) is now good enough that the situation is approaching that in the "semiconductor region"; that is, the existing capabilities establish a threshold of performance that is high enough that a new, less mature approach can dominate a performance niche but have difficulty getting established for general applications and thereby attracting the resources to develop it to a higher level.

7.5 Problems

7.1 (a) Compute the velocity of Cooper pairs in superconducting aluminum. Assume that every aluminum atom yields all three of its valence electrons that can be bound into a Cooper pair. There is a limiting current of about 1.3×10^7 A cm^{-1}. This current is set by mechanisms that cause the Cooper pairs to break apart at velocities higher than the one you have been asked to compute.

(b) In aluminum at room temperature, there is a limiting current of about 2×10^5 A cm^{-1}, i.e., 100 times smaller. This limit is set by resistive heating of the conductor. Obviously, one expects the electron drift velocity to be two orders of magnitude lower, but confirm this by using equation 2.6 with the mobility of aluminum, ~ 13 cm^2 (V s)$^{-1}$.

(c) Compare these maximum drift velocities at 300 K for single electrons and at 1 K for paired ones with the rms thermal velocities given by

$$v_{rms} = \sqrt{\frac{3kT}{m_e}}.$$

7.2 Prove that the resistance of a square of metal film depends only on the thickness of the film and not on its area.

7.3 Calculate the kinetic inductance of a trace of 50 nm thick NbN ($\lambda_L(0) = 200$ nm, $T_c = 12$ K for this thickness) that is 2 cm long and 3 μm wide. Assume that the trace is sunk to a thermal bath of liquid helium under reduced pressure (through pumping) to a temperature of 1.8 K (it would be 4.2 K at 1 atmosphere pressure) and is at 0.3 K above the bath temperature due to parasitic heat leaks.

7.4 Assume the trace from problem 7.3 is used in a MKID pixel similar to that in Figure 7.7. The pixel is 1 mm on a side. There are five capacitor fingers that are 400 μm long and 20 μm wide, with 20 μm gaps and the substrate is 2 μm thick silicon. What is the capacitance? What is the resonant frequency with the inductance from Problem 7.3 (ignoring the magnetic inductance)? If the resonator has a resistance of 1 mΩ, what is the quality factor? Use equation 7.28 to estimate the full width at half maximum of the resonance. Hint: Use the following to estimate the capacitance:

$$C = (\kappa_0 + 1)l[(N - 3) * A_1 + A_2]$$

with

$$A_1 = 4.409 \times tanh\left[0.55 \times \left(\frac{h}{w}\right)^{0.45} \times 10^{-6}\right],$$

$$A_2 = 9.92 \times tanh\left[0.52 \times \left(\frac{h}{w}\right)^{0.5} \times 10^{-6}\right],$$

where h is the thickness of the substrate and w is the width of the N capacitor fingers, assuming that the space between fingers is equal to their width (from Beeresha et al. 2016).

7.5 Suppose one wants to distribute the outputs of 1000 MKID pixels over a total band of 2 GHz and to avoid crosstalk the resonances have to be separated by ten times their FWHMs (taking the results from Problem 7.4). Assume that there are random variations in the 50 nm thickness of the NbN in the inductor. What tolerance must be maintained if the array is specified to have no pixels that exceed this ten times rule to avoid crosstalk?

7.6 In Chapter 4, we discussed very large-scale detector arrays that use CMOS circuitry to provide an amplifier for each pixel and then to multiplex the signals from these amplifiers to a small number of array outputs. Typical permissible data rates correspond to one signal emerging from an array output every 10 μs. In the text, we discuss JFET readouts for STJs, which have significant disadvantages such as inability to operate near the detector operating temperatures. CMOS array readouts can operate that cold. Discuss why they do not offer the potential for large STJ arrays. Assume the array is used for imaging on an 8-meter aperture telescope with pixels set to 0.1 arcsec on a side and that the full optical band from 0.3 to 1 μm is put onto the detectors with a net efficiency of 50% due to optical losses in the telescope and instrument. Take the sky to emit the equivalent of a 21st magnitude (at V) star per square arcsec, corresponding to 1.5×10^{-31} W (m^2 Hz)$^{-1}$.

7.6 Further Reading

Baselmans (2012) – short and readable description of MKIDs

Berggren et al. (2013) – Good overview of all the superconducting detector types discussed in this chapter

Clarke and Braginski (2006) – overview of SQUID-based electronics

Eisenhauer and Raab (2018) – review of imaging spectroscopy instrumentation, including STJs and MKIDs, including comparisons with integral field unit spectrometers

Martin and Verhoeve (2013) – review of all aspects of superconductng tunnel junctions

Mauskopf (2018) – rather technical overview of MKIDs (and transition edge bolometers)

Pobell and Luth (1996) – modern overview of methods for achieving low temperatures

Ruggiero and Rudman (2013) – overview of superconducting electronics

Tinkham (1996) – classic textbook on superconductivity, brought up to date in a second edition

Zmuidzinas (2012) – overview of MKID principles

8

Bolometers

We now turn to thermal detectors, the second major class of detector listed in Chapter 1. Unlike all detector types described so far, these devices do *not* detect photons by the excitation of charge carriers directly. They instead absorb the photons and convert their energy to heat, which is detected by a very sensitive thermometer. The energy that the photons deposit is important to this process; the wavelength is irrelevant, that is, the detector responds identically to signals at any wavelength so long as the number of photons in the signal is adjusted to keep the absorbed energy the same. Thus, the wavelength dependence of responsivity is flat and as broad as the photon absorbing material will allow. Because the absorber is decoupled from the detection process, it can be optimized fully for the wavelength range of interest. Bolometers based on semiconductor or superconductor temperature sensors are the most highly developed form of thermal detector for low light levels and are the detector of choice for many applications, especially in the submillimeter spectral range and as microcalorimeters in the X-ray. For the highest possible performance, such detectors need to be cooled to below 1 K. Bolometers manufactured by etching miniature structures in silicon and silicon nitride provide high performance with large pixels and also in detector arrays. Similar etching technologies are used to manufacture imaging arrays that operate at room temperature, and high-temperature superconductors are making possible a new class of high-performance bolometer operating at intermediate temperature.

8.1 Basic Operation

To illustrate the operation of a bolometer, consider the simple thermal model in Figure 8.1. A detector is connected by a weak thermal link of thermal conductance G (in units of watts per kelvin) to a heat sink at temperature T_0. Assume there are no additional paths for heat loss. The detector is absorbing a constant power P_0,

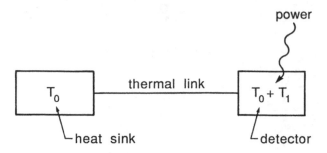

Figure 8.1 Thermal model of a bolometer.

which raises its temperature by an amount T_1 above the heat sink. The definition of G is then

$$G = \frac{P_0}{T_1}. \tag{8.1}$$

Now we introduce an additional, variable power component, $P_v(t)$, which we will assume is deposited by absorbed radiation. The temperature of the detector element will change following the time dependence of $P_v(t)$:

$$\eta P_v(t) = \frac{dQ}{dt} = C\frac{dT_1}{dt}, \tag{8.2}$$

where η is the fraction of power absorbed by the detector (that is, the quantum efficiency), Q is the energy absorbed, and C is the heat capacity, which is defined by $dQ = C dT_1$ and has units of joules per kelvin. The total power, $P_T(t)$, being absorbed by the detector is

$$P_T(t) = P_0 + \eta P_v(t) = GT_1 + C\frac{dT_1}{dt}. \tag{8.3}$$

Suppose $P_v(t)$ is a step function at $t = 0$ such that $P_v = 0$ for $t < 0$ and $P_v = P_1$ for $t \geqslant 0$. The solution to equation 8.3 is then

$$T_1(t) = \left[\begin{array}{ll} \dfrac{P_0}{G}, & t < 0, \\[2ex] \dfrac{P_0}{G} + \dfrac{P_1}{G}\left(1 - e^{-\frac{t}{(C/G)}}\right), & t \geq 0. \end{array}\right. \tag{8.4}$$

By inspection, the thermal time constant of the detector is therefore

$$\tau_T = \frac{C}{G}. \tag{8.5}$$

For times long compared to τ_T, T_1 is proportional to $(P_0 + P_1)$. If T_1 can be measured, this device can measure the input power.

We will develop bolometer theory in more detail later in this chapter when we discuss high-performance cryogenically cooled devices. However, we can begin by discussing room temperature bolometers without any more background.

8.2 Room Temperature Thermal Detectors

Room temperature bolometer arrays are widely used as imagers in the 10 μm spectral region, where the atmosphere is transparent. The thermal radiation of ambient temperature objects around this wavelength is high, making the modest per-pixel sensitivity of these devices well matched to the achievable performance in many applications. Figure 8.2 is a conceptual diagram of two pixels in such a device, which we assume is electrically and physically connected to a readout integrated circuit (ROIC) as part of a detector array. The detectors for these arrays are electrically simple, but mechanically complex as indicated in the figure. They are fabricated of silicon (with a thin layer, typically VO_x or amorphous silicon, as the absorber/resistor) in a complex series of steps that involve bonding the bolometer material to the ROIC wafer, patterning the resulting sandwich with a photoresist and etching to delineate the bolometer pixels, then etching holes and metallizing vias for electrical connection to the input pads of the ROIC, and finally etching away the material under the bolometer sensitive area to suspend it by the vias. The reflective layer on the ROIC is placed one quarter wave underneath the resulting suspended pixel to create a resonant optical cavity which optimizes the absorption of any light that penetrates the pixel into this space. The performance of these devices can be enhanced significantly by sealed packaging that allows them to operate in a vacuum, eliminating the effects of convecting air in introducing thermal noise and acting as a heat loss mechanism. Large-format arrays are available, such as the 2048 × 1536 pixel device with 10 μm pixel pitch from FLIR Systems, the 1920 × 1200 pixel device with 12 μm pixel pitch offered by

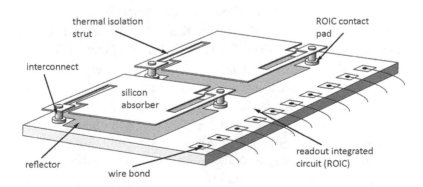

Figure 8.2 Two pixels in a microbolometer array.

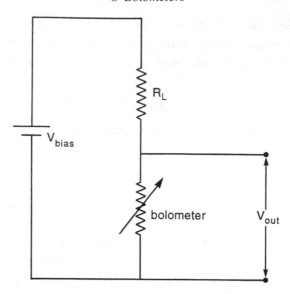

Figure 8.3 Bolometer biasing and readout circuit.

Sierra Olympic, or the 1024×768, 17 µm pixel pitch devices from Leonardo DRS, SemiConductor Devices (SCD), and a number of other suppliers (see, e.g., Mizrahi et al. 2013).

The bolometer unit cell absorber is material with a temperature-dependent resistance and the current through it can be provided with a simple load resistor circuit such as in Section 4.2, reproduced in Figure 8.3. The most common bolometer resistor element is vanadium oxide, indicated as VO_x. There are many phases of vanadium oxide, e.g., VO_2, V_2O_3, V_2O_5; the material in microbolometers is a mixture that is effectively of order $VO_{1.8}$ (Lamsal 2014). The resulting material is a semiconductor with a band gap that produces a strong temperature coefficient of resistance at room temperature. The VO_x is deposited in a sandwich with silicon nitride to enhance and level out the absorption across the 8 to 14 µm range, with achievable values $\gtrsim 70\%$ (Lamsal 2014). The temperature coefficient of resistance is about $2\%/°C$. The ROIC is a CMOS circuit operated similarly to those for hybrid infrared arrays; however, rather than supplying an amplifier for each pixel it is common to provide a high-gain amplifier for each column, with rows selected by sequential biasing. So long as the frequency of applying the bias and interrogating a pixel is fast compared with the thermal time constant of the bolometers, little information is lost due to this intermittent readout. The performance of these arrays is typically quoted in noise equivalent temperature difference (NETD), usually defined as the amount of signal temperature change that is needed to match the internal noise of the detector, i.e., that yields a ratio of signal to noise of one. Typical values are a few tens of mK. In practice this specification can be misleading, since it depends on parameters such as the temperature of the source.

These arrays also respond to energy in the THz region, but with quantum efficiencies of only $\sim 0.2\%$ (Simoens 2013). However, the basic design can be optimized for this spectral range by adding an absorber to the pixels in the form of a metal film with the appropriate ohms/square (Section 7.3.3) and increasing the spacing between the back of the pixel and the reflective layer on the ROIC; quantum efficiency improvement by about $\times 50$ can be achieved (Simoens 2013).

There are two other specialized kinds of temperature sensor for room temperature bolometers: (1) thermopiles and (2) pyroelectric devices. Generally, these approaches are not used for large-format arrays, but rather for single detectors or very small arrays of individual devices. We begin their discussion with thermopiles. When two dissimilar materials are contacted to make a thermocouple, their Fermi levels will adjust to match as we have found is generally the case. However, the structures of the energy levels in the two materials above the Fermi levels will generally be different (a fact we have usually ignored in our previous discussions). This difference can be summarized to first order by saying that the materials have different work functions, defined as the minimum energies above the different Fermi levels required for electrons to escape. Under thermal excitation, the mismatch in work functions results in a tendency for electrons to flow preferentially from one of the materials into the other, setting up a temperature-dependent electromotive force across the joint. If a similar junction is held at a reference temperature, a circuit connecting the two junctions will develop a voltage:

$$\Delta V = \alpha \Delta T. \tag{8.6}$$

Here, α is the Seebeck coefficient. For metal pairs, it is of order 50 $\mu V \, {}^{\circ}C^{-1}$; suitable semiconductors typically have Seebeck coefficients an order of magnitude greater than that for metal pairs. ΔT is the temperature difference between the two joints, and ΔV is the voltage. The response can be increased by connecting a number of thermocouples in series to make a thermopile detector. For more information, consult Stevens (1970), Budde (1983), Pollock (1985), or Teranishi (1997).

Pyroelectric detectors are based on certain specialized materials (for example, triglycerine sulfate (TGS), strontium barium niobate, lithium tantalate, lead scandium tantalate). To create a thermometer, a sample of one of these materials is placed between two electrical contacts in a specific orientation. The molecules of these materials have a charge asymmetry that can be aligned by applying an electric field to the contacts at elevated temperatures. The alignment is retained as a net electrical polarization after the sample has been cooled below its Curie temperature, even after the field has been removed. The detector material is "poled" by putting it through this series of steps. Afterwards, when the crystal changes temperature, a change in the electrical polarization occurs which changes the charge on the contacts. As a result, a current will flow in an external circuit:

$$I = p(T) A \frac{dT}{dt},$$
(8.7)

where A is the thermometer area and $p(T)$ is the pyroelectric coefficient (which can be of the order of 1×10^{-8} C cm^{-2} K^{-1}). Since pyroelectric thermometers are sensitive only to changes in temperature, detectors using them are generally operated with choppers to interrupt the signal reaching them, although they can also be read out in a pulsed mode where their charge is sensed by placing them in an alternating electric field (Hanson 1997). Further information on pyroelectric detectors can be found in Putley (1970) and Hanson (1997).

A Golay cell is another thermal detector consisting of a hermetically sealed container filled with gas and arranged so that expansion of the gas under heating by a photon signal deflects a mirror or bends a membrane. Modern versions provide very broadband detection from 0.1 to 20 THz (i.e., 300–15 μm). They are rugged enclosed units in which an infrared absorber is placed within the cell containing the gas; one side of this cell is a membrane that deforms under pressure. The shape of the membrane is sensed by reflecting a beam of light off it to a photo-diode. An example of the application of these detectors can be found in Kaufmann et al. (2014). An effort to miniaturize and modernize these devices is described by Kenny (1997).

8.3 Cryogenic Semiconductor Bolometers

Bolometers operated at very low temperatures provide state-of-the-art performance in the far infrared to millimeter-wave. Originally, their construction was almost exclusively based on semiconductor temperature sensors, so they had some familial resemblance to the room temperature devices just discussed. The development of low-temperature bolometry was revolutionized by Low's invention of a sensitive germanium-based device (Low 1961), semiconductor bolometers played a central role in the Herschel mission in 2009–2013, and development still continues on such devices (e.g., Adami et al. 2018; Bounissou et al. 2018). In addition, they illustrate many principles of operation that apply to other high-performance bolometer types.

However, there is now strong competition from superconductor temperature sensors, as we will discuss in Section 8.4.

8.3.1 Electrical Properties

The performance of a semiconductor bolometer is based on the temperature dependence of its electrical properties. This dependence is described by the temperature coefficient of resistance,

$$\alpha(T) = \frac{1}{R}\frac{dR}{dT}. \tag{8.8}$$

To obtain appropriate electrical properties at very low temperatures (< 5 K), the semiconductor material must be doped sufficiently heavily that the dominant conductivity mode is hopping, which occurs with the assistance of excited states in the quantum mechanical modes of vibration of the crystal, called phonons and which are often treated as quasiparticles. At the temperatures of interest, high energy phonons are scarce and hopping has to occur to an unoccupied site close to the same energy, thus producing a large and temperature-dependent resistance. This is usually the regime of interest for low-temperature thermometry, and the process is called phonon-assisted variable range hopping (vrh). The resistance is given by an expression of the form:

$$R = R_0 \, e^{[T_0/T]^{1/2}} \tag{8.9}$$

(from McCammon 2005). Equation 8.9 is only applicable for $T \ll T_0$, where T_0 is a temperature of the order of 4–20 K. This relationship gives

$$\alpha(T) = -\frac{1}{2}\left(\frac{\Delta}{T^3}\right)^{-1/2}. \tag{8.10}$$

Temperature sensors are made of germanium or silicon. Very uniform devices can be made of neutron transmutation doped (NTD) germanium; neutron capture (in a nuclear reactor) converts ^{70}Ge to ^{71}Ga (an acceptor impurity) or ^{74}Ge to ^{75}As (a donor) (Palaio et al. 1983). The naturally occurring isotope ratio yields Ge:Ga compensated by 32% with Ge:As. The high uniformity of the process results in excellent reproducibility of the temperature sensors; it has the accompanying disadvantage that the neutron penetration is so large that the technique cannot be used in complex structures – rather, the sensors must be diced from a larger sample and mounted on the bolometer. The alternative is to dope by ion implantation in silicon, which allows small thermal sensors to be fabricated as part of larger structures and hence is useful for bolometer arrays. The disadvantage is that the silicon thermometers can be noisier at low frequencies.

An alternative expression to equation 8.9, applicable at intermediate values of T, is determined from an empirical fit to the bolometer resistance and is of the form

$$R = R_0 \left(\frac{T}{T_0} \right)^{-A}, \tag{8.11}$$

where typically $A \approx 4$. We then have

$$\alpha(T) = -\frac{A}{T}. \tag{8.12}$$

In both of these cases, $\alpha < 0$ and has strong temperature dependence.

Again, the circuit in Figure 8.3 is used to determine the bolometer resistance. We have already discussed the shortcomings of this circuit that arise from the need to operate photodetectors at very large resistance to minimize the Johnson noise. Since there are compelling reasons to operate bolometers at extremely low temperatures, Johnson noise for bolometers is sufficiently small even with modest detector resistance, and the previous arguments no longer apply.

The analysis of the circuit in Figure 8.3 is simplified by setting the load resistance, R_L, to be much larger than the detector resistance so the current through the detector is, to first order, independent of the detector resistance. If R_L is too small and the bolometer controls the current through the circuit, the circuit is subject to a runaway instability in which additional current drives down the bolometer resistance (because of the large negative coefficient of resistance), which allows even more current to flow. This arrangement frequently results in the demise of the bolometer through overheating. For protection, a large value of R_L is almost always used, and an analysis under that assumption does not suffer any serious loss of generality in terms of practical applications. The general treatment can be found in Mather (1982).

8.3.2 Time Response

In this section and the following two, we will derive the time response, responsivity, and noise characteristics of a bolometer in the circuit in Figure 8.3, making the simplifying assumption just discussed. Let $P_I = I^2 R(T)$ be the electrical power dissipated in the detector resistance $R(T)$ by the current I. Because R is a function of temperature, we cannot just add a fixed power of $I^2 R$ to P_0, but have to correct for the change in P_I caused by the change in temperature resulting from the I.

We therefore modify equation 8.1 to $P_I + (dP_I/dT)\, T_1 = GT_1$, and equation 8.3 becomes

$$P_T(t) = \left(G - \frac{dP_I}{dT}\right) T_1 + C \frac{dT_1}{dt}. \tag{8.13}$$

However,

$$\frac{dP_I}{dT} = I^2 \frac{dR}{dT} = \alpha I^2 R = \alpha P_I. \tag{8.14}$$

Substituting this expression into equation 8.13 and rearranging, we obtain

$$P_T(t) = (G - \alpha P_I)T_1 + C \frac{dT_1}{dt}. \tag{8.15}$$

By comparing this result with equation 8.3 and its solution, we find that the time response is again exponential, but with an electrical time constant of

$$\tau_e = \frac{C}{G - \alpha P_I}. \tag{8.16}$$

Since $\alpha < 0$, the electrical time constant for a semiconductor bolometer is shorter than the thermal time constant τ_T defined in equation 8.5. This result is due to electrothermal feedback, which is discussed by Jones (1953) and Mather (1982). We will discuss it in more detail when we derive the noise performance of these detectors.

Because the time response is exponential, the response of the bolometer can be described as in equations 1.37 through 1.42. In particular, the frequency response is obtained by substituting τ_e for τ_{RC} in 1.40, that is,

$$S(f) = \frac{S(0)}{[1 + (2\pi f \tau_e)^2]^{1/2}}, \tag{8.17}$$

where $S(0)$ is the low-frequency responsivity and both it and $S(f)$ have the units of volts (output signal) per watt (input photon power).

8.3.3 Responsivity

To derive the responsivity, let dR, dT, and dV be the changes in the resistance, temperature, and voltage, respectively, across the bolometer that are caused by a change in absorbed power, dP. Then, continuing to assume that R_L is large, we find that

$$dV = I\,dR = I(\alpha R\,dT) = \alpha V\,dT. \tag{8.18}$$

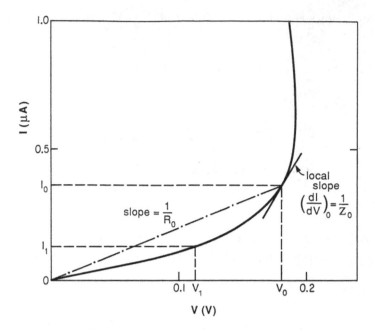

Figure 8.4 Load curve.

At low frequency ($t \gg \tau_e$), we obtain $dT = dP/(G - \alpha P_I)$ from equations 8.3, 8.4, and 8.15. Substituting this expression into equation 8.18, we get $dV = \alpha V dP/(G - \alpha P_I)$, yielding a responsivity in volts per watt:

$$S_E = \frac{dV}{dP} = \frac{\alpha V}{G - \alpha P_I}. \tag{8.19}$$

This responsivity can be determined entirely from the electrical properties of the detector and is therefore termed the electrical responsivity.

The electrical properties of a bolometer, however, are not always available in terms of α and G, and dismantling the detector to determine these parameters usually voids the warranty. What can be determined for an operating detector is an I–V curve (or load curve), which has the form shown in Figure 8.4 for a bolometer with $\alpha < 0$. The load curve is determined by using a high impedance voltmeter (possibly in series with a FET input DC-coupled amplifier) to measure the voltage across the detector as a function of the current through it. The parameters governing the bolometer performance can be determined from a series of measurements centered on the load curve.

The circuit in Figure 8.3 is used to establish an operating point for the bolometer on the I–V curve by adjusting R_L and the bias voltage to establish the desired current through the detector. A generic example is indicated by the voltage V_0 in

Figure 8.4. We are interested in determining the responsivity (and other parameters) of the detector at this operating point. The bolometer resistance is $R_0 = V_0/I_0$. We designate the slope of the load curve at any operating point as $1/\mathbf{Z}$, or

$$\mathbf{Z} = \frac{dV}{dI}, \tag{8.20}$$

which is different from the resistance because of the nonlinearity of the load curve. We can rewrite $\mathbf{Z}$ as follows:

$$\mathbf{Z} = R\frac{d(logV)}{d(logI)} = R\frac{\left[\frac{d(logP)}{d(logR)} + 1\right]}{\left[\frac{d(logP)}{d(logR)} - 1\right]}. \tag{8.21}$$

This expression can be simplified by defining the parameter $H = G/\alpha P$ and noting from equations 8.1 and 8.8,

$$\frac{d(logP)}{d(logR)} = \frac{1}{I^2}\frac{dP}{dR} = \frac{1}{I^2}\frac{dP}{dT}\frac{dT}{dR} = \frac{G}{\alpha I^2 R} = \frac{G}{\alpha P} = H. \tag{8.22}$$

Substituting this result into equation 8.21 and solving for H, we have

$$H = \frac{\mathbf{Z} + R}{\mathbf{Z} - R}, \tag{8.23}$$

which can be determined readily from measurements of the load curve.

If we substitute $G = \alpha H P$ (from equation 8.22) into equation 8.19 and then use the form of H obtained in equation 8.23, the electrical responsivity takes the form

$$S_E = \frac{V}{P(H-1)} = \frac{1}{2I}\left(\frac{\mathbf{Z}}{R} - 1\right). \tag{8.24}$$

The power, P, in equation 8.24 is the electrical energy dissipated in the detector at the operating point, and the other quantities are also measured at this point.

If the temperature coefficient of resistance for the bolometer material is known, the combination of equations 8.22 and 8.23 permits the determination of the thermal conductance, G, from the load curve. Rewriting equation 8.16 in terms of the load curve parameters, we obtain

$$\tau_e = \frac{\tau_T}{1 - \frac{1}{H}} = \tau_T\left[\frac{\mathbf{Z} + R}{2R}\right]. \tag{8.25}$$

Mather's (1982) general derivation allowing any value of R_L shows that there is a correction factor of $(R_L + R)/(R_L + \mathbf{Z})$ to the time constant in equation 8.25, that is,

$$\tau_e = \tau_T\left[\frac{\mathbf{Z} + R}{2R}\right]\left(\frac{R_L + R}{R_L + \mathbf{Z}}\right), \tag{8.26}$$

which can be solved for τ_T ($= C/G$) to determine the heat capacity of the detector in terms of τ_e. (The value of τ_e is obtained by determining the bolometer response to a varying signal; refer to equation 8.17.)

To convert S_E to a responsivity to incident radiation, or "radiant responsivity," we must allow for the possibility that only a fraction, η, of the incident energy is absorbed by the bolometer, where η is the quantum efficiency. We then have

$$S_R = \frac{\eta}{2I}\left(\frac{\mathbf{Z}}{R} - 1\right), \tag{8.27}$$

again in volts per watt. Unlike photoconductors and photodiodes, the bolometer response is independent of the wavelength of operation (as long as the quantum efficiency is independent of wavelength).

8.3.4 Noise and Noise Equivalent Power (NEP)

The measures of performance for photodetectors, such as read noise, do not all translate to bolometers. The fundamental measure for them is the noise equivalent power (*NEP*), defined as the power received by the detector that would yield a signal to noise (rms) ratio of one when operated with a 1 Hz electronic bandwidth. A derivation of the bolometer noise must account for the electrothermal feedback mechanism already encountered in discussing the time response. The bolometer is, of course, subject to Johnson noise characterized by a noise voltage $V_J = \langle I_J^2 \rangle^{1/2} R$, where $\langle I_J^2 \rangle = 4kT\, df/R$ from equation 2.25. The Johnson noise dissipates power in the detector. When V_J is added to the bias voltage, the power dissipated in the detector increases, and, because of the negative temperature coefficient of resistance, the detector resistance decreases, thus reducing the net change in voltage across the detector. Similarly, when V_J opposes the bias voltage, the power dissipated in the detector is decreased, and the resistance increases, decreasing the net voltage change. Since the detector response opposes the ohmic voltage changes resulting from Johnson noise, the observed noise should be less than predicted by equation 2.25 with a fixed value of resistance.

For a bolometer with no signal and no noise, we can take the output voltage to be

$$V_0 = IR + I^2 R\, S_E, \tag{8.28}$$

where the second term represents the response of the detector to the power dissipated by the current established by the bias voltage. Adding Johnson noise and substituting for the responsivity from equation 8.24, it can be shown that

$$\langle V_N{}^2 \rangle^{1/2} \approx \langle I_J{}^2 \rangle^{1/2} R + \frac{\langle I_J{}^2 \rangle^{1/2} R}{2}\left(\frac{\mathbf{Z}}{R} - 1\right) \approx \left(\frac{R + \mathbf{Z}}{2}\right)\left(\frac{4kT\, df}{R}\right)^{1/2}. \tag{8.29}$$

For example, if $\mathbf{Z} = 0$, electrothermal feedback reduces the noise by a factor of two relative to Johnson noise for a fixed resistance, whereas a detector operated with a small bias voltage (for example, V_1 in Figure 8.4 will have $\mathbf{Z} \approx R$ and nearly the full Johnson noise for a fixed resistance.)

The *NEP* for a bolometer under Johnson-noise-limited conditions is

$$NEP_J = \langle V_N^2 \rangle / (S_R (df)^{1/2}). \tag{8.30}$$

Using equation 8.29, it can be written as

$$NEP_J = (4kTR)^{1/2} \frac{I}{\eta} \left| \frac{\mathbf{Z}+R}{\mathbf{Z}-R} \right|$$

$$= (4kTP)^{1/2} \frac{|H|}{\eta}$$

$$= \left(\frac{4kT}{P} \right)^{1/2} \frac{G}{\eta |\alpha|}, \tag{8.31}$$

where we have used $H = G/\alpha P$. We have disregarded the frequency dependence of *NEP*; to include this effect, equation 8.31 should be multiplied by the denominator in equation 8.17. The temperature dependence of NEP_J can be obtained by substituting for α from equation 8.10 or 8.12:

$$NEP_J \sim \begin{cases} GT^2 \text{ for } \alpha \sim T^{-3/2}, \\ GT^{3/2} \text{ for } \alpha \sim T^{-1}. \end{cases} \tag{8.32}$$

This expression makes clear the advantages in operating the bolometer at very low temperatures. We will soon show that the temperature dependence of G can increase these advantages. In addition, equation 8.31 implies that if possible the detector should be biased such that $\mathbf{Z}$ is negative. Such values can be obtained, but it is usually found that excess noise is produced by the relatively large current that must flow through the detector to maintain the operating point. In addition to Johnson noise, a second type of fundamental noise for a bolometer is thermal noise due to fluctuations of entropy across the thermal link that connects the detector to the heat sink. More detailed derivations can be found in van der Ziel (1976), and Mather (1982); here we will determine the magnitude of this noise through an analogy with the treatment of Johnson noise in Section 2.2.5.

We construct a thermal circuit diagram of the bolometer as shown in Figure 8.5, which is analogous to the electrical circuit used for the discussion of Johnson noise (see Figure 2.10). Here the thermal link is replaced by an equivalent resistance, R_T, and the ability of the bolometer to store thermal energy is represented by a capacitance, C_T. As in the case of Johnson noise, the energy that is stored fluctuates under thermodynamic equilibrium. The resulting fluctuating temperature of the

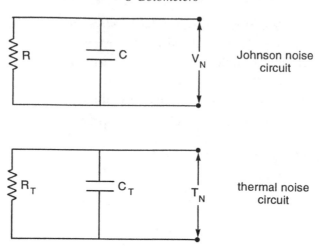

Figure 8.5 Analogy between Johnson noise and thermal noise.

bolometer is represented by T_N (analogous to V_N). By analogy with the electrical case (equation 2.24), the energy E_T stored in C_T should be

$$E_T = \frac{1}{2}C_T T_N{}^2,\tag{8.33}$$

and the dissipation of energy in R_T occurs at a rate

$$P_T = \frac{T_N{}^2}{R_T}.\tag{8.34}$$

From these relationships, we see that C_T has units of joules per kelvin squared and R_T has units of kelvin squared per watt. We also expect the circuit to have exponential time response with a time constant of $R_T C_T$. From equation 8.5, this relationship gives us

$$R_T C_T = \frac{C}{G},\tag{8.35}$$

where C is the heat capacity of the bolometer and G is the thermal conductance to the bath. Applying the dimensional constraints stated just after equation 8.34 (and being careful not to confuse thermal capacitance C_T with heat capacity C), we find that

$$C_T = \frac{C}{T}\tag{8.36}$$

and

$$R_T = \frac{T}{G}.\tag{8.37}$$

From equation 2.25, we know that $\langle V_J^2 \rangle = (4kTR)df$, or in the present case by substituting $\langle T_N^2 \rangle$ for $\langle V_J^2 \rangle$ and R_T for R and using equation 8.37,

$$\langle T_N^2 \rangle = \frac{4kT^2\,df}{G}. \tag{8.38}$$

Now let a signal power, P_v, fall on the bolometer; it will produce a temperature change of

$$\Delta T_S = \frac{\eta P_v}{G}. \tag{8.39}$$

We compute the value of P_v that will give unity signal to noise against thermal fluctuations by setting $\Delta T_S = \langle T_N^2 \rangle^{1/2}$, yielding

$$P_v = \frac{(4kT^2Gdf)^{1/2}}{\eta}, \tag{8.40}$$

and

$$NEP_T = \frac{(4kT^2G)^{1/2}}{\eta}. \tag{8.41}$$

Again, there will be a frequency dependence as discussed after the derivation of NEP_J in equation 8.31.

The third fundamental performance limit for a bolometer is photon noise. Bolometers are not subject to G–R noise, so the noise current can be calculated analogously to equations 2.19 through 2.23 with the appropriate reduction by $\sqrt{2}$:

$$\langle I_N^2 \rangle = \frac{2\,(hc)^2\,\varphi\eta\,df\,S_E^2}{(\lambda R)^2}. \tag{8.42}$$

In addition,

$$S(AW^{-1}) = \frac{\eta S_E}{R}, \tag{8.43}$$

so the *NEP* is

$$NEP_{ph} = \frac{hc}{\lambda}\left(\frac{2\varphi}{\eta}\right)^{1/2}, \tag{8.44}$$

where φ is the photons s^{-1} falling on the detector. The *NEP* will be a function of the frequency of operation of the bolometer. The frequency dependence of the detector responsivity is given in equation 8.17, and the parameters that control it are discussed in detail in Section 8.5. The corresponding modifications in the expressions for *NEP* – equations 8.31. 8.41, and 8.44 – are straightforward.

As is usual, the total *NEP* of a bolometer is given by the quadratic combination of the *NEP*s from various noise sources, for example:

$$NEP = (NEP_J{}^2 + NEP_T{}^2 + NEP_{ph}{}^2 + \cdots)^{1/2}. \tag{8.45}$$

8.3.5 Readout

Semiconductor temperature sensors operate at very high impedance and hence match well to JFET-based readout amplifiers. Since JFETs must operate at relatively high temperatures (≥ 60 K), the thermal mismatch to the sub-kelvin detectors is an engineering challenge that limits the use of this technology in truly large-format detector arrays. Nonetheless, it has been used in high-performance moderate-format arrays of order 100 pixels. An alternative approach to solving this problem and expanding to larger arrays is described next.

8.4 Superconducting Bolometers

8.4.1 Temperature Sensing

Superconductivity was introduced in Chapter 7; Section 7.1.2 in particular describes the change of resistance of a superconducting film near the critical temperature, T_c. Recalling equation 7.8,

$$\frac{R}{R_n} = \frac{1}{2}\left\{ 1 + \left[1 - \left(\frac{T_c - T}{\delta T_c}\right)^2 \right]^{1/2} \right\}, \quad T_c - \delta T_c \leq T \leq T_c, \tag{8.46}$$

where R_n is the normal resistance and $\delta T_c = I(dI_c/dT)^{-1}$ and I_c is the maximum current at which the entire wire can be superconducting. As shown in Figure 7.3, the transition from normal resistance to full superconductivity ("zero" resistance) occurs over about 0.1 K of temperature change, making possible extremely sensitive temperature sensing. Devices using this behavior are termed "transition edge sensors" (TESs). For a typical germanium thermometer, the temperature coefficient of resistance, α in equation 8.10 or 8.12 is -20 K^{-1}. For a low-temperature superconducting thermometer the absolute magnitude of α can be 10–50 times greater, that is, up to ~ 1000.

The usefulness of TESs is enhanced by the ability to tune T_c to bring it to a value that is convenient to achieve with a cooling system. When a superconducting metal is placed in contact with a non-superconducting one, the electron states in the two materials cannot change completely abruptly at the interface. Consequently the Cooper-paired state is carried over to the normal metal, until the pairing is destroyed by scattering. The electron order in the normal metal is also carried over to the superconductor, resulting in a reduction of the superconducting gap near the interface. Thus, T_c is reduced and there are signs of weak superconductivity over tiny distances in the normal metal.

Near T_c, the resistance of a superconductor increases with increasing temperature, just the opposite of semiconductor behavior. Consequently, α is greater than zero. If the bolometer is biased as we have discussed for semiconductor bolometers, at constant current, then equation 8.19 shows that the responsivity can become formally infinite when $G - \alpha I^2 R = 0$, which can readily be achieved for $\alpha > 0$. This condition – huge output power for infinitesmal input power – is unacceptable, since it corresponds to a thermal runaway of the device. Instead, a TES is biased at constant voltage and the current through it is sensed as a measure of its temperature. Figure 7.4, here repeated in a MUX in Figure 8.6, shows a circuit that provides constant voltage bias. The bolometer is operated at a resistance of a few ohms, so the bias voltage is controlled by the 16 mΩ shunt resistor.

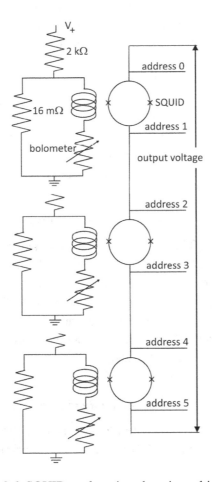

Figure 8.6 SQUID readout time domain multiplexing.

The electrical power dissipated in the device is then

$$P_{elec} = V_{bias}^2 / R. \tag{8.47}$$

Recall the discussion of negative electrothermal feedback in a semiconductor bolometer (Section 8.3.4). In analogy, if the input power to a constant voltage biased superconducting bolometer is increased, the value of R increases and the electrical power decreases from equation 8.47. That is, the constant voltage biased superconducting bolometer has the same property that the constant current biased semiconducting one had: the change in electrical power dissipation partially compensates for the changes in input power. That is, the device behaves as if it had a negative value for α in our previous discussion. As a result, the behavior in this operating mode is parallel to the derivations for a semiconductor thermometer. For example, the response time of the bolometer is reduced compared with the thermal time constant. Given the large values possible for α at the superconducting transition, the boost in speed can be much greater than that for semiconducting detectors, so superconducting bolometers can be substantially faster (by an order of magnitude) than semiconducting ones with similar thermal conductance values and heat capacities.

The feedback provided by the large and effectively negative value of α is also valuable in extending the dynamic range of the device. When it absorbs an increased power level and its temperature increases, the increase in resistance decreases the bias current and hence the bias power, tending to counteract the net temperature rise. However, should the absorbed power be large enough to drive the temperature above T_c, the device becomes inoperative.

8.4.2 SQUID Readouts

A suitable readout for a TES is a superconducting quantum interference device (SQUID), as discussed in Section 7.2. To use a SQUID readout, the bolometer current is passed through a coil that excites currents in the loop as shown in Figure 7.4; we illustrate the same process in Figure 8.6. If the SQUID is biased close to $2I_0$, small changes in the current through the coil are readily detected on the SQUID output. Thus, the coil/SQUID unit acts as a low input impedance amplifier for the current conducted through the bolometer.

8.4.3 Array Readouts

There are various methods to multiplex signals from TESs in detector arrays, reviewed by Ullom and Bennett (2015). Here we describe two of the most mature approaches.

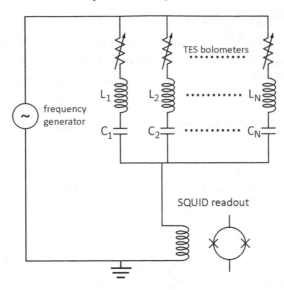

Figure 8.7 SQUID readout frequency domain multiplexing.

8.4.3.1 Time Domain Multiplexing

TES bolometer arrays can use SQUIDs for readout functions analogous to those we have discussed for photodiode and IBC detector arrays. The operation of a simple SQUID time domain multiplexer is illustrated in Figure 8.6 (Chervenak et al. 1999; Benford et al. 2000). The biases across the SQUIDs are controlled by the address lines. Each SQUID can be switched from an operational state to a superconducting one if it is biased to carry about 100 μA. The address lines are set so all the SQUIDs in series are superconducting except one, and then only that one contributes to the output voltage. By a suitable series of bias settings, each SQUID amplifier can be read out in turn.

8.4.3.2 Frequency Domain Multiplexing

An arrangement for frequency domain multiplexing is shown in Figure 8.7. The parallel circuit in Figure 8.6 has been replaced with a series one, the shunt resistor with a capacitor and an inductor, and the voltage source with a modulated one so the TESs are AC-biased, each with a unique frequency (typically between 300 kHz and 1 MHz). A tuned LC filter is in series with each bolometer, to (1) simplify the biasing, allowing a single line carrying a "comb" of frequencies with one for each bolometer; and (2) limit the bandwidth for current noise just around the bias frequency for that bolometer. All the pixels are connected to a common SQUID readout but since each has a distinct value for its bias frequency, each has a different resonant frequency and its signal will emerge at that frequency. The signals are demodulated with room-temperature electronics to retrieve the signals for each TES. In principle, the use of a single SQUID reduces noise aliasing and helps retain the signal to noise that would apply without multiplexing. This style of multiplexing requires fewer SQUIDs and lines to them and can simplify system design.

8.5 Bolometer Construction Components

The preceding theoretical analysis and discussion give an accurate description of the bolometer performance achieved in practice – that is, bolometers whose construction and operating parameters are known accurately perform near to theoretical expectations. Bolometer development therefore concentrates on construction techniques, use of improved materials, measurement of material parameters, and utilization of low temperatures. These issues are illustrated in the following discussion, which will concentrate on bolometers designed for high-sensitivity operation at very low temperatures. The review by Richards (1994) is a more advanced overview of the same issues.

The *NEP* at a given temperature is improved by reducing the thermal conductance, as shown by equations 8.31 and 8.41. The time response of the bolometer is given by equations 8.5 and 8.16 and is improved by reducing the heat capacity and increasing the thermal conductance of the detector. Therefore, both the heat capacity and the thermal conductance are critical in bolometer optimization. Heat capacity can be minimized by the choice of construction materials as well as by minimizing the volume of materials with high specific heat. Construction techniques have been developed that allow control of the thermal conductance over a very wide range. We will find that the temperature dependencies of the heat capacities and thermal conductances can significantly increase the benefits of operating the bolometer at very low temperatures.

8.5.1 Heat Capacity

Three distinct types of material can contribute to the heat capacity of a low-temperature bolometer: (1) semiconductor (and insulator) components; (2) metals; and (3) superconductors.

For pure semiconductor material at low temperature, an approximation to the specific heat (heat capacity per unit volume) is given by the Debye theory for heat capacity of a crystal lattice:

$$c_v{}^{lat} = \frac{12\pi^4 N_A k}{5}\left(\frac{T}{\Theta_D}\right)^3, \tag{8.48}$$

where N_A is Avogadro's number and Θ_D is the Debye temperature. (For a discussion of this relationship and others to come, see, for example, Goldsmid (1965)). As examples, for germanium, $\Theta_D = 374$ K and $c_v \sim 7 \times 10^{-6}T^3$ J K^{-4} cm^{-3}, while for silicon, $\Theta_D = 645$ K and c_v is a factor of 5 times less when compared with germanium.

Free electrons in a metal provide another contribution to the specific heat:

$$c_v^e = \gamma_e T = \gamma_e T \ (\rho/\text{mole}), \tag{8.49}$$

where the parameter γ_e depends on the metal but is of order 10^{-3} J mole^{-1} K^{-2}, and the second equality converts the units of γ_e from mole^{-1} to volume^{-1} by multiplying by the moles per unit volume, that is, ρ mole^{-1}. For brass ($\rho = 8.5$ g cm^{-3}, 1 mole $= 64$ g), $\gamma_e \sim 1.3 \times 10^{-4}$ J cm^{-3} K^{-2}.

In general for both metals and semiconductors, c_v includes contributions both from the lattice and from free electrons:

$$c_v = DT^3 + \gamma_e T \rightarrow \gamma_e T \ \text{as} \ T \rightarrow 0, \tag{8.50}$$

where D is a constant from, for example, equation 8.48. At temperatures < 1 K, the electronic contribution to the specific heat even of the semiconductor components will become important, as indicated by condition 8.50. For example, silicon thermometers are preferred over germanium for bolometers operating above about 1 K because of their reduced heat capacity. For low-temperature (that is, 0.3 K or 0.1 K) detectors, the electronic contribution dominates and the performance of the two semiconductors is similar.

The heat capacity of low-temperature superconductors is dominated by the electronic contribution above the critical temperature. Just at T_c, the specific heat increases by a factor of 2.43 from the value in equation 8.49. Below T_c, the specific heat drops in proportion to T^3, analogous to the Debye relation. For a transition edge bolometer, however, the temperature is held close to T_c, so the specific heat of the thermometer can be taken to be about 2.4 times the metallic value.

Although much of the structure of the bolometer chip is of material that follows the Debye relation, there is usually metal, heavily doped semiconductor, and/or superconductor in the leads, the contacts, the absorber, and/or the thermometer. Thus,

$$C = c_v^{lat} V_{lat} + c_v^e V_e + c_v^{sup} V_{sup}. \tag{8.51}$$

For $T > 1$ K, V_{lat} is approximately the volume of semiconductor and V_e is the effective volume of the metal in the leads and contacts. The effect of the heat capacity of the leads is reduced because they do not experience the full temperature variation seen by the thermometer chip. Nonetheless, it is important to minimize their heat capacity by making them of dielectric material carrying thin metal (or superconducting) films or with ion implants to provide for electrical conduction. Much of the progress in bolometer performance over the past three decades has centered on reducing the volume of materials that contribute to the free electronic heat capacity as well as minimizing the total volume of material in the detector.

8.5.2 Thermal Conductance

The time response, responsivity, and *NEP* depend on the conductance of the thermal link, G. We assume that this parameter is dominated by the behavior of the metal or other electrical conductors in the leads, although G may also be dominated by dielectrics in the leads or by additional supports used to hold the bolometer in place. The thermal conductivity of metals is given roughly by the Wiedemann–Franz relation:

$$k_e \approx \left[\frac{\pi^2 k^2}{3q^2}\right] \sigma T, \tag{8.52}$$

where the expression in brackets is the Lorentz number ($\approx 2.45 \times 10^{-8}$ W Ω K^{-2}) and σ is the electrical conductivity. At low temperatures, the value of the equivalent Lorentz number tends to decrease, in some cases by as much as an order of magnitude. This behavior can be allowed for by taking $k_e \sim T^a$, $a \geqslant 1$. Assuming that the bolometer is mounted on two leads of length L and cross-sectional area A, the conductance is

$$G = 2\frac{A}{L}k_e. \tag{8.53}$$

Equation 8.52 shows that there is a proportionality between thermal and electrical conductivity, implying that low-resistance leads would tend to be accompanied by high values of G. For proper bolometer operation, the leads must have much lower resistance than the thermometer. This requirement poses no problem for semiconductor bolometers, given their high thermometer resistances; however, superconducting thermometers have resistances of only a few ohms and could not achieve low values of G with normal metal leads. However, since superconductors depend on a different mechanism for electrical conduction, they need not obey the Wiedemann–Franz relation, making them useful for cryogenic electrical leads in many applications. Superconducting leads are critical for use with superconducting thermometers, since they provide good thermal isolation with very low electrical resistance.

Thermal conductances useful with bolometers operating at 1 to 4 K range from about 10^{-7} to 10^{-3} W K^{-1}, whereas bolometers at 0.1 K utilize predominantly dielectric leads with $G < 10^{-10}$ W K^{-1}. This wide range of thermal conductances permits the frequency response and *NEP* of the bolometer to be tailored for a variety of applications.

With further understanding of the behavior of C and G, we can revisit the frequency response of bolometers. Substituting equations 8.51 and 8.53 into equation 8.5, we find that the time constant $\tau \sim T^2$ as long as the semiconductor lattice dominates the heat capacity of the bolometer, and τ is roughly independent of T when the metallic parts dominate.

Where the heat capacity is dominated by the lattice, bolometers built for differing operating temperatures in which G is adjusted to keep the same frequency response will have $G \propto T^2$. Substituting in equation 8.31, the Johnson-noise-limited *NEP* goes as $T^{7/2}$ for a given bolometer with $\alpha = -A/T$ as in equation 8.12; with the same assumptions the thermal noise limited *NEP* goes as T^2. If the heat capacity is dominated by free electrons, the frequency response is unaffected by reducing the temperature and the *NEP*s go as in equations 8.31 and 8.41.

As we have seen, the theoretically predicted improvement of *NEP* with reduced temperature varies considerably depending on the construction of the detector. As an empirical example, typical silicon bolometers with similar sensitive areas (~ 1000 μm^2) and time response (~ 6 ms) have *NEP*s of $\sim 5 \times 10^{-15}$ W Hz$^{-1/2}$, $\sim 3 \times 10^{-16}$ W Hz$^{-1/2}$, and 0.5–1×10^{-17} W Hz$^{-1/2}$ respectively at 1.5, 0.3, and 0.1 K. These values suggest *NEP* $\propto T^{(2\text{ to }2.5)}$; a scaling as $T^{5/2}$ is often used as an approximate relation.

8.5.3 Quantum Efficiency

In some cases, a semiconductor thermometer chip can also be a perfectly adequate absorber of the incident radiation, such as the microbolometer illustrated in Figure 8.2. However, high-performance bolometers almost always find it advantageous to separate the two functions. A semiconductor or superconducting thermometer of adequate size to be used also as an absorber in the far infrared and submillimeter regions would contribute too much heat capacity. Instead, grids are fabricated with member spacing well under the wavelength of interest. Typically the members are metallized to create a net impedance equal that of free space, which maximizes the absorption in the absence of interference effects (Section 7.3.3). In many cases, the absorption is maximized by placing the bolometer within a resonant cavity and further concentrating energy by placing a reflective surface behind it (a "backshort"), again spaced to maximize interference effects that encourage absorption. The temperature sensor can then be a tiny volume compared with the absorbing area of the detector. Some of these approaches can be seen in the examples of Planck HFI, Herschel PACS, and SCUBA-2 bolometers illustrated below.

Bolometers are not terribly choosy about what kinds of radiation they detect. Consequently, devices very similar to those used for submm- and mm-wave detection but with suitable absorbers are used as microcalorimeters in the X-ray, as discussed in Section 8.7.

8.6 Examples of Bolometer Construction

Unlike many of the detector types discussed so far, which are basically planar electronic devices, bolometers are marvelously sophisticated miniature three-dimensional mechanical structures. The possibilities have been revolutionized by precision etching techniques in silicon and silicon nitride that allow construction of precise, complex, and miniature structures. The manufacture of ultraminiature (and very thin) structures by etching allows very complex optimized bolometers with heat capacity minimized by drastically reducing the volume of the material in the detector. We discuss three examples here.

8.6.1 Planck High Frequency Instrument

The focal plane for the Planck High Frequency Instrument (HFI) is basically a set of individually optimized pixels mounted together and fed by feedhorns. There is a total of six frequency bands, with four bolometers in each band. The instrument is optimized for measuring the diffuse background radiation, not for individual source detection. The bolometers are constructed as spider webs (Figure 8.8), with silicon nitride "beams" 5–10 μm wide and 1 μm thick (Yun et al. 2003). These elements are mounted in feedhorn-fed cavities formed by the back face of the feedhorn and a backshort reflector placed 1/4 wave behind the absorber. The absorber diameter is set to 2λ, where λ is the mid-wavelength of the corresponding spectral band. Temperature sensing is by NTD germanium, and the entire detector is operated at 0.1 K. The bolometers are connected in a simple load-resistor circuit and read out by a JFET output amplifier operated at 120 K and shielded from the detectors so its emission does not compromise their performance. Further details about these detectors are given in the example at the end of this chapter.

Construction of these bolometers depends on the robustness of very thin Si_3N_4 membranes, as well as the use of etches that attack silicon much faster than silicon nitride. This process of silicon "micromachining" yields complex miniature structures that enable a broad variety of bolometer concepts. An abbreviated assembly flow for the HFI bolometers is: (1) deposit Si_3N_4 onto Si wafer; (2) mask the wafer and prepare windows for metallization; (3) metallize the Si_3N_4 for both the absorber and the thermometer leads; (4) mask and etch the wafer to establish the spider web pattern in the Si_3N_4; (5) mask and evaporate indium bumps; (6) bump bond

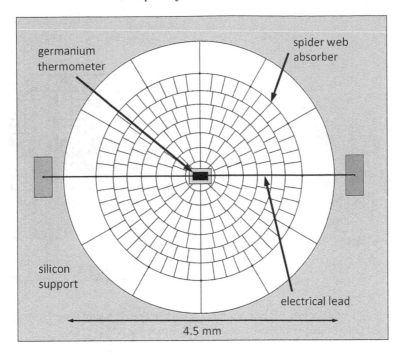

Figure 8.8 A Planck HFI "spider web" bolometer for the 100 GHz channel. The instrument had higher-frequency channels, up to 857 GHz. To first order, the bolometer dimensions were scaled inversely with the frequency down to about 40% of this size and then held constant from construction considerations. After Yun et al. (2003).

a neutron-transmutation-doped germanium thermometer to the unit; (7) mask the back of the wafer and etch away silicon from the back up to the Si_3N_4 membrane to complete the free standing bolometer. All of these steps except number (6) are conducted in standard wafer processing and avoid the delicate hand construction that was required for traditional bolometers. Even step (6) can be conducted during wafer processing, for example, if a superconducting thermometer is used in place of NTD germanium (Gildemeister et al. 1999). Similar steps are used for the two other bolometer types described below, but we will not describe them in comparable detail.

8.6.2 Herschel PACS Bolometer Array

One channel of the Herschel/PACS instrument uses a 2048 pixel array of bolometers (Billot et al. 2006). The architecture of this array is vaguely similar to the direct hybrid arrays for the near- and mid-infrared. One silicon wafer is patterned with bolometers, each in the form of a silicon mesh, as shown in Figure 8.9. A second

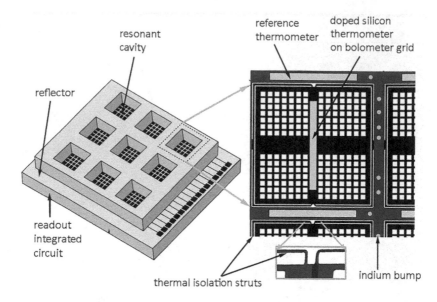

Figure 8.9 Construction of the bolometer array for the Herschel PACS instrument. The pixel (right image) is approximately 750 μm square.

silicon wafer is used to fabricate the MOSFET-based readouts, and the two are joined by indium bump bonding. The delicate construction of the PACS detectors, as shown in Figure 8.9, shows off the ability to etch exquisitely complex miniature structures in silicon. In this instance, the absorbers are fine grids of silicon (grid spacing ≪ the wavelength of operation) to minimize their mass and heat capacity. The silicon mechanical structure around the grid region provides the heat sink; the grid is connected to but isolated from it with thin and long silicon rods. The rods and grid both need to be designed to achieve appropriate response and time constant characteristics. Each grid is blackened with a thin layer of titanium nitride. Quarter-wave resonant structures surrounding it tune the absorption to higher values over limited spectral bands. For each bolometer, a silicon-based thermometer doped by ion implantation to have appropriate temperature-sensitive resistance lies at the center of the grid with a reference thermometer on the frame. Large resistance values are used so the fundamental noise is large enough to utilize MOSFET read-out amplifiers. To minimize thermal noise and optimize the material properties, the bolometer array is operated at 0.3 K. Further details are in Billot et al. (2006).

8.6.3 SCUBA-2

The SCUBA-2 bolometer arrays utilize TESs and time-division SQUID-based MUXs (Irwin and Hilton 2005; Holland et al. (2013)). The transition edge sensors

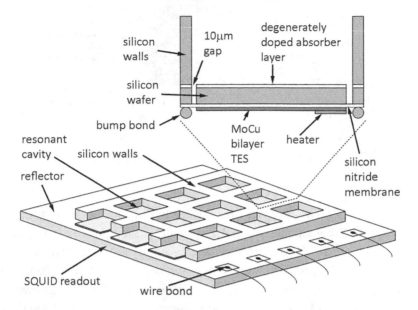

Figure 8.10 A SCUBA-2 focal plane. The pixel cross-section (upper right) shows the pixel walls designed to enhance absorption through interference effects, the doped top layer of the pixel that is responsible for absorption, the thermal isolation around the edges of the pixel through the thin (0.5 μm thick) SiN material, the Mo/Cu bilayer TES, and the indium bump that connects the pixel to the readout wafer.

are formed from a molybdenum (transition temperature 0.90 K) /copper(non-superconducting) (Mo/Cu) bilayer of material, with the relative thicknesses used through the proximity effect to tune the transition temperatures of the TESs. The full focal plane is divided into sub-arrays of 32 columns by 40 rows of bolometers. The sub-arrays are fabricated as a detector and a readout wafer and joined by indium bump bonding. The pixel upper surface is ion-implanted with phosphorus to absorb the photons and the Mo/Cu layer not only serves as the TES, but acts as a backshort. An underside silicon nitride membrane supports each bolometer and provides a weak thermal link to the cold bath. A heater is placed in a thin-line geometry around the edge of each pixel. These devices are illustrated in Figure 8.10.

8.7 Other Thermal Detectors

8.7.1 X-ray Calorimeters

TES bolometers are also used as energy-resolving detector arrays (or single detectors) for the X- and gamma-ray ranges, in which application they are termed

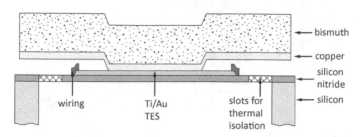

Figure 8.11 A schematic of a microcalorimeter.

"microcalorimeters." These devices operate according to the same principles as the mm-wave TES bolometers, but not surprisingly, the optimization of detectors for this application differs significantly (see Figure 8.11). The absorber needs to combine low specific heat with high atomic number: bismuth (Z = 83) combines both of these attributes, and gold (Z = 79) is also used. Bismuth, however, has low thermal conductivity, which can lead to unequal distribution of the deposited energy in the absorber and less precise thermometry. To mitigate this issue, a layer of copper is added between absorber and thermometer. The thermal isolation also differs from that for mm-wave TESs. The calorimeter must operate in a pulse counting mode; spatial and energy information would be mixed if the absorbed photons were not detected individually. Consequently, the thermal isolation is optimized for fast response, but not so fast that the signal from an absorbed photon decays before it can be intercepted by the readout electronics. Some degree of thermal isolation is also desirable to be sure that the absorbed energy is distributed approximately uniformly before leaking to the outside world. The device in Figure 8.11 is isolated by mounting on a silicon nitride membrane, as with the mm-wave devices, but with higher conduction to the bath; the conduction is tuned to an optimum level by adjusting the membrane thickness and by adding empty slots to it. The TESs are read out in a multiplexed array as described in Section 8.4.3. In the most common version of a time domain multiplexer, the speed of an array is limited by the time to turn on a SQUID of about 320 ns (Ullom and Bennett 2015). Sample observed energy resolutions are $R = E/\Delta E \gtrsim 2000$ between 1.5 and 5 keV (Ullom and Bennett 2015).

8.7.2 Hot Electron Bolometers

A bolometer for the submillimeter- and mm-wave spectral regions can be implemented quite differently from those described so far, by making use of properties of n-type InSb. In this material the impurity levels are merged with the conduction band even at very low temperatures; as a result, the ionization energy can be very small, of order 0.001 eV, and the bolometer has a sea of free electrons. Photons incident on the detector are absorbed by the free electrons, which are thereby raised above their thermal equilibrium energies and become hot electrons. Eventually, an equilibrium will be reached between de-excitation and the photo-induced effects. Since the hot electrons in InSb do not interact strongly with the lattice to de-excite, they can accumulate to a significant density. For InSb, the hot electrons strongly affect the mobility and hence the conductivity (equation 2.5); the photon signal can be monitored by measuring the detector resistance.

The parameters of these bolometers are largely set by the properties of the InSb (for example, Kinch and Rollin 1963; Putley 1977); nonetheless, they are well suited for certain applications. Voltage responsivities are 100 to 1000 V W^{-1}, the thermal conductance, G, is about 5×10^{-5} W K^{-1}, and the heat capacity of the electron sea $C \sim 3/2\, nkv$, where n is the carrier concentration and v is the volume of the detector. A typical detector might be 2 mm on a side ($v \sim 10^{-2}$ cm^3) and have $n \sim 5 \times 10^{13}$ cm^{-3}; then $C \sim 10^{-11}$ J K^{-1} and the detector time constant (from equation 9.5) is 2×10^{-7} s. In this example operating at 4 K, the thermally limited *NEP* is 2×10^{-13} W Hz$^{-1/2}$, from equation 8.41 and assuming unity quantum efficiency. The detector is fast (response time $\lesssim$ μs) and of only modest sensitivity compared with the bolometers we have discussed previously.

The quantum efficiency of these detectors depends on the absorption efficiency of the free electrons. At low frequencies, the absorption is large and wavelength independent, and the quantum efficiency is near unity; at high frequencies, the absorption efficiency goes as λ^2. The crossover between these behaviors is at $\lambda \sim 1.6$ mm; at 1 mm, the absorption coefficient $a = 22$ cm^{-1}, and at $\lambda = 100$ μm, $a = 0.3$ cm^{-1}. Consequently, the detectors become inefficient at wavelengths short of about 300 μm.

A substantial advance in speed is achieved by generating the hot electrons in superconducting material. Similarly to a TES bolometer, a superconducting film is biased to operate near its transition temperature. At this temperature, the super-conducting energy gap is suppressed and the concentration of Cooper pairs is very small, so most electrons are unpaired. These electrons do not show superconduc-tivity, but approximate the behavior of normal electrons with a Fermi energy distri-bution. When one of these electrons absorbs a photon, the energy is shared with other electrons to produce a tiny temperature change that causes the resistance of the superconducting film to change substantially. This film is deposited on a substrate that acts as the heat sink/bath; the boundary layer between them acts as the thermal link to the low temperature. Thus, this arrangement has all of the components described for more conventional bolometers already discussed.

The process is illustrated schematically in Figure 8.12 with indicative time constants. The input photon is absorbed by an electron in the superconductor.

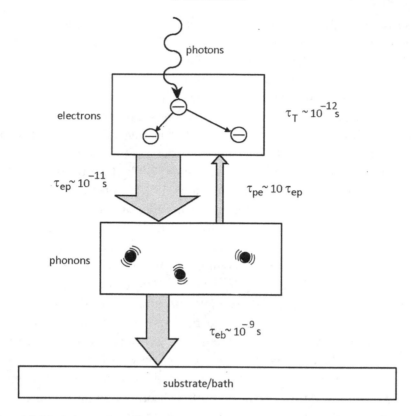

Figure 8.12 Schematic operation of a superconducting hot electron bolometer, with approximate time constants indicated. After Semenov et al. (2002).

The absorbed energy is quickly shared by the initial electron to others in the superconductor, with a thermalization time of order 10^{-12} s, resulting in a small increase in the electron temperature that results in the resistance change. They in turn pass the energy to vibrations in the crystal lattice; the vibrational states are quantized and described in terms of phonons. (The electron temperature can relax via diffusion or phonon emission; we focus on phonon-cooled devices in the following.) The time for electron energy relaxation through interactions with phonons is τ_{ep}, for typical materials $0.1 - 1 \times 10^{-11}$ s. The energy then flows into the substrate of the detector, which is thermally sunk to the low-temperature bath; describing this process in terms of phonons, τ_{eb} is the time for phonon escape to the temperature bath, of order $0.4-4 \times 10^{-9}$ s. Time τ_{pe} characterizes phonon flow into the electron sea; to cool the electrons requires that it be significantly longer than the time constant for the flow in the opposite direction, τ_{ep}. It is typically of order 10^{-10} s. To be more specific, when the energy flow is balanced and in equilibrium, $\tau_{pe} = (C_p/C_e)\,\tau_{ep}$, where C_p and C_e are the phonon and electron specific heats, respectively.

These devices are much smaller than the wavelengths at which they are used; a practical bolometer consists of a tiny strip of metal (e.g., 1 μm^3 in volume). Energy is coupled into the electrons in the microbolometer by integrating it with an antenna structure. Thermal isolation of the microbolometer strip can be achieved by a combination of thermal boundary resistance and reflection effects at the interface between the strip and superconducting structures to which it is attached. The electron sea in the tiny microbolometer metal strip can have thermal conductances to the lattice of $G = 10^{-11}$–10^{-13} W K^{-1} for $T = 0.3$–0.1 K. Because of the very small volume, the time constants of such microbolometers are very short, $\sim$ns. Responsivities can be as high as 10^9 V W^{-1}, which combined with the low G values provide very low *NEP*s, estimated at 3×10^{-19} W Hz$^{-1/2}$ at 0.1 K (e.g., Nahum et al. 1993). These performance levels are very attractive in situations where the limitations of antenna coupling are acceptable, such as those discussed in Chapter 10.

Alternately to this example, the thermal link may be dominated by diffusion of the hot electrons out of the bolometer. The latter mechanism occurs with a time constant of

$$\tau_{diff} = \frac{L^2}{\pi^2 D}, \tag{8.54}$$

where L is the length of the bolometer and D is the diffusion constant. This expression can be derived as for equation 5.9, recalling that electrons can diffuse out of both ends of the hot electron bolometer microstrip. As an example, devices of superconducting aluminum have $D \sim 2.5$–6 cm^2 s^{-1} and with $L \leq 0.6$ μm can have $\tau_{diff} < 10^{-10}$ s (e.g., Burke et al. 1996). The thermal conductance of devices longer than $\sim$0.6 μm is dominated by phonon cooling.

8.8 Example: Design of a Bolometer

We consider the design of a spider web bolometer such as the Planck HFI one illustrated in Figure 8.8. Design details have been taken from Yun et al. (2003) and Holmes et al. (2008). The Planck telescope scanned the sky as the spacecraft rotated at 1 rpm. The smallest instrument field of view was 5′, which a source transited in 14 ms, necessitating a time response for the bolometers in the few ms range. So that the performance would be sky-photon-limited rather than limited by bolometer noise, the *NEP*s were required to be $< 9 \times 10^{-18}$ W Hz$^{-1/2}$.

We first consider the NTD germanium temperature sensor, which is 30 μm thick, 100 μm wide, and 300 μm long. We compute its temperature-dependent resistance from equation 8.9. The material has an effective doping density of 5.6×10^{16} Ga cm^{-3}, which yields $T_0 = 18$ K and a conductivity of 10 (Ω cm)$^{-1}$ at this temperature. From equations 2.1 and 8.9, at T = 0.1 K we find a resistance of 6.7 MΩ. This allows use in the circuit of Figure 8.3 with $R_L = 20$ MΩ and a readout amplifier with a JFET first stage, which will have an input impedance well above the bolometer resistance. A low-noise JFET can have significant input capacitance, but the load resistance is small enough to support time response in the ms range.

The germanium chip is mounted on a Si$_3$N$_4$ grid, constructed by micromachining as discussed in Section 8.6.1. The thermal conductance of this material at 0.1 K is 1.7×10^{-4} W K^{-1} m^{-1} (Woodcraft et al. 2000). A grid element is < 1 μm thick and 5–10 μm wide by ∼5 mm long (depending on the band, since the size is scaled to the wavelength), and there are 12 of them. They therefore provide a heat path of about 2×10^{-12} W K^{-1}. We will find later in the discussion that this heat path is small compared with others. The grid has a spacing of ∼320 μm, much finer than the operating wavelength of 3000 μm, and a thin (∼12 nm thick) film of gold (with a titanium buffer layer) is deposited on it adjusted to match the impedance of free space. It therefore provides good absorption of the signal, as described in Section 7.3.3. The absorption is enhanced by placing a 1/4-wave backshort behind the absorber.

The strength of the thermal link to the 0.1 K bath, G, is set from a combination of requirements, i.e., that it be small enough to achieve the *NEP* needed to be photon-limited on the background, but large enough to meet the requirements for frequency response (and to satisfy saturation limits). For the 100 GHz channel, these considerations led to $G = 7 \times 10^{-11}$ W K^{-1}. The strength of the thermal link is controlled by adjusting the length, width, and thickness of the gold electrical leads deposited onto opposite legs of the Si$_3$N$_4$ absorber, according to the Wiedemann–Franz relation, equation 8.52. They are connected to boron-implanted contacts on the NTD germanium chip.

The *NEP* is determined by: (1) phonon (thermal) noise, equation 8.41; (2) Johnson noise in the germanium chip, equation 8.31; (3) photon noise, equation 8.44; and (4) any excess noise component. These terms add quadratically. For the HFI bolometer, the first one is dominant, with a value of 6×10^{-18} W Hz$^{-1/2}$ for unity absorption efficiency (and ignoring a correction for thermal gradients along the thermal link). The response time is given by equation 8.5 as modified for electrothermal feedback in equation 8.16: $\tau \sim 0.6\ C/G$. The heat capacity of $\sim 1.2 \times 10^{-12}$ J K^{-1} includes contributions by the germanium chip, the metallic components, and contaminants left from the processing during construction. The result is $\tau = 10$ ms.

8.9 Problems

8.1 *NEP* is unpopular in some quarters because it gets smaller the better the detector performance − bigger is worse, not better. Also many detectors have noise currents proportional to detector area, A, and hence noise proportional to the square root of area. These considerations have led to the invention of $D^* = A^{1/2}/NEP$. Discuss the applicability of D^* as a figure of merit for bolometer performance.

8.2 Discuss the advantages and disadvantages of operating a bolometer with a transimpedance amplifier.

8.3 Consider a dewar where liquid helium cools apparatus that absorbs the energy passing from the room (at $20\,°C$) through a 3 cm diameter hole (with full view of the room from one side). How long will it take to boil away one liter of helium? (The cooling capacity of liquid helium is 2562 J L^{-1}.)

8.4 Estimate the gallium density above which tunneling/hopping conductivity becomes significant in germanium. Hint: Use the result of equation 6.42 and the discussion following it, along with the excitation energy derived from λ_c in Table 2.4.

8.5 Because the Planck HFI instrument was designed to measure diffuse backgrounds, it put a premium on detector stability. Why would this have argued against using transition edge temperature sensors instead of semiconductor ones?

8.6 It is uncommon to cool a bolometer in a spacecraft to 0.1 K; because it is easier to achieve, a temperature of 0.3 K from a helium-3 refrigerator is more popular. Suppose a bolometer like the one in the example were operated at 0.3 K (with the NTD germanium chip selected to have the same resistance). Could it achieve similar performance? Assume the heat capacity behaves as a metal (the most favorable case) and the thermal link is also dominated by metals. What adjustments would be needed to recover the performance achieved at 0.1 K? Are they feasible?

8.10 Further Reading

Clarke and Braginski (2006) – overview of SQUID-based electronics
Kruse (1997) – shows how thermal detector principles apply to room-temperature devices. The remaining articles in this volume provide in-depth discussions of room-temperature thermal detectors and arrays.

McCammon (2005) – very thorough discussion of cryogenic semiconductor thermistors
Mauskopf (2018) – rather technical overview of MKIDs and transition edge bolometers
Pobell and Luth (1996) – overview of methods for achieving low temperatures
Richards (1994) – detailed and thorough review of high-performance bolometer technologies
Simoens (2013) – comprehensive review of bolometers
Ullom and Bennett (2015) – up to date review of microcalorimeters

9

Visible and Infrared Coherent Receivers

Coherent receivers are the third and last general category of detector to discuss. These devices mix the electromagnetic field of the incoming photons with a local oscillating field to produce a signal at the difference, or beat, frequency. In the optical and near-to-mid-infrared, the oscillations of the photon field are of such high frequency that they cannot be captured in electronics, and the mixing must be conducted with the incoming photons and a locally generated stream of photons of nearly the same frequency. In the submillimeter through the radio, the photon field can be converted into an oscillating current, providing much more freedom in receiver design (next chapter). In either case, unlike the output from the incoherent detectors discussed so far, this signal directly encodes the spectrum of the incoming signal over a range of input frequencies and also retains information about the phase of the incoming wavefront. Coherent receivers monopolize radio applications; in this chapter we discuss two specialized applications in the optical and infrared. They are not used as widely at these wavelengths because of their narrow spectral bandwidths, small fields of view, and inability to be constructed in simple, large-format spatial arrays.

9.1 Basic Operation

In general, any device that measures the field strength of the incoming photon – that is, that has the potential to measure and preserve phase information directly – is a coherent receiver. Heterodyne receivers are the most important class of such device. They function by mixing signals of different frequency; if two such signals are added together, they "beat" against each other. The resulting signal contains frequencies only from the original two signals, but its *amplitude* is modulated at the difference, or beat, frequency (see Figure 9.1). Heterodyne receivers measure this amplitude. This chapter not only discusses such receivers for the optical

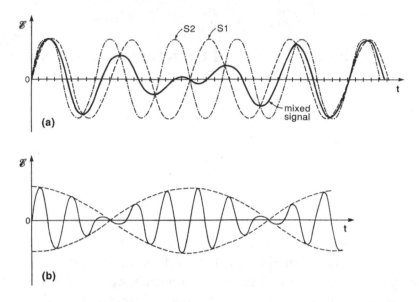

Figure 9.1 (a) Mixing of two sinusoidal signals, S_1 and S_2, with the result shown as the heavy line. (b) The mixed signal on a compressed timescale to illustrate the beating at the difference frequency.

and infrared, but also derives some general performance attributes that will be needed in the following chapter where we discuss heterodyne submm- and mm-wave receivers.

In Figure 9.1, two oscillating fields have been mixed, and the resulting field is indicated by the solid line. If this field is measured by a linear device, two results are possible: (1) if the time constant of the device is short, the output will simply follow the solid curve in Figure 9.1, which contains power only at the original frequencies; and (2) if the time constant is long, the device will not respond because on average the output signal contains equal positive and negative excursions. In neither case will the output contain any component of power at the beat frequency. For heterodyne operation, the mixed field must be passed through a nonlinear circuit element or *mixer* that converts power from the original frequencies to the beat frequency. In the submm- and mm-wave (and radio) region this element is a diode or other nonlinear electrical circuit component. For visible and infrared operation, the nonlinear element is a photon detector, sometimes also termed a photomixer.

The action of the nonlinear element in converting the power to the beat frequency is illustrated in Figure 9.2. We have already argued that if the element has a linear *I–V* curve (see Figure 9.2(a)), then the conversion efficiency is zero. Similarly, any mixer having a characteristic curve that is an odd function of voltage around the

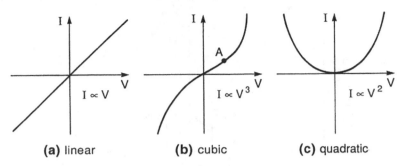

Figure 9.2 *I–V* curves of three hypothetical mixer elements.

origin will have zero conversion efficiency if operated at zero bias; the cubic curve in Figure 9.2(b) is an example. In this case, some conversion will occur if the mixer is biased above zero. For example, if the operating point is set at A, the average change in current will be greater for positive than for negative voltage swings. Much greater efficiency is achieved, however, with a characteristic curve that is an even function of voltage – for example the quadratic curve in Figure 9.2(c). Moreover, if $I \propto V^2$, the output current is proportional to the square of the signal amplitude. The amplitude is proportional to the field strength in the incoming photon, and since the signal power goes as the square of the field strength, a quadratic *I–V* curve for the mixer produces an output proportional to power, which is usually what we want to measure (that is, $I \propto V^2 \propto \mathcal{E}^2 \propto P$, where $\mathcal{E}$ is the strength of the electric field of the signal photons). Hence, we prefer to use "square law" devices as fundamental mixers because their output is linear with input power.

Even if we cannot find an electronic device with an *I–V* curve exactly like the one shown in Figure 9.2(c), we can assume that our mixer has a characteristic curve that can be represented by a Taylor series:

$$I(V) = I(V_0) + \left(\frac{dI}{dV}\right)_{V=V_0} dV + \frac{1}{2!}\left(\frac{d^2I}{dV^2}\right)_{V=V_0} dV^2$$

$$+ \frac{1}{3!}\left(\frac{d^3I}{dV^3}\right)_{V=V_0} dV^3 + \cdots, \tag{9.1}$$

where V_0 is the voltage at the operating point. As we have seen, the first two terms are of no importance for fundamental mixers ($I(V_0)$ is a DC current and the second term gives a net response of zero). The fourth and higher terms can usually be ignored as long as $dV (= V - V_0)$ is small; the device is then a good square law mixer as long as d^2I/dV^2 is reasonably large.

Assume that we are using a square law mixer, and let it be illuminated with a mixture of two sources of power, one a signal at frequency ω_S and the other at ω_{LO}.

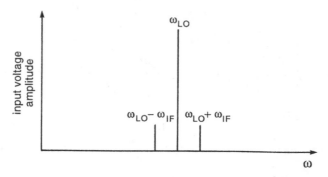

Figure 9.3   The ambiguity in the frequency of the input signal after heterodyne conversion to the intermediate frequency.

Let the power at ω_{LO} originate from a source within the instrument called the local oscillator (LO), and assume it is much stronger than the weaker, unknown signal at ω_S. We also specify that $\omega_S > \omega_{LO}$. Then the mixed signal will be amplitude modulated at the intermediate frequency $\omega_{IF} = |\omega_S - \omega_{LO}|$. The signal at ω_{IF} contains spectral and phase information about the signal at ω_S as long as the LO signal at ω_{LO} is steady in frequency and phase, and if the difference between ω_S and ω_{LO} is neither so large that it exceeds the maximum frequency response of our equipment nor so small that it falls below its low-frequency cutoff. The signal has been downconverted to a much lower frequency than ω_S or ω_{LO}, in the frequency range where it can be processed by conventional electronics. If the LO is absolutely steady in frequency and phase, this transformation retains without distortion the relative frequencies and phases of the input signal ranging around ω_S. As a result, it is relatively easy to extract the spectrum of the source over this range or to measure the phase of the incoming photons.

There is no easy way of telling in the mixed signal whether $\omega_S > \omega_{LO}$ or $\omega_{LO} > \omega_S$. Because we have lost the initial information regarding the relative values of ω_S and ω_{LO}, many of the derivations of receiver performance will assume that the input signal contains two components of equal strength, one above and the other below the LO frequency ω_{LO} (see Figure 9.3). Since the signal at ω_{IF} can arise from a combination of true inputs at $\omega_{LO} + \omega_{IF}$ and $\omega_{LO} - \omega_{IF}$, it is referred to as a double sideband signal. When observing continuum sources, the ambiguity in the frequency of the input signal is usually a minor inconvenience. When observing spectral lines, however, the unwanted "image" frequency signal can result in serious complications in calibration.[1] In general, visible or infrared receivers operate double sideband.

---

[1] Submillimeter- and millimeter-wave receivers can be operated single sideband if the image is suppressed by tuning the mixer or with a narrow bandpass "image rejection" filter that is placed in front of the receiver; this

The output of the mixer is passed to an intermediate frequency amplifier and then either directly to a rectifying and smoothing circuit called a "detector" or, more commonly, to a "backend" spectrometer. This terminology differs from that used for incoherent devices, where the detector is the device that receives the photons and converts them to an electrical signal. The backend spectrometer divides the output into the respective frequencies to create the spectrum of the source.

9.2 Visible and Infrared Heterodyne

To provide a specific basis, we discuss heterodyne components in two applications: (1) near 1 μm in LIDAR (or LADAR) systems; and (2) for ultra-high-resolution spectroscopy in the 10 μm region. LIDAR (light detection and ranging) and LADAR (laser detection and ranging) are two names (and acronyms) for the same thing: a laser is shined on a target and the return signal from scattering is detected, with the range to the scattering surface determined from the timing of the return. Coherent LIDAR uses a heterodyne detector system allowing in addition measurement of Doppler shifts and hence motions of the target toward and away from the instrument. Heterodyne systems in the 10 μm range are used to study planetary atmospheres (including our own) at spectral resolutions of $\lambda / \Delta \lambda \sim 10^7$, a resolution virtually unattainable by conventional dispersive optical spectrometers of reasonable size.

9.2.1 Local Oscillator and Mixer: Principles of Operation

The basic components of a high-frequency heterodyne receiver are shown in Figure 9.4. The highest frequency heterodyne receivers, i.e., those in the optical and mid-infrared discussed in this chapter, use a continuous wave (CW) laser as the local oscillator. The laser light and the signal are combined by a beam splitter, sometimes called a diplexer. The output is mixed in a photon detector; because such a photomixer responds to power, or field strength squared, it is a square law mixer.

The operation of potential visible and infrared heterodyne photomixers is described in previous chapters; we will assume the mixer is a photodiode. Because the mixer must follow the heterodyne signal, it must respond to far higher frequency (to 1 GHz or more) than we considered in the earlier discussions. Compared with typical operation as incoherent detectors, where time constants of milliseconds to

filter blocks the photons at image frequencies before they are mixed with the *LO* signal. A better if more complex approach is to provide a second *LO* signal shifted 90° in phase into a second mixer and deliver the signal to both mixers. The resulting phase difference between the mixer outputs can be used to separate the signals from the two frequencies. Further discussion will be provided in the next chapter.

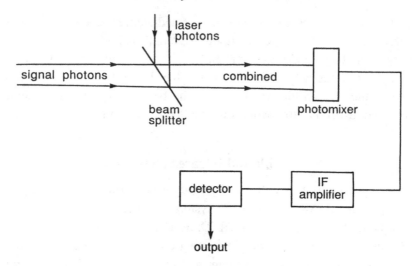

Figure 9.4 An infrared or visible heterodyne receiver.

seconds are acceptable, heterodyne use demands careful optimization for high-frequency response. PIN or avalanche photodiodes are preferred for this reason, as discussed in Sections 3.1.4.1 and 3.1.4.2.

Assume we have such a mixer arranged as in Figure 9.4, that the two photon beams falling on the mixer are well collimated, and that both are linearly polarized in the same direction. The combined electric field in that direction is

$$\mathcal{E}_T(t) = \mathcal{E}_{LO}(t) + \mathcal{E}_S(t), \tag{9.2}$$

where $\mathcal{E}_{LO}(t)$ and $\mathcal{E}_S(t)$ are the electric fields of the laser and source, respectively.

The electric and magnetic fields of any electromagnetic wave can be written as

$$\mathcal{E}(t) = \mathcal{E}_0 e^{-j\omega t} = \mathcal{E}_0 \cos \omega t - j \mathcal{E}_0 \sin \omega t \tag{9.3}$$

and

$$\mathcal{H}(t) = \mathcal{H}_0 e^{-j\omega t} = \mathcal{H}_0 \cos \omega t - j \mathcal{H}_0 \sin \omega t, \tag{9.4}$$

where $\mathcal{E}_0$ and $\mathcal{H}_0$ are the complex amplitudes of the fields and $\omega = 2\pi \nu$, with ν the frequency of the wave, and where j is $\sqrt{(-1)}$. The electric and magnetic field amplitudes are related by

$$\mathcal{E}_0 = \left(\frac{\mu}{\varepsilon}\right)^{1/2} \mathcal{H}_0, \tag{9.5}$$

where μ is the magnetic permeability and ε is the dielectric permittivity of the medium.

We want to determine the current that is generated in the mixer as a result of the electric field $\mathcal{E}_T(t)$. The rate at which energy passes through a unit area normal to the propagation direction of the wave is given by the real part of the Poynting vector:

$$S(t) = Re[\mathcal{E}(t)] \, Re[\mathcal{H}(t)]\hat{i}. \tag{9.6}$$

The direction of S is given by the unit vector $\hat{i}$ and is perpendicular to the directions of both $\mathcal{E}(t)$ and $\mathcal{H}(t)$. The time average of the Poynting vector is[2]

$$\langle S \rangle_t = \langle Re\,[\mathcal{E}\,(t)] \, Re\,[\mathcal{H}\,(t)] \rangle_t \, \hat{i} = \frac{1}{2} Re\,[\mathcal{E}\,(t)\,\mathcal{H}^*\,(t)]\,\hat{i}. \tag{9.7}$$

$\mathcal{H}^*$ is the complex conjugate of $\mathcal{H}$, obtained by substituting $-j$ for j. Using equation 9.4 to substitute for $\mathcal{H}^*(t)$ in equation 9.7, the power falling on a mixer of area A (normal to the propagation direction of the radiation) and averaged in time over a cycle of the input frequencies is

$$\langle P\,(t) \rangle = \frac{A}{2} \left(\frac{\varepsilon}{\mu}\right)^{1/2} |\mathcal{E}\,(t)|^2. \tag{9.8}$$

To express the time dependence explicitly, we replace $\mathcal{E}(t)$ and $\mathcal{E}^*(t)$ in $|\mathcal{E}(t)|$ in the expression above with $\mathcal{E}_T(t)$ and $\mathcal{E}_T^*(t)$ from equations 9.2 and 9.3 to get

$$\langle P\,(t) \rangle = \frac{A}{2} \left(\frac{\varepsilon}{\mu}\right)^{1/2} [(\mathcal{E}_{0,LO}\, e^{-j\omega_{LO}t} + \mathcal{E}_{0,s}\, e^{-j\omega_s t})$$
$$\times (\mathcal{E}_{0,LO}^*\, e^{j\omega_{LO}t} + \mathcal{E}_{0,s}\, e^{j\omega_s t})]$$

[2] The second equality can be easily demonstrated. Referring to equations 9.3 and 9.4, and letting $\mathcal{E}_0 = a + jb$ and $\mathcal{H}_0 = c + jd$,

$$Re[\mathcal{E}(t)]Re[\mathcal{H}(t)] = (a\cos\omega t + b\sin\omega t)(c\cos\omega t + d\sin\omega t)$$
$$= (ac)\cos^2\omega t + (bd)\sin^2\omega t$$
$$+ (ad + bc)\cos\omega t \sin\omega t. \tag{9.6a}$$

Because $\langle\cos\omega t \sin\omega t\rangle_t = 0$ and $\langle\sin^2\omega t\rangle_t = \langle\cos^2\omega t\rangle_t = 1/2$, we get

$$\langle Re[\mathcal{E}(t)]Re[\mathcal{H}(t)]\rangle_t = \frac{1}{2}(ac + bd). \tag{9.6b}$$

Now we calculate the quantity in the second form of equation 9.7; it is

$$\frac{1}{2}Re[\mathcal{E}(t)\mathcal{H}^*(t)] = \frac{1}{2}Re[(a + jb)(\cos\omega t - j\sin\omega t)$$
$$\times (c - jd)(\cos\omega t + j\sin\omega t]$$
$$= \frac{1}{2}(ac + bd), \tag{9.6c}$$

which is equivalent to the result in equation 9.7.

$$= \frac{A}{2}\left(\frac{\varepsilon}{\mu}\right)^{1/2}\left[|\mathcal{E}_{0,LO}|^2 + |\mathcal{E}_{0,S}|^2\right.$$

$$\left. + \mathcal{E}_{0,S}\mathcal{E}_{0,LO}^* \, e^{-j(\omega_S - \omega_{LO})t} + \mathcal{E}_{0,S}^*\mathcal{E}_{0,LO} \, e^{j(\omega_S - \omega_{LO})t}\right]. \qquad (9.9)$$

The photocurrent, $I(t)$, that is generated by this power can be obtained from equation 3.1:

$$I(t) = \frac{\eta q}{h\nu} P(t), \qquad (9.10)$$

where η is the quantum efficiency of the mixer and $\nu = \omega/2\pi$ is the frequency of the incoming photons. Substituting for $P(t)$ from equation 9.9, we obtain

$$I(t) = I_{LO} + I_S + 2(I_{LO}I_S)^{1/2}\cos\left[(\omega_S - \omega_{LO})t + \phi\right], \qquad (9.11)$$

where the DC current from the laser is

$$I_{LO} = \frac{\eta q A}{2h\nu}\left(\frac{\varepsilon}{\mu}\right)^{1/2}|\mathcal{E}_{0,LO}|^2 \qquad (9.12)$$

and a similar expression holds for the current from the source, I_S. The relative phase between $\mathcal{E}_S$ and $\mathcal{E}_L$ is

$$\phi = arctan\left[\frac{Re[\mathcal{E}_{0,S}]Im[\mathcal{E}_{0,LO}] - Im[\mathcal{E}_{0,S}]Re[\mathcal{E}_{0,LO}]}{Re[\mathcal{E}_{0,S}]Re[\mathcal{E}_{0,LO}] + Im[\mathcal{E}_{0,S}]Im[\mathcal{E}_{0,LO}]}\right]. \qquad (9.13)$$

Thus, the photocurrent contains a component oscillating at the intermediate frequency, $\omega_{IF} = |\omega_S - \omega_{LO}|$. In principle, this current also contains components at ω_S and ω_{LO}, but since in this case these frequencies may be 10^{12} to 10^{14} Hz, these components appear as the cycle-averaged currents I_S and I_{LO}, which are DC in nature. The IF current (the third term in equation 9.11) is the heterodyne signal and has a mean-square-amplitude of

$$\langle I_{IF}^2\rangle_t = 2\,I_{LO}\,I_S; \qquad (9.14)$$

recall that $\langle cos^2\omega t\rangle = 1/2$.

An important result is that the signal strength in equation 9.14 depends on the *LO* power. As a result, many forms of noise can be overcome by increasing the output of the local oscillator (see Blaney (1975) for additional discussion of the noise attributes of heterodyne receivers). The ability to provide an increase in power while downconverting the input signal frequency is characteristic of quantum mixers, such as the photomixers discussed here. The conversion gain is defined as the *IF* output power that can be delivered by the mixer to the next stage of electronics (sometimes called the exchangeable power) divided by the input signal power. In practice, the gain in output power with increasing *LO* power is limited by saturation in the mixer. Nonetheless, the gain of quantum mixers can be useful in overcoming the noise contributions of the post-mixer electronics and in providing high signal to noise from the receiver. The gain provided by quantum mixers can be contrasted with the behavior of classical mixers, which downconvert the frequency but provide no increase in power. Chapter 10 provides further discussion of this distinction.

As shown in equation 9.13, the phase, ϕ, in the heterodyne signal is a measure of the phase of the input signal (assuming that the phase of the local oscillator – the laser in this case – is stable). Thus, the heterodyne signal can be used directly with another, similar signal to reconstruct information about the wavefront. If the signal also has good phase stability, as applies in the radio regime and into the mm-wave, this attribute allows signals from two different telescopes to be combined coherently, as if the telescopes were part of a single large instrument, making it possible to do wide-baseline spatial interferometry. In the optical and infrared, transmission through the atmosphere results in too much phase instability to achieve this type of operation routinely.

9.2.2 Local Oscillator

At the high frequencies of the near-infrared and visible spectral regions, the only form of local oscillator with reasonably high power output is a continuous wave laser. Laser frequencies/wavelengths are selected according to the application of the LIDAR system. A concern with these systems is that eyes can be damaged by strong light in the 0.4–1.5 µm range, and precautions are required to prevent exposure (damage can occur outside this range also, but at higher thresholds). Thus, many systems operate at 1.6 µm (Er-doped solid state lasers) or 2.0 µm (Tm and Ho-doped lasers).

The frequency range of the heterodyne spectrum is limited to the *IF* bandwidth around the laser frequency, i.e., a tiny fraction of the laser frequency. As a result, heterodyne receivers are generally not utilized as general-purpose spectrometers in the visible range. However, in LIDAR, the laser output that is used for ranging to the target can also be used as the *LO*. So that return signals with no frequency shift can be included in the *IF*, the *LO* signal is offset using an acousto-optical modulator (AOM) (Figure 9.5) operating at high frequency, up to about a GHz. In this device, the piezoelectric transducer injects compressional waves (termed "sound waves" although the frequency is far, far higher than human hearing) into a suitable optical medium (e.g., a crystal such as lithium niobate or water) that result in moving alternative compressive and rarefactive zones, with a corresponding modulation of the refractive index of the material. This arrangement is called a Bragg cell. Incident light is diffracted by this pattern of refractive index; if the light is at the first-order angle, it will efficiently produce an emerging first-order output. However, because

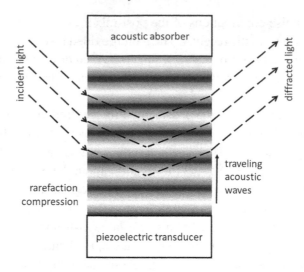

Figure 9.5 Acousto-optical modulator. The piezoelectric transducer inputs high-frequency vibrational waves into a suitable optical medium.

the diffraction occurs off a moving refractive index modulation, the emergent light will be Doppler shifted by the frequency of the sound pattern (if a higher order, m, is used the shift is m times the sound frequency).

Heterodyne receivers used as very high resolution spectrometers in the mid-infrared (particularly around 10 μm) traditionally have used carbon dioxide lasers as LOs. By selecting the isotopic composition of the carbon dioxide gas and utilizing optical elements to select wavelengths of specific rotational lines in a CO_2 vibrational transition, it is possible to access a bit more than 1 μm of spectral range. Much more flexibility in operational wavelengths is now provided with quantum cascade lasers (QCLs), which can be tuned to the desired frequency. These lasers make use of multiple quantum well structures, called superlattices. As we saw in Section 6.4, this type of structure leads to each potential well having a number of discrete energy sub-bands. The layer thicknesses can be designed to produce a population inversion between two sub-bands, and hence to support laser emission. The frequency of this emission is determined primarily by the layer thicknesses, which can be adjusted to tune the emission wavelength of QCLs over a wide range. Tunable diode lasers are available operating from ∼3 to 30 μm. The tradeoff is poorer sensitivity, due to a number of issues with single-mode diode lasers such as limited output power.

9.2.3 Post-Mixer Electronics

The heterodyne signal derived above is a low-level, high-frequency AC current; to be compatible with simple post-mixer electronics, it needs to be amplified and

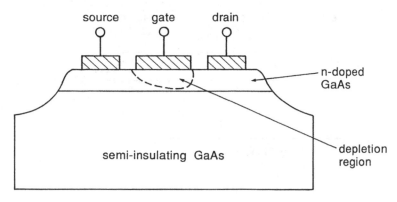

Figure 9.6   A metal-semiconductor field effect transistor (MESFET).

converted into a slowly varying voltage that is proportional to the time-averaged input signal power.

9.2.3.1 IF Amplifier

Because of the low power in the *IF* signal, a critical component in a high-performance receiver is the amplifier that follows the mixer and boosts the *IF* signal power. High-speed amplifiers can be constructed with MOSFETs, but the best performance in this application is usually obtained with high electron mobility transistors (HEMTs) built on GaAs or InP (Streetman and Banerjee 2014). These devices are made extremely fast by a combination of maximizing the speed of the charge carriers passing across them and minimizing the distance these carriers must travel.

The HEMT is based on the metal–semiconductor field effect transistor (MES-FET), which is a cross between the JFET and MOSFET discussed in Chapter 4. Fabrication of a MESFET can start with a substrate of GaAs doped with chromium. The Fermi level is then close to the middle of the bandgap, giving the material relatively high resistance and leading to its being called a "semi-insulating layer." An n-doped layer is grown on this substrate to form the channel, with contacts for the source and drain and a gate formed as a Schottky diode[3] between them on this layer (see Figure 9.6). The electron flow between source and drain in this channel can be regulated by the reverse bias on the gate. As with the JFET and MOSFET, with an adequately large reverse bias a region depleted of charge carriers grows to the semi-insulating layer and pinches off the current. Because this structure is very simple, MESFETs can be made extremely small, which reduces the electron transit time between the source and drain and increases the response speed.

---

[3] Schottky diodes result from the asymmetry to current flow at a metal–semiconductor interface. They are discussed in a different context in the next chapter, Section 10.2.1.

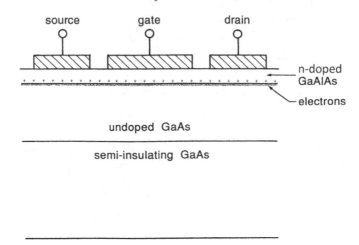

Figure 9.7 A high electron mobility transistor (HEMT).

For a given field the electron drift velocity between drain and source is proportional to the mobility of the material in the channel:

$$\mu_n = -\frac{\langle v_x \rangle}{\mathcal{E}_x}, \tag{9.15}$$

where μ_n is the electron mobility and $\langle v_x \rangle$ is the drift velocity (see equation 2.6). High-speed MESFETs are built in GaAs or InP rather than in silicon because of the higher electron mobility of the former materials. However, as the size of the transistor is decreased, the n-type doping in the channel must be increased leading to decreased mobility due to impurity scattering.

This problem can be overcome by forming a heterojunction; refer to the discussion of quantum well detectors in Section 6.4, particularly to Figure 6.14. If the MESFET is grown on a thin layer of GaAlAs heavily doped n-type and deposited over an undoped layer of GaAs, the Fermi level in the GaAsAl can lie above the bottom of the conduction band in the GaAs. Consequently, electrons diffuse into the GaAs; they are held close to the interface by the positive space charge retained in the GaAlAs as shown in Figure 9.7. The thin layer of free electrons is described as a two-dimensional electron gas (2DEG). Because the GaAs is undoped, the mobility is high and the 2DEG can carry large currents. Scattering can be further reduced by adding a very thin layer of undoped GaAsAl between the doped layer and the GaAs. At room temperature, the mobility is dominated by lattice scattering, so it can be increased further by cooling the device: values of $\geq 2.5 \times 10^5$ cm^2 (V s)$^{-1}$ and $\geq 2 \times 10^6$ cm^2 (V s)$^{-1}$ can be achieved at 77 K and 4 K, respectively. The high-frequency response can be further improved by minimizing the size of the gate to

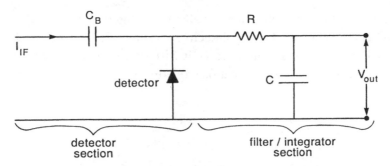

Figure 9.8 The detector stage.

reduce parasitic capacitances. Gates smaller than 0.1 μm are used for extremely high-frequency operation.

Epitaxial growth can be used to produce more complex versions of HEMTs than in our simple example. The processing starts with a substrate of either GaAs or InP, which have similar bandgap properties. A number of thin layers can be grown on this substrate to enhance lattice matching, optimize the space between active layers, control heterojunction bandgaps, and for physical protection of the device. The control offered by doping, layer thickness, and variable bandgap materials allows for a variety of device architectures, and the degrees of freedom are further increased because very thin layers can tolerate slight lattice mismatches between materials (allowing construction of "pseudomorphic" HEMTs). There are some relevant generalizations, however, connected with the substrate material. Use of GaAs is a relatively mature technology, allowing lower-cost manufacturing. The InP-based devices provide better high-frequency behavior (to well beyond 10^{11} Hz) and usually lower noise. The AlGaN/GaN devices are well suited for high-power applications. As will be discussed in the next chapter, up to $\sim 1.2 \times 10^{11}$ Hz, HEMTs have sufficiently low noise that they can even be used to amplify the signal prior to the mixer, boosting receiver performance (Wilson 2018). The very lowest noise is achieved with InP HEMTs operated up to $\gtrsim 8$ GHz, providing high-performance amplification in *IF* stages over this wide a frequency band.

9.2.3.2 Detector Stage

The conversion to a slowly varying output can be done by a "detector" stage that rectifies the signal and sends it through a low-pass filter. We would like the circuit to act as a square law detector because $\langle I_{IF}^2 \rangle$ is proportional to I_S (see equation 9.14), which in turn is proportional to the power in the incoming signal.

A suitable circuit for this purpose is shown in Figure 9.8. Here C_B is used to block DC components of the signal such as I_{LO} and I_S in equation 9.11. If the diode is held near zero bias with $I \ll I_0$, it will act as a good square law detector. To demonstrate, solve the diode equation (equation 3.24) for the voltage and expand in terms of I/I_0:

$$V = \frac{kT}{q} ln \left(1 + \frac{I}{I_0} \right)$$

$$= \frac{kT}{q} \left[\frac{I}{I_0} - \frac{1}{2} \left(\frac{I}{I_0} \right)^2 + \frac{1}{3} \left(\frac{I}{I_0} \right)^3 - \frac{1}{4} \left(\frac{I}{I_0} \right)^4 + \cdots \right]. \quad (9.16)$$

As previously discussed, the first term in the expansion will have zero conversion efficiency, and, near zero bias, the third term will have negligible efficiency. For $I \ll I_0$, the fourth and higher terms are also negligible. Thus, the dominant behavior is square law as long as the current is kept small and the operating point is held near zero bias.

The RC circuit on the output of the detection stage acts to integrate the output with a time constant $\tau = RC$. It therefore smooths the output to provide V_{out} in a form that is easily handled by the circuitry that follows. Particularly if a filter is used to restrict the range of IF frequencies passed to this circuit, it serves as a "total power detector," that is, it provides a measure of the total power from the source into the filter bandpass. A slightly more complex arrangement can recover both polarities of the IF signal for greater efficiency.

However, it is usually desirable to carry out a variety of operations with the IF signal itself before smoothing it in the detector stage. The down-converted IF signal could be sent to a bank of narrow bandpass fixed-width electronic filters that divide the IF band into small frequency intervals, each of which maps back to a unique difference from the LO frequency, i.e., to a unique input frequency to the receiver from the target source. A detector stage can then be put at the output of each filter, so the outputs are proportional to the power at a sequence of input frequencies; that is, they provide a spectrum. In this manner the total IF bandwidth is divided into a spectrum, even though only a single observation with a single receiver has been made; the process is called spectral multiplexing. However, although it is conceptually simple, a high-performance, hardware-based filter bank can be an engineering challenge. The individual filters need to have closely matched properties and be robust against drift of those properties due to effects like temperature changes. A filter bank is also inflexible in use; the resolution must be set during design and construction. Finally, these devices are complex electronically and expensive to build if many channels are required. For these reasons,

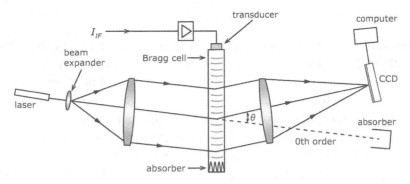

Figure 9.9 An acousto-optical spectrometer.

although hardware filter banks were once widely used, alternative approaches are now the norm.

We will discuss two software-based types of backend spectrometer in the next chapter. However, a more hardware-oriented approach, an acousto-optical spectrometer (AOS), is used as the backend for many of the optical and infrared heterodyne systems discussed in this chapter. An AOS is a close relative of the acousto-optical modulator we encountered a few pages back. In the case of the AOS (see Figure 9.9), the IF signal is used to excite the piezoelectric transducer attached to the Bragg cell (again, a transparent volume containing either a crystal such as lithium niobate or water). The vibrations produce ultrasonic waves that propagate through the Bragg cell and produce periodic density variations. The resulting variations in the index of refraction make the cell act like a diffraction grating; when light from a near-infrared laser diode passes through the cell, it is deflected accordingly. An absorber is provided for the zero-order path, and the first-order path is focused onto a CCD by camera optics. The efficiency of the diffraction depends on the strength of the ultrasound waves, so the intensity of the diffracted light is proportional to the IF power injected into the Bragg cell, while the deflection angle and hence position on the CCD is determined by the ultrasound wavelength, i.e., the IF frequency. The output signal is basically the Fourier transform of the IF signal. An AOS is capable of resolving the IF signal into more than 2000 spectral channels.

9.3 Heterodyne Performance Attributes

This section discusses some general aspects of *all* heterodyne receivers, including those for the submm- and mm-wave described in the next chapter and even for longer-wavelength systems in the radio regime.

9.3.1 Bandwidth

The spectral bandwidth of a heterodyne receiver is determined by the achievable bandwidth at the intermediate frequency. The *IF* bandwidth, Δf_{IF}, can be limited either by the frequency response of the photomixer or by that of any circuitry that amplifies or filters the heterodyne signal. Some forms of infrared mixer impose severe limitations on the bandwidth. The *RC* time constants of photoconductor mixers are usually short because of the reduction in *R* by the *LO* power, but recombination times can limit the bandwidth. For example, the long recombination times in germanium limit the frequency response of far-infrared germanium photoconductors to $< 10^8$ Hz. InSb hot electron bolometers can be used as mixers in the submillimeter region, but their thermal time constants limit the achievable bandpasses to $\sim 10^6$ Hz. In more favorable cases, such as photodiode mixers, current technical limitations often hold Δf_{IF} to no more than a few times 10^9 Hz due to the *RC* time constant of the mixer. In the infrared, the bandwidth so defined is very small compared with the frequency of the + signal photons; at 10 μm, $\nu = 3 \times 10^{13}$ Hz, and, even assuming a heterodyne bandwidth of 3×10^9 Hz, the spectral bandwidth is 0.01% of the operating frequency. Therefore, heterodyne receivers operating in this wavelength region have poor signal to noise on continuum sources. As we have discussed, they are used primarily for measurements of spectral lines and/or Doppler shifts at extremely high resolution. In the submm- and mm-wave ranges, heterodyne systems become more competitive in observing continuum sources (at 3 mm (10^{11} Hz) $\Delta f_{IF} = 3 \times 10^9$ Hz corresponds to a 6% bandwidth for a double sideband receiver).

9.3.2 Time Response

The time response of a heterodyne receiver can be as fast as the period of the *IF* signal, $1/f_{IF}$; that is, a few nanoseconds. It may also be limited by the time response of the mixer (see above) or of the detector stage.

9.3.3 Throughput

The phases of the *LO* and signal beams must be very well matched and stable, or the fringes will not be steady, washing out the *IF* signal. In the submillimeter and longer wavelengths, this behavior can be achieved electronically since receivers can respond to the oscillations of the input photon electric fields. However, in the optical and infrared, the matching must be achieved with the input beams optically. The two beams must be coincident, their diameters should be equal, and the wavefronts must have the same curvature. They must also be identically polarized, so that their electric vectors will coincide.

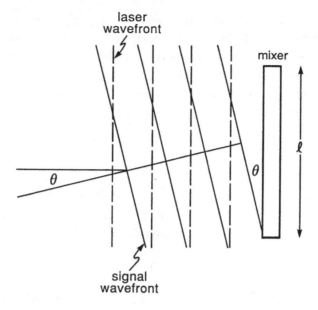

Figure 9.10 Required alignment of the local oscillator and signal wavefronts on the photomixer.

However, the signal photons cannot be concentrated onto the mixer in a parallel beam; even for a point source, they will strike it over a range of angles. The requirement that interference occurs at the mixer between the laser and the signal photons sets a requirement on the useful range of acceptance angle for the heterodyne receiver.

We will assume that the *LO* photon stream is perfectly collimated and strikes the mixer perpendicular to its face. Consider Figure 9.10, which shows the general case of a signal wavefront striking the mixer face at a nonzero angle θ (θ is measured from the normal to the mixer face). If $\theta \neq 0$, there will be a phase shift across the mixer between the signal and laser fields; thus, the *IF* signals from different parts of the mixer will not add and may even begin to cancel. Full cancellation will occur when

$$\ell \sin \theta_{max} = \lambda \approx \ell \theta_{max}, \tag{9.17}$$

where ℓ is the length of the mixer (assumed square) and λ is the wavelength of observation. Taking the central ray of the signal to strike the mixer perpendicularly, from equation 1.10 in the case of small θ, we get $\theta_{max}^2 \approx \Omega$, the solid angle over which signal interferes efficiently with the *LO*. We also have $\ell^2 = A$, the area of the mixer. Thus, we can rewrite equation 9.17 as

$$A\Omega \approx \lambda^2. \tag{9.18}$$

$A\Omega$, frequently called the etendue, is invariant in any aberration-free system. That is, optical elements can be placed in the beam to magnify or demagnify any focal plane of the system, but perfect geometric optics leave $A\Omega$ unchanged, and imperfect optics can only increase it. Therefore, equation 9.18 is also a constraint on the beam that can be accepted by a telescope or by any other optical system that concentrates the signal on the heterodyne mixer; this condition can be alternately expressed as

$$\Phi \approx \frac{\lambda}{D}, \tag{9.19}$$

where D is the diameter of the telescope aperture and Φ is the angular diameter of the field of view on the sky. This relationship defines the condition for diffraction-limited imaging; compare it to the Rayleigh criterion, which is just

$$\Phi = 1.22\frac{\lambda}{D}. \tag{9.20}$$

Thus, a coherent receiver should operate at the diffraction limit of the telescope (if the receiver only accepts a resolution element smaller than the diffraction limit of the telescope, there is a significant loss of energy from the source).

At infrared wavelengths, atmospheric turbulence can cause significant image motion that, if uncompensated, broadens the net image flux distribution beyond the diffraction limit. These effects are even more serious at shorter wavelengths. The constraint on etendue can result in loss of signal because the receiver accepts only a fraction of the broadened image. The constraint is a much less serious shortcoming at longer wavelengths such as the submillimeter range.

A second restriction, already mentioned, is that the interference that produces a heterodyne signal only occurs for components of the source photon electric field vector that are parallel to the electric field vector of the laser power. Since the laser light is polarized, only a single polarization of the source emission produces any signal.

The relation between wavelength and etendue in equation 9.19 and the constraints on polarization are manifestations of the "antenna theorem," which applies to all heterodyne receivers. The combination of the constraint derived in equation 9.19 and the sensitivity to only one plane of polarization is sometimes collectively described by saying that heterodyne receivers are single-mode detectors, or that they are sensitive to only one transverse mode of the radiation field.

9.3.4 Signal to Noise and Fundamental Detection Limits

The derivation of the detection limits of a heterodyne receiver is significantly different from similar derivations for incoherent detectors. In fact, it is usually expressed in terms of "noise temperature," but that parameter does not relate well to the specifications for the detectors discussed up to now. We therefore derive expressions for NEP in this chapter and then convert the results to noise temperature.

As pointed out after the derivation of equation 9.14, we need to distinguish between types of noise: (1) those that are independent of the LO-generated current, I_{LO}, and (2) those that depend on I_{LO}. In principle, the first category can be eliminated at least for quantum mixers by using a local oscillator with sufficient power to raise the signal strength out of the noise (there are, however, frequently practical problems in doing so). Thus, the second category alone contains the fundamental noise limits for heterodyne receivers. Two types of fundamental noise need to be considered: (1) noise in the mixer from the generation of charge carriers by the LO power; and (2) noise from thermal background detected by the system.

In considering the photon (shot) noise from the LO, we assume that the LO power has been increased to provide $I_{LO} \gg I_S$ and $I_{LO} \gg I_B$, where I_B is the signal in the mixer from background photons. From equation 9.10, the current generated by the LO in a photodiode mixer is

$$I_{LO} = \frac{\eta q P_{LO}}{h\nu}, \tag{9.21}$$

where P_{LO} and ν are the LO power and frequency, respectively, and η is the quantum efficiency of the mixer. We define the frequency bandwidth over which the mixer operates to be Δf_{IF}, the IF bandwidth. The noise current is then, from equation 3.3,

$$\langle I_S^2 \rangle = 2q\, I_{LO}\, \Delta f_{IF} = \frac{2q^2 \eta\, P_{LO}\, \Delta f_{IF}}{h\nu}. \tag{9.22}$$

In considering the noise in the thermal background, we note from equation 9.14 that the background noise current is

$$\langle I_B^2 \rangle = 2\, I_{LO}\, I_B, \tag{9.23}$$

where I_{LO} is from equation 9.21 and similarly

$$I_B = \frac{\eta q P_B}{h\nu}. \tag{9.24}$$

Here the background power is

$$P_B = \frac{1}{2} L_\nu(\varepsilon, T_B)\, A\, \Omega\, (2\, \Delta f_{IF}), \tag{9.25}$$

where L_ν is spectral radiance for an effective emissivity ε and a temperature T_B. The factor 1/2 accounts for the heterodyne receiver being sensitive to only one polarization direction, A is the sensitive area of the receiver, and Ω is the solid angle over which it is illuminated. We have assumed a double sideband receiver, so background power is taken to be detected in two bands of width Δf_{IF} placed, respectively, above and below the *LO* frequency (see Figure 9.3); that is, the bandwidth $\Delta \nu$ for signals is $2\Delta f_{IF}$. This bandwidth is twice the frequency bandwidth used in computing the shot noise (equation 9.22) because the two sidebands are merged into a single range of *IF* frequency.

As just discussed, we can take the etendue, $A\Omega \approx \lambda^2 = c^2/\nu^2$ (see equation 9.18 and the surrounding discussion), for heterodyne detection. Substituting into equation 9.23 from equations 9.21, 9.24, and 9.25, the noise current from the background signal is

$$\langle I_B^2 \rangle = \frac{4\eta^2 q^2 \varepsilon \, P_{LO} \, \Delta f_{IF}}{h\nu \left(e^{h\nu/kT_B} - 1\right)}. \tag{9.26}$$

Equations 9.22 and 9.26 show how to calculate the two fundamental noise contributions. We next consider how these two noise mechanisms affect the signal-to-noise ratio achieved by the receiver. We continue to assume that we can make I_{LO} arbitrarily large so that all noise mechanisms not dependent on I_{LO} are negligible compared with the two discussed above. Then, since the output power goes as I^2, the instantaneous signal-to-noise ratio at the output of the *IF* stage and at the *IF* bandwidth is

$$\left(\frac{S}{N}\right)_{IF} = \frac{\langle I_{IF}^2 \rangle}{\langle I_S^2 \rangle + \langle I_B^2 \rangle}, \tag{9.27}$$

where the signal and noise currents are from equations 9.14, 9.22, and 9.26. Substituting for these currents and then for I_{LO} from equation 9.21 and I_S from a similar relation, we obtain

$$\left(\frac{S}{N}\right)_{IF} = \frac{\eta P_S}{h\nu \, \Delta f_{IF} \left[1 + \frac{2\eta\varepsilon}{e^{h\nu/kT_B} - 1}\right]}. \tag{9.28}$$

The signal to noise derived in equation 9.28 is independent of P_{LO} and can therefore be treated as the fundamental performance limit of a heterodyne receiver system at the *IF* output. This equation has two limiting cases. The first case, known as the quantum limit, occurs when the shot noise from the *LO* power dominates; the second case is called the thermal limit, for which noise from the background emission dominates. The dividing point between these limits can be derived by setting the two terms in square brackets in equation 9.28 equal to each other and solving to obtain:

$$\frac{h\nu}{kT_B} = ln\,(1 + 2\,\eta\,\varepsilon). \tag{9.29}$$

Since η and ε are all less than, but of the order of, one, the logarithmic term is of order unity. Therefore, the division between the two regimes can often be simplified by stating that for $h\nu \gg kT_B$ the quantum limit holds, while for $h\nu \ll kT_B$ we get the thermal limit.

A quantity useful for high-speed systems such as those used in communications is the minimum detectable power, MDP_{IF}, required to give $S/N = 1$ at the IF bandwidth:

$$MDP_{IF} = \frac{h\nu \, \Delta f_{IF}}{\eta}\left[1 + \frac{2\eta\varepsilon}{e^{h\nu/kT_B} - 1}\right]. \tag{9.30}$$

However, in most applications a detector stage is used to rectify and smooth the IF signal and we need to include this stage in the performance evaluation. As with other detection systems, the instantaneous signal-to-noise ratio is improved in proportion to the square root of the effective integration time. When a smoothing stage is used (such as the filter/integrator circuit shown in Figure 9.8), the S/N increases in proportion to the square root of the ratio of effective integration times of the IF and smoothing circuits, that is,

$$\left(\frac{S}{N}\right)_{out} = \left(\frac{S}{N}\right)_{IF}\left(\frac{\tau_{RC}}{\tau_{IF}}\right)^{1/2}. \tag{9.31}$$

To convert integration times to frequency bandwidths, we should use the procedures in Chapter 2, for example, equations 2.21 or 2.22. We can usually assume that the output of the IF stage is sampled uniformly over a time interval τ_{IF}, so $\tau_{IF} = 1/2\Delta f_{IF}$, and that the integrator is a single-stage RC circuit (as shown in Figure 9.8), so $\tau_{RC} = 1/4\Delta f_{RC}$. The noise equivalent power, NEP, is the signal that can be detected at a signal-to-noise ratio of one in unity frequency bandwidth. Therefore, setting $\Delta f_{RC} = 1$ Hz, the NEP in the quantum limit for a heterodyne receiver system, NEP_H, is

$$\begin{aligned} NEP_H &= \frac{h\nu}{\eta}\,\Delta f_{IF}\left[\frac{\tau_{IF}}{\tau_{RC}(1\text{Hz})}\right]^{1/2} \\ &= \frac{h\nu}{\eta}\,\Delta f_{IF}\left[\frac{2\Delta f_{RC}(1\text{Hz})}{\Delta f_{IF}}\right]^{1/2} \\ &= \frac{h\nu}{\eta}(2\Delta f_{IF})^{1/2}. \end{aligned} \tag{9.32}$$

In the thermal limit,

$$NEP_H = \frac{2h\nu\varepsilon}{e^{h\nu/kT_B} - 1}(2\Delta f_{IF})^{1/2}. \tag{9.33}$$

Although the result is obscured by our normalization of the NEP to unity bandwidth, inspection of equation 9.32 also shows the general result that the effective noise bandwidth of the receiver is proportional to the geometric average of the predetector and postdetector bandwidths, that is, to $(\Delta f_{IF} \, \Delta f_{RC})^{1/2}$ (for example, see Robinson (1962)).

9.3.5 Noise Temperature

Although equations 9.32 and 9.33 are useful for describing the theoretical performance of heterodyne receivers in a given situation, they are inadequate for making general comparisons between heterodyne and incoherent detector response to continuum sources. If the *LO* power can be increased without limit, these equations imply that the *NEP* of heterodyne receivers is reduced (that is, the signal-to-noise ratio is increased) by narrowing the *IF* bandwidth. This behavior arises because the equations assume that all the source power continues to fall within this bandwidth, which clearly is a physically unreasonable assumption for a continuum source. To overcome this limitation, a thermal continuum source is introduced through a noise temperature, T_N, defined such that a matched blackbody at the receiver input at a temperature T_N produces $S/N = 1$. The lower T_N, the fainter a source gives $S/N = 1$, and the better is the performance of the receiver.

In the thermal limit, if the effective source emissivity $\varepsilon = 1$, then by definition $T_N = T_B$. More generally, by substituting

$$P_S = L_\nu(T_N) A\Omega \, \Delta f_{IF} \tag{9.34}$$

into equation 9.28 and setting $S/N = 1$, it can be shown that the double sideband noise temperature in the thermal limit at the output of the *IF* stage is

$$T_N = \frac{h\nu}{k} \frac{1}{ln(\varepsilon - 1 + e^{h\nu/kT_B}) - ln\,\varepsilon}. \tag{9.35}$$

To derive a noise temperature in the quantum limit, we again start from equations 9.28 and 9.34. For the double sideband case, the quantum limit is

$$T_N = \frac{h\nu}{k\,ln\,(1 + 2\eta)}. \tag{9.36}$$

In an ideal case, we set η to 1 to obtain

$$T_N \approx \frac{h\nu}{k}. \tag{9.37}$$

The quantum limit expressed in equation 9.37 can be justified in terms of the Heisenberg uncertainty principle, which states that the uncertainty, ΔP, in a measurement of power will be

$$\Delta P = \frac{h\nu}{\Delta t}, \tag{9.38}$$

where Δt is the observation time. As is conventional in discussing receiver performance, we represent the noise by that of an equivalent ideal resistor. Starting with
equation 2.25 for Johnson noise in a resistor within a frequency bandwidth df and
converting to power noise within a time interval Δt,

$$\langle P \rangle \Delta t = kT_N. \tag{9.39}$$

Setting $\Delta P \approx \langle P \rangle$, we obtain

$$T_N \sim \frac{h\nu}{k}. \tag{9.40}$$

Although we have concentrated on fundamental noise sources, noise that is independent of the LO power is also present. For example, the mixer detects the total
background power, including that at frequencies outside the bandpasses determined
by Δf_{IF}. This signal can contribute noise, particularly if the background is fluctuating rapidly. A second source of noise is Johnson noise in the mixer; it can
normally be eliminated by cooling the mixer. A third possibility is that the local
oscillator contributes noise through phase or amplitude instability or both. Fourth,
the amplifiers used in the signal chain may dominate the noise. We will consider
characterization of these nonfundamental noise sources for submillimeter receivers
in Section 10.5. We will then work through an example to characterize a heterodyne
receiver working at 10 μm.

9.4 Problems

9.1 Consider a square mixer of width ℓ illuminated by plane-parallel signal and
LO waves. Assume that the LO is perfectly aligned on the mixer and the signal
strikes it at an angle θ, in a direction parallel to the x-axis of the mixer. Show
that the current from a surface element of the mixer is

$$dI(t) \approx \frac{dxdy}{\ell^2} \left[I_{LO} + 2(I_S I_{LO})^{1/2} \cos\left(\omega t + \frac{2\pi \sin\theta}{\lambda}x + \phi'' \right) \right]$$

where ϕ'' is a fixed phase shift. Use this result to prove the result of
equation 9.17.

9.2 Compute the "quantum limit" for an ideal incoherent detector. That is, for a
photodiode or photomultiplier show that the minimum detectable power when
dark current and Johnson noise can be neglected is

$$P_{min} = \frac{h\nu \, df}{\eta},$$

where ν is the frequency of the signal photons, df is the detector electronic bandwidth, and η is the quantum efficiency. Compare with the heterodyne detector.

9.3 Suppose we could suspend reality for a while and there was a heterodyne receiver available working in the 3 μm atmospheric window with an IF bandwidth of 10^{13} Hz. Discuss its potential performance advantages and disadvantages relative to HgCdTe photodiodes, particularly for signal to noise on faint source continua. Is a company making this device a promising investment opportunity?

9.5 Further Reading

Aadit et al. (2017) – comprehensive discussion of HEMTs
Keyes and Quist (1970) – dated but still useful article on infrared heterodyne
Nakagawa et al. (2016) – detailed description of a modern heterodyne spectrometer
National Research Council (2014) – broad overview of LIDAR
Schmülling et al. (1998) – description of a heterodyne spectrometer
Sonnabend et al. (2008) – description of a heterodyne spectrometer
Teich (1970) – advanced review of infrared heterodyne, dated but useful for theory; same volume as Keyes and Quist (1970)

10

Submillimeter- and Millimeter-Wave Heterodyne Receivers

"Heterodyne" is the process of combining signals at two nearly equal frequencies to downconvert to a lower one, and is not linked to any specific implementation of this process.[1] As a result, the general principles derived in Chapter 9 are equally valid for submm- and mm-wave (and longer-wavelength radio) heterodyne receivers even though some of their operations differ significantly. That is, they are still subject to limitations such those expressed in the antenna theorem. However, some performance attributes that limit the general usefulness of infrared heterodyne receivers, such as limited bandpass, cease to be serious limitations as the wavelength of operation increases. Heterodyne receivers are therefore the preferred approach for high-resolution spectroscopy in the submillimeter spectral region, and their usefulness is expanded as the wavelength increases into the millimeter regime and beyond. At wavelengths longer than a few millimeters, they are used to the exclusion of all other kinds of detectors. Not only are these receivers easily adapted for spectroscopy, their outputs can be combined to reconstruct the incoming wavefront, making interferometry between different receivers possible. A dramatic application of this latter capability is the use of intercontinental-baseline radio-telescope interferometers to achieve milliarcsecond resolution in astronomy.

10.1 Basic Operation

10.1.1 Radio Receivers in General

The operational principles of heterodyne receivers were described in Section 9.1. At centimeter and longer wavelengths (frequencies $\lesssim$ 30 GHz), mixers take the form of diodes or transistors. The performance can be improved by adding a low-noise

---

[1] "Superheterodyne" has the same meaning; it is a contraction of "supersonic heterodyne" which refers to the frequencies involved being above the range of human hearing.

amplifier ahead of the mixer to boost the signal. In the mm-wave, for performance close to the quantum limit, amplifiers with sufficiently low noise are not available and the signal must be delivered directly to the mixer, which for the best performance needs to be a highly specialized device. This wavelength dividing line is not fixed (Wilson 2018), and amplifiers with sufficiently low noise to begin approaching fundamental limits up to 200 GHz are under development (e.g., White et al. 2019; Yagoubov et al. 2019). In addition, when performance approaching fundamental limits is not required, diode (or transistor) mixers are used into the mm-wave range, such as uncooled versions of the Schottky diodes discussed in this chapter.

The operation of the components that follow the mixer in a radio-, submm- or mm-wave receiver is similar to the systems discussed in Chapter 9. Such components can be used for amplification, frequency conversion, and detection. In simple cases, e.g., broadcast amplitude modulated (AM) radio, a tunable filter plus a simple smoothing circuit can be adequate (although somewhat more complex approaches are also used). However, to take full advantage of the heterodyne capabilities, much more complex "backend" spectrometers are needed. Often, much of the expense in a heterodyne receiver system is in these spectrometers and in other equipment that processes the *IF* signal. Because these items can be identical from one system to another, sometimes a single set is used with different receiver "frontends" that together can operate over a broad range of signal frequencies. In any case, the approaches to backends are similar independent of the receiver type.

10.1.2 Submillimeter- and Millimeter-Wave Receivers

In this chapter, we will focus on the receiver components that must be changed from the optical and infrared devices in Chapter 9 for operation in the submillimeter and millimeter, that is, on mixers and local oscillators; we will also discuss additional types of backend that are commonly used in the mm-wave and radio.

There are three underlying reasons for the differences between the mixer design for submm- or mm-wave and infrared receivers. The first is that high-quality, fast photon detectors are not available at wavelengths longer than the infrared. As discussed in Chapter 3, no high-performance photodiodes are available with response at wavelengths longer than ~15 μm. (Although we have not discussed the issue, the photoconductors and other detectors that operate at longer infrared wavelengths have significant handicaps as mixers, e.g., relatively long recombination times and hence poor frequency response.) A second reason is that, in contrast to photon detectors, it is possible to manufacture mixers with response extending to sufficiently high frequencies to respond directly to the electric field of the photons.

Superconductors play a large role in such mixers, since, as we saw in Problem 7.1, their speed of electrical response is orders of magnitude faster than for materials with normal conductivity.

A third reason is a more general result of the requirement for high-frequency response. Efficient absorption of the energy of the incoming photons by a photo-mixer requires the mixer to have dimensions at least comparable to the photon wavelength. As the electronic devices used as mixers are made larger, their frequency response generally becomes worse (or their capacitances become larger with equally deleterious effects). Consequently, as we approach longer wavelengths it becomes desirable to couple the energy into a mixer much smaller than the wavelength of the photons.

Submillimeter- and mm-wave mixers are therefore designed as optimized electrical components. The photon stream from the source is collected by a telescope or "primary antenna" and concentrated onto a secondary antenna. The electric field of the photon stream creates an oscillating current in the secondary antenna, which conducts the current to the mixer. When the signal and *LO* power are combined in the mixer, the oscillating current will beat in amplitude at their difference frequency as illustrated in Figure 9.1. The mixer is a suitably nonlinear electrical element that converts this amplitude beating into the *IF* signal. A block diagram of this arrangement is shown in Figure 10.1.

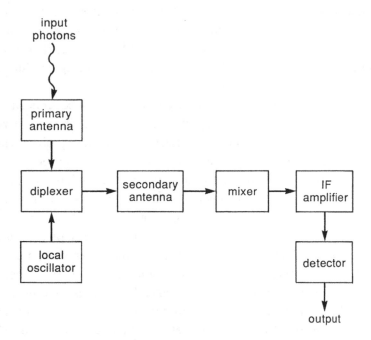

Figure 10.1 Block diagram of a mm- or submm-wave heterodyne receiver.

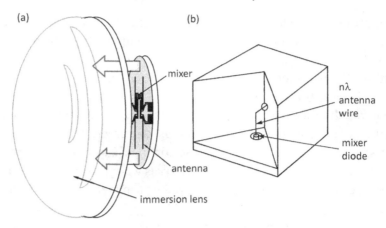

Figure 10.2 Quasi-optical receiver arrangements: (a) lens/planar antenna mixer mount; (b) corner reflector mixer mount.

The secondary antenna can take the form of a wire or a similar structure defined by lithography arranged within the receiver optics. Because of the restrictions expressed in the antenna theorem (see Section 9.3.3), the secondary antenna should be adjusted to receive only one mode of the photon stream because the fields due to the others will differ in phase and will tend to interfere with the heterodyne process. Since the bulk of the energy is in the longest wavelength, or fundamental mode, the antennas and their surrounding structures are designed to operate at this mode. This behavior accounts for the term "single-mode detector" introduced in the preceding chapter. The overall performance of the receiver depends critically upon the efficiency of transfer of power to the mixer, requiring careful impedance matching of the antenna to the mixer.

In addition, the signal energy from free space must be concentrated efficiently onto the antenna wire to improve the efficiency and directivity of its antenna pattern. Conventional lithography can overcome these problems to produce structures consisting of a planar antenna, a mixer, and appropriate impedance matching devices integrated on a dielectric substrate, although in the submillimeter fabricating structures on a small enough scale for good single mode coupling becomes challenging. We show two examples of means to focus the energy onto a secondary antenna: (1) a lens (Figure 10.2(a) feeding a planar antenna); or (2) a corner cube reflector (Figure 10.2(b)). In the latter case, the secondary antenna contacts the mixer through a hole in the bottom of the corner reflector. Because of the heavy reliance on standard optical techniques and devices (lenses and mirrors), the arrangements discussed in this paragraph are called quasi-optical feeds. They are described by Goldsmith et al. (1989) and Goldsmith (1998). In principle, such feeds can be highly efficient (Filipovic et al. 1993).

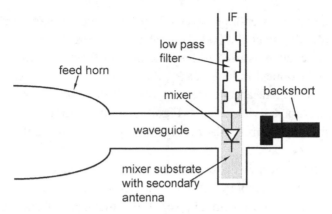

Figure 10.3 A waveguide-based mixer mount.

At frequencies below ∼800 GHz, the signal can be coupled to the mixer in a different way, illustrated in Figure 10.3. The signal flux is directed into a feed horn that concentrates the energy into a waveguide. The mixer and secondary antenna are mounted across the waveguide, with the antenna in the direction of the electric field. Waveguides can carry photon streams in a variety of modes corresponding to different oscillation patterns in the guide; most of the power is in the fundamental mode. To concentrate the fundamental mode onto the secondary antenna, the waveguide is terminated with a conducting backshort (see Figure 10.3), designed to provide a resonant cavity to maximize the absorption. In principle, the transfer of power is maximized when the length of waveguide section between mixer and backshort is adjusted to $(m/4)\lambda$, where m is an odd integer. Sometimes the backshort is made adjustable so the efficiency can be optimized empirically. The *IF* signal is brought out through another channel, which contains a low-pass filter to keep the input signal power from escaping through it. Further details are given in Christiansen and Högbom (1985), Kraus (1986), Bahl (1989), and Walker et al. (1992).[2] A suitably optimized waveguide structure usually couples energy to the secondary antenna more efficiently than does a quasi-optical feed, but with computer optimization it has been possible to improve the performance of the latter type of feed substantially (e.g., Carlstrom and Zmuidzinas 1996; Gonzalez 2016; Joint et al. 2019).

At microwave and radio frequencies, waveguides can carry signals substantial distances without significant losses. The required fabrication tolerances in the submm-wave region are very stringent to achieve this advantage, but there is

---

[2] In contrast to a waveguide mixer feed, a lithographically defined quasi-optical feed usually cannot have adjustable tuning elements, requiring that the original elements be designed well and implemented very accurately (Goldsmith 1998).

continued progress in extending waveguide performance to very high frequencies (e.g., Drouet d'Aubigny et al. 2000; Boussaha et al. 2012; Davis et al. 2017).

A variety of schemes is used to feed the energy from both the signal and the *LO* into the mixer optics. In many cases, the *LO* power is reflected off a diplexer that transmits the signal, an arrangement very similar to that used with infrared heterodyne receivers and illustrated in Figure 9.4. Other arrangements can be used with waveguides; for example, the *LO* signal may be brought to the mixer through a second waveguide.

Despite the differences in mixers and *LO*s, the performance attributes discussed in Section 9.3 are applicable to submm- and mm-wave receivers as well as to infrared ones. As pointed out in Chapter 9, the maximum frequency bandwidth of heterodyne receivers is often limited by the *IF* bandwidth, which is similar from one receiver to another; therefore, the fractional bandwidth changes inversely with operating frequency (assuming that the bandwidth is not limited by the mixer). Hence, although heterodyne receivers are well adapted to high-resolution spectroscopy at all frequencies, they can be used more effectively for low-resolution spectroscopy and continuum detection as the operational frequency decreases from the infrared to the submm- to the mm-wave spectral region. With decreasing frequency, it also becomes increasingly feasible for a receiver to be fed efficiently with a diffraction-limited beam; the tolerances on optical components relax in proportion to wavelength, and the effects of atmospheric turbulence are decreased relative to the achievable angular resolution. The only infrared performance limitation discussed in Chapter 9 (besides the fundamental ones like the quantum limit and antenna theorem) that remains as an equally serious inconvenience at longer wavelengths is the difficulty in making large-scale spatial arrays of heterodyne receivers.

10.2 Mixers

One measure of the performance of a mixer is the magnitude of the quadratic deviation of its $I-V$ curve from linearity at the operating point. Recalling equation 9.1, the $I-V$ curve of a diode can be expanded in a Taylor series:

$$I(V) = I(V_0) + \left(\frac{dI}{dV}\right)_{V=V_0} dV + \frac{1}{2!}\left(\frac{d^2I}{dV^2}\right)_{V=V_0} dV^2$$

$$+ \frac{1}{3!}\left(\frac{d^3I}{dV^3}\right)_{V=V_0} dV^3 + \cdots . \tag{10.1}$$

A similar expansion is applicable to any other potential type of mixer. The mixing process is centered on the third term; a figure of merit is then

$$\Psi = \frac{d^2 I/dV^2}{2 \, dI/dV}.$$ (10.2)

Therefore, strongly nonlinear circuit elements are used as mixers. The three most widely applied classes of device are semiconductor diodes, superconductor–insulator–superconductor (SIS) junctions, and hot electron bolometers (HEBs), as described below.

10.2.1 Diode Mixers

In theory, a junction diode similar to those described in Chapter 3 could be used as a mixer. However, junction diodes have frequency response limited to ≤ 1 GHz by the recombination time required for charge carriers that have crossed the junction (for example, Sze 2000). For submm- and mm-wave receivers, it is necessary to use a device that has been optimized for high frequencies by restricting the junction area and removing the surrounding semiconductor through which charge must diffuse and recombine. Nonetheless, many of the parameters of diode mixers can be described in terms of the derivations in Chapter 3.

Suitable high-frequency diodes can be produced at a contact between a metal and a semiconductor: that is, Schottky diodes. A possible physical arrangement is shown in Figure 10.4. Schottky diodes were originally constructed by pressing a pointed metal wire, called a cat's whisker, against a piece of doped semiconductor. Modern versions of these devices are called "point-contact diodes."

More consistent performance is obtained if the contact is made by depositing metal on the semiconductor in a carefully defined geometry, using photolithographic techniques. The epitaxial layer is covered with a layer of SiO_2 insulator, a hole is etched into the insulator to expose the epitaxial layer, and a metal layer deposited over the insulator makes contact through the hole, see Figure 10.5. These devices are more easily and reliably constructed and more rugged physically than point-contact diodes, and they can have well-matched characteristics and can be fabricated together to implement balanced circuits involving two or more devices. The point-contact diodes remain relevant for high-performance, high-frequency operation because they can have lower parasitic shunt capacitance and finger inductance.

The critical doping concentration for the Schottky diode need only be maintained in a thin layer of the semiconductor below the contact. The remainder of the semiconductor can be doped to minimize the resistance in series with the diode junction. These features are illustrated in Figure 10.4, where an n-doped layer has been epitaxially grown on a heavily doped substrate and is in turn covered with an oxide layer. The metallization has been applied through a hole in the oxide layer. In the closely related Mott diode, the thickness of the doped layer

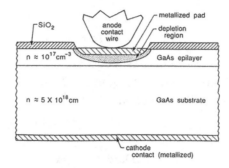

Figure 10.4 Cross-section of a point-contact Schottky diode. The wire contact is made through a metallized region on a GaAs epitaxially deposited layer.

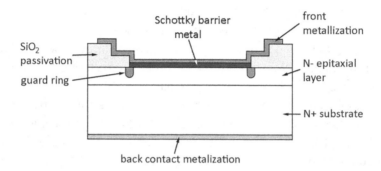

Figure 10.5 Cross-section of a Schottky diode constructed entirely through photolithography. A SiO₂ oxide layer is deposited over the epitaxial layer to isolate most of it electrically.

is adjusted so that the depletion region will just reach the conducting substrate at zero bias. Consequently, the contribution to the series resistance from the relatively low conductivity epitaxially grown layer is minimized.

Figure 10.6 shows the band diagram of such a contact, where the work function of the metal, W_m, is larger than the electron affinity of the semiconductor, χ; the semiconductor is assumed to be doped n-type. Surface charges will accumulate at the contact to equalize the Fermi levels while maintaining the contact potential. The behavior of this device under forward and reverse biases is illustrated in Figure 10.7. The height of the potential barrier seen from the semiconductor side changes with the direction of the bias, whereas the height is virtually independent of bias as viewed from the metal side. This asymmetry produces the nonlinear $I–V$ curve of the diode. A Schottky diode can also be made with a p-type semiconductor as long as $W_m < \chi + E_g$.

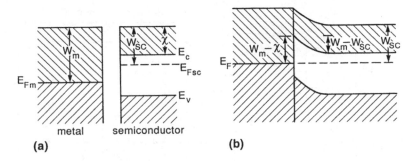

Figure 10.6 Band diagram for a Schottky diode: (a) shows the bands before contact and (b) shows the bands after contact and establishment of equilibrium.

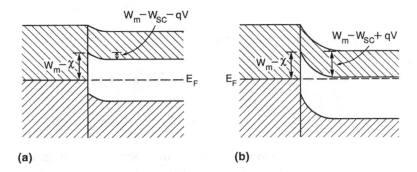

Figure 10.7 The diode action of a Schottky junction: (a) forward-biased; (b) back-biased.

The $I-V$ curve of a Schottky diode is described approximately by the diode equation (equation 3.24); any deviations from this behavior can frequently be treated by rewriting the expression as in equation 3.27:

$$I = I_0 \left(e^{qV/mkT} - 1 \right), \tag{10.3}$$

where m is called the slope parameter or "ideality factor" and I_0 is the saturation current. In most realistic situations, the calculation of the saturation current is complex; however, in the ideal case where it arises purely through thermal emission, it is given by

$$I_0 = AA^{**}T^2 e^{-\frac{q\phi_b}{kT}}, \tag{10.4}$$

where A^{**} is a constant called the modified Richardson constant, A is the diode area, and ϕ_b is the barrier height as seen from the metal side.

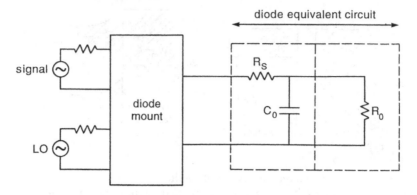

Figure 10.8　Simplified equivalent circuit for a diode mixer.

A large value of m reduces the sharpness of the knee in the diode curve, thus reducing the response of the mixer to the input field. For a diode described by equation 10.3, the figure of merit in equation 10.2 is

$$\Psi = q/2mkT. \tag{10.5}$$

An increase in m is equivalent to an increase in the operating temperature and corresponding decrease in the figure of merit (for the same I_0). Cooling the diode can improve its performance significantly as long as the ideality factor does not change. The predicted improvement is seen down to about 70 K. Below this temperature, the ideality factor begins to increase, and below about 30 K there is little performance improvement with reduced temperature. One reason for this behavior is that quantum mechanical tunneling through the diode potential barrier has less temperature dependence than the thermally driven current over the barrier and therefore dominates the behavior at low temperatures, causing the ideality factor to increase.

Complex models are used to study mixer behavior (see Note at end of chapter), but some basic properties can be derived by considering a simple equivalent circuit, such as in Figure 10.8. Here C_0 is the junction capacitance, R_S is the series resistance of the semiconductor substrate on which the diode is grown, and R_0 is the nonlinear equivalent resistance of the diode. The mixing process in a Schottky diode occurs in R_0; R_S and C_0 are parasitic elements that interfere with the operation of the mixer. As the circuit is drawn, it emphasizes that R_S and C_0 act as a low-pass filter on the mixer element and hence control the high-frequency limit of the receiver. The cutoff frequency of the mixer is

$$f_c = \frac{1}{2\pi R_S C_0}, \tag{10.6}$$

which shows that we want to minimize both R_S and C_0. Particularly for a cooled mixer, we can take $R_S \ll R_0$. From equation 10.3,

$$R_0 = \left(\frac{dI}{dV}\right)^{-1} = \frac{mkT}{qI_0\left(e^{\frac{qV}{mkT}}\right)} \approx \frac{mkT}{qI}. \tag{10.7}$$

High-performance diode mixers can have $R_0 \sim 10^{14}$ Ω, although lower values are more typical.

The capacitance, C_0, can be derived from our previous discussion of junction diodes by assuming that the metal has extremely large N_A if the semiconductor is n-type or extremely large N_D if the semiconductor is p-type. Then, from equation 3.33, the thickness of the potential barrier is

$$w = \left[\frac{2\varepsilon}{qN}(V_0 - V_b)\right]^{1/2}, \tag{10.8}$$

where N is the dopant concentration in the semiconductor, ε is its dielectric permittivity, and V_0 is the contact potential. The diode capacitance can be obtained from equations 3.34 and 10.8 to be

$$C_0 = A\left[\frac{qN\varepsilon}{2(V_0 - V_b)}\right]^{1/2}. \tag{10.9}$$

To minimize R_S, we would like to increase the doping concentration N. From equation 10.9, however, it is apparent that doing so increases the capacitance. Thus, the best we can do to control R_S is to select a semiconductor with high mobility (see equation 2.5). The mobility of GaAs is over six times that of silicon for equivalent doping levels (see Table 2.3); hence, high-frequency diodes are normally fabricated on GaAs (InP also has high mobility and is a useful alternative material to GaAs). Typical values for R_S in cooled diodes are a few hundred ohms, but values as low as ~ 20 Ω have been achieved. When Schottky diodes are not cooled, the thermal generation of charge carriers can reduce R_S sufficiently that the frequency response is limited by the electron transit time across the junction, rather than by the RC time constant of the mixer.

Given the doping concentration, C_0 can be minimized by reducing the diode area. However, from equations 3.24 and 3.25, R_0 increases as the diode area, A, decreases. A compromise that also takes fabrication difficulties into account is to make diodes with diameters of about 2 μm, although smaller diodes (down to ~ 0.25 μm) are used in the submillimeter. C_0 can be very small, $\sim 3 \times 10^{-15}$ F μm^{-1}. The resulting values of f_c are typically a few times 10^{11} Hz but can be an order of magnitude higher for devices optimized for very high frequency operation. Above f_c, the performance of diode mixers is limited by their parasitic capacitances

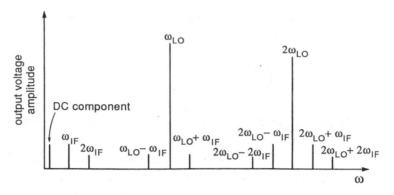

Figure 10.9 Signals present in a mixer.

and the resulting RC time constants. Thus, Schottky mixers are useful up to $\sim 3 \times 10^{12}$ Hz.

A diode mixer is an example of a classical device. The electric field of the signal drives an oscillating current through the diode, which behaves like a voltage-controlled switch that conducts whenever this current produces a voltage across it that exceeds some threshold. Examination of Figure 9.1(b) demonstrates that the action of such a switch can generate a heterodyne signal suitable for detection and smoothing, but that the conversion gain can be no more than one, or equivalently the conversion in these devices occurs with a loss in power.

To understand the diode mixer conversion loss, it is necessary to review basic mixer performance more carefully than previously. In Chapter 9, we discussed the simple situation where the output contains the input frequencies and the beat frequency. However, harmonics of the local oscillator, $n\omega_{LO}$ (where $n = 1, 2, 3, \ldots$), will also be present as well as the "sum frequency" $\omega_{LO} + \omega_S$. All of these signals can beat with each other, producing a highly complex output signal as indicated in Figure 10.9. Only the *IF* signal is useful for the detection process; power from the incoming photon stream that is used in generating signals at the other frequencies is lost and reduces the conversion efficiency of the mixer. At high operating frequencies, the mixer capacitance and other stray capacitances will tend to short-circuit the higher harmonics. The simple model in Figure 10.8 includes a low-pass filter in the form of R_S and C_0. Therefore, by far the most important unwanted signal is the image sideband at $\omega_I = \omega_{LO} - \omega_{IF}$. This signal is generated as a result of the beat of the signal at ω_S with the second harmonic of the *LO* at $2\omega_{LO}$ as well as the beat between ω_{IF} and ω_{LO}. In general, half the power is converted to the image frequency (where it is of no use in the detection process) along with half to the *IF*. Because of this behavior, diode mixers generally deliver at least a factor of 2 *less* power to the *IF* stage than the signal power into them. Expressing this in the usual way as 10 times the log of the loss factor, i.e., expressed as dB, they have a conversion loss of at least 3 dB. A more detailed derivation of this result is given in the Note at the end of this chapter. Diode mixers are useful only because they reduce the signal frequency to a range where very low noise amplifiers and other circuitry can be used.

This behavior contrasts with that of the photomixers discussed in Chapter 9. The photomixers are an example of quantum devices, where absorption of a photon frees a charge carrier. After its release, electric fields can do work on the charge carrier, making possible a mixer with conversion gain greater than unity, that is, the output power can exceed the power in signal photons incident on the mixer.

The performance of Schottky mixers is limited primarily by the modest nonlinearity in the I–V curve (or equivalently the modest value of the figure of merit, Ψ). This behavior is a fundamental result of the relatively large bandgap in the semiconductor. The consequence is that diode mixers require a large LO power to operate at low noise. Mixer types based on superconductors have much sharper nonlinearities, resulting in much larger values of the figure of merit, Ψ. Consequently, they require much lower LO powers and generally outperform Schottky mixers. However, superconducting mixers are only useful where the performance requirements demand the system-level complexities of very low temperature operation. In many applications, Schottky diode mixers are preferred because they provide good performance at conveniently accessible temperatures (including room temperature).

10.2.2 Superconductor–Insulator–Superconductor (SIS) Mixers

The second important class of mixer uses the properties of superconductors to produce a strongly nonlinear circuit element. When two superconductors are separated by an insulator thinner than the binding distance of a Cooper pair, we have an SIS, or superconductor–insulator–superconductor structure. An example is shown schematically in Figure 10.10 (and we encountered another example in the superconducting tunnel junctions discussed in Chapter 7). The most common materials for SIS junctions are niobium superconductor layers with Al_2O_3 as the insulator. The layers are deposited in a vacuum. Standard photolithographic techniques are used to confine the overlap (junction) area to be small so that the frequency response of the finished device is acceptable. The mixers are mounted in configurations defined by photolithography that include the feed antenna, as well as various features to improve the efficiency of absorption and the power transfer to the IF feed. Additional discussion of the fabrication of these devices can be found in Zmuidzinas and Richards (2004).

The current that can be carried by the SIS structure can be understood in terms of the band structure of the superconductor and the process of normal electrons

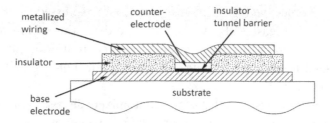

Figure 10.10 A SIS tunnel junction. In this example, the layers are sputtered onto a substrate of silicon, sapphire, or quartz. The base electrode is 200 nm of niobium, the insulating barrier 5 nm of Al_2O_3, and the counter-electrode is 100 nm of niobium. The junction area is defined by photolithography on the counterelectrode, followed by depositing ~ 200 nm of SiO or SiO_2 as an electrical insulator. A metal layer, in this example of niobium, provides electrical contact to the junction. After Zmuidzinas and Richards (2004).

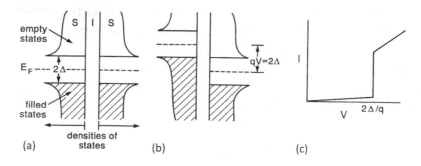

Figure 10.11 SIS junction operation: (a) shows the band diagram with no applied voltage; (b) shows the band diagram with just enough applied voltage to align the highest energy filled states on one side with the lowest energy empty ones on the other; (c) is the $I-V$ curve.

tunneling through an insulator. Referring to Figure 10.11, for simplicity we assume that the superconductors are close enough to absolute zero that the ground states are full and the excited states are empty. We are interested in the flow of electrons from left to right as the voltage across the device is increased (for example, from that in Figure 10.11(a) to that in Figure 10.11(b)). With no voltage applied across the junction, Cooper pairs can flow from one superconductor to the other to carry very small currents (the Josephson effect) because the two superconductors share the same energy states. An applied voltage will shift the energy states, but the insulator will block the flow of any additional current as long as the voltage difference is smaller than the energy gap. When the voltage just exceeds the gap, and if the insulator is thin enough, a significant current suddenly becomes possible via quasiparticle tunneling through the insulator. The current is proportional to {the density of filled states in the left side superconductor} times {the density of accessible empty states in the right side superconductor} times {the tunneling probability, P_T}. In our discussion of the superconducting tunnel junction in Section 7.4.1, we provided an expanded description of the tunneling through the insulator.

In the simplest model, $P_T = 0$ for $V < 2\Delta/q$. For $V \geqslant 2\Delta/q$, P_T can be estimated from a simple one-dimensional quantum mechanical calculation, such as the one in Section 6.4. The height of the potential barrier is determined by the bandgap of the insulating material and is typically of the order of 1 eV. The calculation then shows that reasonably large tunneling probabilities ($P_T \sim 10^{-6}$) require barrier widths of $w \sim 10^{-9}$ m, i.e., in the nm range. The derivation also shows that P_T is not steeply dependent on the particle energy as long as $E \ll U_2$ (that is, the energy of the particle is well short of allowing it to get over the barrier without tunneling). Therefore, the rate of tunneling will be controlled largely by the density-of-state terms – referring to Figure 10.11, the density of filled "valence" states in the left side superconductor and of empty "conduction" states in the right side superconductor. In particular, because the densities of permitted states tend to infinity at the edges of the bands, the tunneling current rises abruptly at the voltage difference across the device that just brings the top of the lower band in the left side superconductor to the energy level of the bottom of the upper band in the right side one. Figure 10.11(c) illustrates this behavior. The sharp inflection in the I–V curve provides the nonlinearity needed for a high-performance mixer.

A number of effects act to round off the abrupt onset of conductivity predicted in the above discussion. For example, as the temperature increases above absolute zero, the bandgap becomes smaller and its effect on the I–V curve becomes less pronounced. In addition, since the probability of excited states follows the Fermi function (equation 2.30), as the temperature increases an increasing number of electrons will be lifted into the upper energy band, leaving vacancies in the lower one. Consequently, the probability increases with temperature that tunneling will occur between states created by thermal excitation, which is possible without imposing sufficient voltage to align the lower band on one side of the junction with the upper band on the other. Finally, various other factors – for example, strain on the material, a dependence of the bandgap on direction in the crystal – can smear out the sharp density-of-state peaks at the superconducting bandgap. Nonetheless, a typical SIS mixer has a far sharper inflection in its I–V curve than does a Schottky diode. Fundamentally, this behavior is possible because the superconducting bandgap is of order 1000 times smaller than the bandgaps in semiconductors, leading to a figure of merit, Ψ, three orders of magnitude larger than for Schottky diodes. Since, to first order, the required *LO* power is inversely proportional to Ψ, SIS mixers can operate

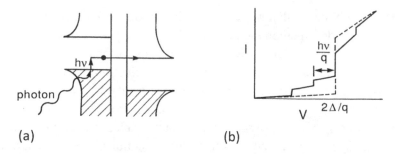

(a)   (b)

Figure 10.12 Quantum-assisted tunneling in a SIS junction: (a) shows the band diagram and (b) shows the *I–V* curve.

with much lower *LO* power than is needed for semiconductor-based devices. This behavior is particularly important at high frequencies, where it is difficult to obtain a large *LO* power from tunable oscillators.

Although the discussion in the preceding paragraph indicates some of the practical limitations, the mixer performance as measured by the figure of merit in equation 10.2 appears to improve without limit as the curvature of the *I–V* curve is made sharper. However, there is a fundamental limit to this behavior; for a sufficiently sharp inflection in the *I–V* curve, a single absorbed photon causes an electron to tunnel through the energy barrier. In this limit,

$$\Psi = \frac{q}{h\nu} \tag{10.10}$$

(Hinken 1989), where q is the charge on the electron and ν is the frequency of the absorbed photon; that is, an input power of $h\nu/\Delta t$ produces a current of $q/\Delta t$. Figure 10.12(a) shows how quantum-assisted tunneling occurs in an SIS junction and the effect on the *I–V* curve is illustrated in Figure 10.12(b). An SIS mixer operating in a regime where quantum-assisted tunneling occurs becomes a quantum device. Therefore, it can produce conversion gains greater than unity (Tucker 1979, 1980). The conditions to achieve this state are discussed by Hinken (1989). However, with the use of quiet HEMT-based *IF* amplifiers, very low noise receivers generally utilize SIS mixers with gain < 1, where more stable operation can be achieved.

The SIS junction is a classical parallel plate capacitor. An undesirable side-effect of the very thin insulator required to allow tunneling is a high capacitance. In the most common case of an Al_2O_3 insulator ($\kappa_0 \sim 10$), assuming a thickness of 3×10^{-9} m, the junction capacitance is 3×10^{-14} F μm^{-1}, an order of magnitude higher than for a Schottky diode mixer. To achieve good efficiency at high frequencies requires that the SIS junction be mounted in an arrangement that uses inductances to cancel most of its capacitance. The tuning inductance is usually

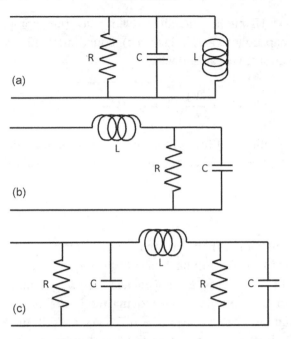

Figure 10.13 Impedance matching stripline equivalent circuits.

provided by a short section of superconducting stripline. Figure 10.13 shows equivalent circuits for some of the possible stripline configurations (for further discussion of these issues, see Shi and Noguchi 1998). The tuning inductance can be placed either in parallel (Figure 10.13(a)) or series (Figure 10.13(b)) with the SIS junction. Taking the case in Figure 10.13a as an example, the impedance looking into the mixer is given by

$$\frac{1}{Z} = \frac{1}{R} + \frac{1}{1/(j\omega C)} + \frac{1}{j\omega L}. \tag{10.11}$$

The maximum power is transferred when the impedances of the capacitor and the inductor cancel, which we have already encountered in the discussion of MKIDs (Chapter 7) to occur when

$$L = \frac{1}{\omega_0^2 C}, \tag{10.12}$$

where ω_0 is the desired angular frequency of operation, $\omega_0 = 2\pi f_0$, where f_0 is the desired electronic frequency.

Because the cancellation of the detrimental effects of the mixer capacitance depends on the resonance properties of the *RLC* circuit, it will be effective over only a limited frequency range. We return to equation 10.11, and use it to compute the real part of the impedance, $|Z| = \sqrt{Z^*Z}$, where Z^* is the complex conjugate of Z

(i.e., with $j = -j$). Using the generalized Ohm's law (i.e., complex representation of the effects of capacitance and inductance), equation 10.12, and a little patience with the manipulations, we can show that

$$|I| = \frac{|V|}{R}\sqrt{1 + Q^2\left[\frac{\omega}{\omega_0} - \frac{\omega_0}{\omega}\right]^2}, \tag{10.13}$$

where $|I|$ is the amplitude of the current. We have already encountered the quality factor $Q = \omega_0 RC$; the resonance becomes sharper as the value of Q increases. It can be taken to be

$$Q \sim \frac{\omega_0}{\Delta\omega}, \tag{10.14}$$

where $\Delta\omega$ is the full width at half maximum of the filter power response. The useful frequency bandwidth of the response of a high-frequency SIS mixer is typically about 100 GHz. In some cases, the impedance seen by the mixer can be adjusted over a modest range by mechanically moving the backshort that forms one wall of a cavity surrounding the mixer (Figure 10.3). As a result, the efficiency peak of the mixer can be adjusted over a modest additional range. Broader bandwidths can also be achieved with distributed circuits, in which the radiation is absorbed as it propagates along a long ($\sim$40 μm) SIS junction due to photon-assisted tunneling (Tong et al. 1995); such devices are not limited by the junction RC product.

    In either the parallel or serial inductance tuning circuits, additional circuit elements are required: termination with a RF short-circuit element in the first case, and an impedance-raising transformer in the second. An alternative that avoids this complication, as shown in Figure 10.13(c), is to use two identical junctions separated by a microstrip tuning inductance. A high-performance mount based on this principle is shown in Figure 10.14.

    The leads of the stripline that guide the energy to the SIS mixer are usually made of superconducting material. The lead conductance, R_S (see Figure 10.8), is therefore much smaller than for a Schottky diode, although we cannot set it quite to zero because there are likely to be some normal conductors in the current path. The small lead conductance is another reason these devices are well adapted to high frequencies as long as the SIS junction area is small enough that it does not have an overly large capacitance.

    However, another mechanism ultimately sets a cutoff to the frequency response. In addition to issues of fabrication and junction size, the superconducting material for the SIS junction must be selected to allow operation at the desired frequency. If an alternating current is applied at such a high frequency, f, that the photon energy, hf, exceeds 2Δ, then the current can break the Cooper pairs; the response of the mixer falls rapidly (as $1/f^2$) with increasing frequency above this limit. This

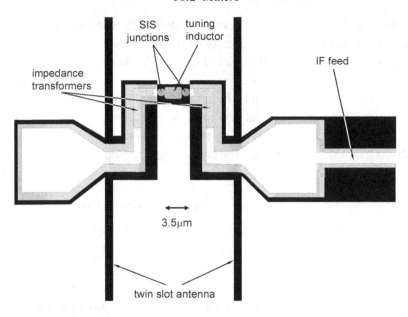

Figure 10.14 High-frequency SIS mixer with impedance matching by means of dual junctions with an inductance between them. After Zmuidzinas et al. (1998).

energy limit is characterized by a frequency $f_u = 2\Delta/h$; for example, for lead ($\Delta = 2.7$ meV), $f_u = 6.5 \times 10^{11}$ Hz. Slightly higher frequencies can be reached with niobium junctions because of their slightly larger bandgap. In principle, very high frequency mixers can operate the SIS junction over a voltage range of $\pm 2\Delta/q$, to achieve upper frequency limits approaching $2f_u$. These devices must be operated in a magnetic field to suppress Josephson tunneling near zero voltage (Walker et al. 1992); otherwise, this effect imposes an unstable operating regime, which dramatically increases the noise. A further issue is that the superconducting leads in the stripline become strongly dissipative above f_u. Improved performance above f_u can be achieved using aluminum for the SIS junction leads, even though it is a normal conductor at the temperatures of interest.

Most mixers avoid the complication of applying magnetic fields and must operate below f_u. The bandwidths of these devices are further reduced by the necessity to avoid the low-voltage regime of Josephson tunneling. High-frequency operation would then require selection of a superconductor with a larger bandgap than that of niobium. NbN and NbTiN have bandgaps corresponding to $f_u \sim 12 \times 10^{11}$ Hz. Although it has proven difficult to make high-performance mixers of the former material (e.g., Dieleman et al. 1997), the Herschel HIFI instrument uses mixers of NbTiN (Jackson et al. 2001). However, at frequencies of a THz and higher, hot electron bolometers provide competitive performance and are the device of choice, as discussed below.

Our discussion of the method of operation of the SIS junction has been brief and nonrigorous. A more complete treatment of the background physics can be found in Solymar (1972), while Hinken (1989) and Zmuidzinas and Richards (2004) provide reviews of the operation as mixers. Further exploration of the theoretical performance of these devices can be found in Tucker (1979, 1980) and Tucker and Feldman (1985). The SIS mixer is now the standard for operation whenever maximum performance in the submm- and mm-wave region (at least up to ~900 GHz) is worth the expense of providing very low temperature operation.

10.2.3 Other Quasiparticle Tunneling Mixers

A variety of other mixers operate by quasiparticle tunneling; their operational principles are therefore similar to those just discussed for SIS junctions. Two examples are superconductor–insulator–normal metal, or SIN, junctions and closely related devices where the metal is replaced by heavily doped semiconductor; these superconductor–degenerate semiconductor junctions are known as super Schottky diodes.

10.2.4 Hot Electron Bolometers

The InSb hot electron bolometer (HEB) described in Section 8.7.2 is a relatively easily constructed detector for the submm- and mm-wave regime that can be adapted for use as a mixer. In this case, the photons are absorbed directly by the mixer element, as was the case for the infrared heterodyne mixers in Chapter 9; in many ways, the bolometer mixer is more closely analogous to those devices than to the ones discussed previously in this chapter. The InSb hot electron bolometer mixers were limited to very narrow IF bandwidths by their response time constants of microseconds.

Response in the GHz range is provided by generating the hot electrons in superconductive material, as also discussed in Section 8.7.2 (refer to that section for a description of the operation of these devices). The advantage of the GHz frequency response is the broad IF bandwidth it makes possible. However, HEBs are typically operated in receivers operating at ~10^{12} Hz and up, and their response cannot follow these frequencies. The signal must therefore be concentrated onto the HEB in a fashion that results in the interference with the *LO* directly. This can be achieved with miniature antenna structures so the HEB itself can be much smaller than the wavelength of the signal. To minimize the volume of the bolometer element further, it can be made long and thin and placed across a waveguide or other feed arrangement, much as a secondary antenna might be for some other mixer type.

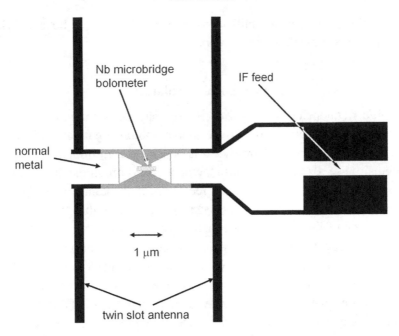

Figure 10.15 Hot electron bolometer mixer. After McGrath et al. (1999).

The good frequency response and high performance levels achieved with metal strip hot electron microbolometers make them interesting as mixers in the very high frequency submillimeter, beyond the range of SIS junctions and Schottky diodes. Figure 10.15 illustrates a typical HEB receiver. The feed arrangement is analogous to that for SIS and Schottky diode devices and automatically imposes the single-mode restrictions applicable to heterodyne operation. In the example in Figure 10.15, energy is coupled from the antenna into a sub-micron long strip of superconducting Nb extending between two pads of normal metal. The thermal link is dominated by electron diffusion cooling to these metal pads. Thermal conductances of $G = 10^{-11}$ to 10^{-13} W K^{-1} can be achieved for $T = 0.3$ to 0.1 K, yielding responsivities up to 10^9 V W^{-1} and *NEP*s estimated at below 10^{-18} W Hz$^{-1/2}$. The low *NEP*s of such hot electron microbolometers have the critical advantage of minimizing the *LO* power requirements (<100 nW) at these high frequencies. The time constants are short due to the small volume of the bolometer element (~ 1 μm^3 in volume) and resulting low capacitance.

Noise temperatures of about ten times the quantum limit have been achieved by HEB receivers uniformly from ~ 500 to 5000 GHz (0.5–5 THz). Useful *IF* bandwidths of $\gtrsim 10$ GHz have been achieved with operating frequencies up to ~ 5 THz (e.g., Novoselov and Cherednichenko 2017). As we show in Section 10.5.1, below

~800 GHz their performance falls short of SIS mixers, but above 1 THz HEB mixers offer the best available performance.

10.3 Local Oscillators

At very high frequencies, the local oscillator may be a continuous wave (CW) laser, whose output is combined with the source photon stream by optical elements and a beamsplitting diplexer. However, it is better at submillimeter and millimeter wavelengths to use a similar feed arrangement based on an electronic *LO*, which can drive a wire antenna or waveguide feed horn that radiates its output into the optical system. Electronic oscillators have the advantage that they can be tuned to the frequency of interest. High-frequency local oscillators can be built around electronic components with negative resistance; the operation of such circuits will be illustrated through the case of the commonly used Gunn oscillator. We then discuss an alternative use of traditional electronic components in a highly optimized millimeter-wave monolithic integrated circuit (MMIC) for an oscillator. In both cases frequency multiplication by electronic circuitry is often used to allow operation of the basic oscillator in an optimum frequency range and then boost its output to the very high frequencies required for submillimeter heterodyne receivers.

10.3.1 Gunn Oscillator

The energy bands in some semiconductors, such as gallium arsenide (GaAs), include a satellite band in addition to the valence and conduction bands, see Figure 10.16. The energy of this third band is higher than the normal conduction band and it is empty unless additional energy is provided to promote electrons into it. This can happen to electrons that start out below the Fermi level and are given the needed energy by applying a forward voltage to generate a strong electric field. In GaAs the mobility (or drift velocity) of the electrons in the satellite band is lower than in the conduction band. As the forward voltage increases, more and more electrons can reach the third band, causing them to move more slowly, and the current through the device decreases (despite the increasing voltage). This is contrary to the usual voltage–current relationship, e.g., $I = V/R$, and is described as the creation of a region of negative differential resistance. That is, if the electric field exceeds a critical value, the differential conductivity of the material, $dJ/d\mathcal{E}$, is negative, as is its differential resistance.

If a space charge perturbation is imposed within a semiconductor, it normally dies out exponentially. We considered just this situation in the discussion of dielectric relaxation in Chapter 2 (see equation 2.17):

$$N_p(t) = N_p(0)e^{-\frac{t}{\tau_d}}, \qquad (10.15)$$

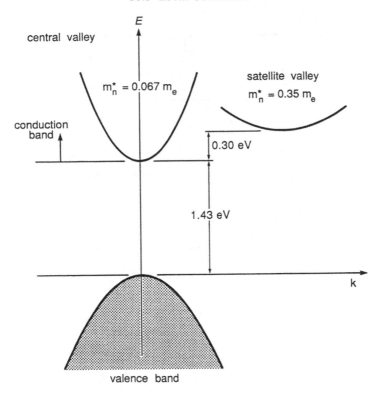

Figure 10.16 Band diagram of GaAs.

where

$$\tau_d = \frac{\varepsilon}{\sigma}. \tag{10.16}$$

From equations 10.15 and 10.16, we see a profound effect resulting from negative conductivity: a space charge fluctuation (such as might occur through random fluctuations in the charge carrier distribution) will *grow* exponentially.

A Gunn effect device is illustrated schematically in Figure 10.17,[3] which lets us follow its operation. A voltage exceeding the critical electric field is placed across the device (see Figure 10.17(a)). A crystal defect (or the negative electrode) provides a nucleation site where a dipole space charge distribution forms (see Figure 10.17(b)). This space charge distribution drifts toward the positive electrode as part of the overall current through the device, and as it drifts it grows in strength as shown in equation 10.15 – the drifting dipole zone is called a domain.

---

[3] These devices are commonly called Gunn diodes, although they consist of a length of suitable material, e.g., GaAs, between two contacts that maintain the bias; they do not contain a junction. The terminology originates in the fact that they are two-terminal devices and distinguishes them from transistors, which require three terminals.

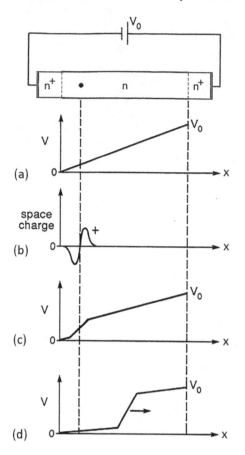

Figure 10.17 The Gunn effect. The physical arrangement is indicated at the top. The heavily doped regions (n^+) provide an ohmic connection to the cathode and anode contacts. The active region is doped n-type so electrons are the majority charge carrier. The behavior is: (a) shows the application of a voltage that exceeds the critical electric field; (b) and (c) illustrate the growth of a space charge dipole at a nucleation site and the resulting voltage profile; and (d) shows the voltage profile as this dipole drifts toward the positive electrode.

Most of the voltage drop across the device occurs at the domain, so the field in the rest of the device is reduced below the critical value, inhibiting the creation of additional domains (see Figures 10.17(c) and 10.17(d)). Once the domain reaches the positive electrode, it is collected as a current pulse and the situation returns to Figure 10.17(a), allowing the sequence to repeat. That is, a microwave oscillator can be created simply by applying a DC voltage to bias the device into its negative resistance region. A detailed description of the Gunn effect can be found in Streetman and Banerjee (2014).

The oscillation frequency is determined partly by the properties of the middle diode layer, but can be tuned by external factors. In practical oscillators, an electronic resonator is usually added to control frequency, in the form of a waveguide, microwave cavity, or YIG sphere (a YIG device uses a yttrium-iron-garnet crystal along with a variable magnetic field to provide a tunable circuit). The diode cancels the loss resistance of the resonator, so it produces oscillations at its resonant frequency. The frequency can be tuned mechanically, by adjusting the size of the cavity, or in the case of YIG spheres by changing the magnetic field. Gallium arsenide Gunn diodes are made for frequencies up to 200 GHz, gallium nitride materials can reach up to 3 THz.

The simple Gunn effect device described above could be used as an oscillator, but it would offer little flexibility in frequency of operation. It is possible to obtain greater control and to extend the performance to higher frequencies by using the negative resistance characteristics provided by this effect in a resonant circuit. In fact, in a common operating mode, *limited space charge accumulation*, the voltage across the Gunn device is varied so rapidly that domains have no opportunity to form, and only the negative resistance is of interest.

Gunn oscillators are convenient *LO* sources over roughly the 30–150 GHz frequency band, where they can produce powers of ∼100 mW. Up to 100 GHz, Gunn devices are fabricated on GaAs; operation at higher frequency can be obtained by making similar devices on InP. This latter material has a band structure like that in GaAs, but it has a shorter time constant for electrons in the central valley of the conduction band to gain or lose energy, resulting in the higher frequency of operation. Because Gunn oscillators can be tuned to the desired frequency and have a long operating life, they are a very common form of *LO*.

10.3.2 IMPATT Diode Local Oscillators

Impact avalanche transit time (IMPATT) diodes (Streetman and Banerjee 2014) can be used in a similar fashion to Gunn diodes in the construction of high-frequency oscillators. IMPATT diode oscillators can operate from roughly 30 to 300 GHz, with power outputs of ∼100 mW.

Two functions are combined in the construction of an IMPATT diode: (1) a junction that can be back-biased to provide large localized field strengths; and (2) a drift region with low doping through which charge carriers travel to be collected at an electrode. The diode is operated with a large DC back-bias, which generates a field just under breakdown across the junction. A rapidly oscillating AC voltage is imposed on top of the bias. When the AC voltage increases the back-bias, the junction breaks down due to avalanching of the charge carriers and a large pulse of charge carriers is injected into the drift region. The finite time required for this pulse to cross to the other side and reach the electrode imposes a delay on the output current pulse. The diode is manufactured with a drift region thickness so that this delay is half the period of the AC voltage oscillation. Thus, the current pulse arrives 180° out of phase with the AC voltage; the device conducts a positive-going current when the AC voltage goes negative. This behavior provides a negative resistance at the oscillating frequency, and the resulting instability can drive the AC voltage as we illustrated for the Gunn device.

Traded against their better high-frequency limit, IMPATT diode oscillators are much noisier than those using Gunn devices because their operation depends on avalanching. In addition, they are more difficult to tune in frequency because of the specific frequency-dependent adjustments that must be made in their design.

10.3.3 MMIC Local Oscillators

The development of low-noise, high-frequency HEMT-based millimeter-wave monolithic integrated circuits (MMICs) with low noise to beyond 100 GHz makes possible a different class of *LO*. Conventional tunable oscillators operate on the general principle that feedback with a phase shift can render a simple *LC* circuit unstable so it oscillates. An example is shown in Figure 10.18; this form was invented by the American engineer Edwin Colpitts. In this circuit, the AC feedback factor (voltage gain of the feedback network) is given in analogy with a resistive divider as

$$B(\omega) = Z(C_1)/(Z(C_1) + Z(L)), \tag{10.17}$$

where we use Z to indicate we are dealing with complex impedances and $\omega = 2\pi f$ where f is the frequency. Similarly, the gain of the amplifier is $A(\omega)$. The condition required for oscillation is

$$|A(\omega)B(\omega)| \geq 1 \tag{10.18}$$

and that the phase shift in the feedback be an integer times 2π, i.e., the phase shift is a multiple of 360° so the fedback signal adds to the intrinsic one – this condition is defined as positive feedback. Resonance occurs at a frequency ω_0 where the impedances within the dashed ellipse in Figure 10.18 cancel, i.e.,

$$Z(C_1) + Z(C_2) + Z(L) = 0, \tag{10.19}$$

so from equation 10.17,

$$B(\omega_0) = -\frac{Z(C_1)}{Z(C_2)} = -\frac{C_1}{C_2}, \tag{10.20}$$

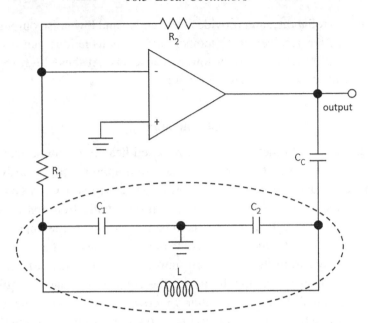

Figure 10.18 Operation of a Colpitts oscillator. The dashed ellipse surrounds a resonant *LC* circuit. The DC gain of the op-amp is controlled by the resistors R_1 and R_2. Resonance occurs at $\omega = 1/\sqrt{(LC_T)}$ where C_T is the series combination of the capacitances of C_1 and C_2.

that is, at resonance $B(\omega)$ is purely negative, equivalent to the feedback network providing 180° of phase shift; since the op-amp also provides this amount of phase shift, the condition for positive feedback is satisfied. Varying the values of the capacitances or the inductance in the resonant feedback circuit changes the oscillation frequency, allowing the *LO* to be tuned to the frequency desired.

10.3.4 Frequency Multiplication

When very high *LO* frequencies are required, the output of a Gunn or MMIC oscillator can be input to high-power diodes. The resulting waveform deviates substantially from a sine wave and hence contains significant overtones at higher frequencies than the fundamental of the oscillator. These overtones can be extracted by filtering to give a higher-frequency signal, with an efficiency of $>25\%$ per frequency doubler stage. Such a circuit can be used at a starting frequency of 10–20 GHz that is increased through multiple stages. MMIC power amplifiers are used at each multiplication stage to maintain the power level, making possible 100 mW of *LO* power around 100 Ghz. High-power diode arrays can boost the *LO* power from that frequency to above 300 GHz. Such *LO*s operate over a broad tuning range, have high reliability, and provide a very stable and low noise output that does not compromise the receiver noise temperature. This technique can also provide power up to $\gtrsim 1000$ GHz, but only low power levels. At these high frequencies, quantum cascade lasers are a useful alternative.

10.4 Backends

Modern backend spectrometers are usually based heavily on numerical methods. One backend approach that retains the conceptual approach of a hardware filter bank is to utilize discrete Fourier transforms (DFTs) to sort the signal into frequency bins. A straightforward use of DFTs in this application raises two issues. First, because the DFT calculation needs to be truncated at some upper frequency, a signal at a single pure frequency will come out with wings on the bin distribution, resulting in leakage from the correct bin into neighboring ones. Mathematically, this issue arises because setting limits on the frequency range is equivalent to convolving the signal with a rectangular window, i.e., $\Pi(x)$ in Table 1.2, which transforms to a sinc function, i.e. $\sin(\pi u)/\pi u$ so there is ringing around the central value. Second, the bins cannot be pure frequency boxes with flat response over their range; as a result, there is loss of energy between bin centers, called "scalloping loss." A "polyphase" filter bank mitigates these issues by weighting the points in the range admitted by the rectangular filter, a process somewhat analogous to apodization used in the imaging domain. Conceptually, this could be accomplished by convolving the signal with an appropriate filter function, then downsampling to limit the data to the minimum needed to represent the filtered product. Mathematically, the nth filtered value is

$$\hat{x}[n] = \sum_i x[i]h[n-i], \tag{10.21}$$

where the original data stream is represented by $x[i]$ and the filter function is $h[i]$. However, this approach involves a large number of redundant calculations. Each value of $h[i]$ in one convolution value is repeated when the calculation is advanced to the next i, except for one that is new. It would seem that downsampling first and then filtering would increase the unwanted artifacts, but it is possible to carry out this process by some reordering followed by application of the desired filter. This approach is used in a polyphase filterbank.

Another important approach is to compute the autocorrelation of the *IF* signal – the integral of the results of multiplying the signal by itself with a sequence of equally spaced delays:

$$R(\tau) = \lim_{(T \to \infty)} \frac{1}{2T} \int_{-T}^{T} U(t)U(t+\tau)dt. \tag{10.22}$$

To measure the autocorrelation, the *IF* signal is first digitized. High-speed performance is required; the Nyquist Theorem says that the digitization rate must be at least twice the IF bandwidth. Therefore, autocorrelators often digitize to only a small number of bits; the loss of information is surprisingly modest if the gains are set optimally. The Fourier transform gives the spectrum, according to the Wiener–Khinchin Theorem (Wilson et al. 2014). Because it is not possible actually to carry out the limit to infinite time in equation 10.22, as we have just discussed for DFTs autocorrelators produce spectra with ringing due to any sharp spectral features. These artifacts can be reduced by filtering the signal (e.g., with a Hanning filter), but with a loss in spectral resolution. Autocorrelators are flexible in operating parameters and very stable, since they work digitally.

With advances in high-speed computer technology, polyphase filter banks and autocorrelators require far fewer resources (both funds and maintenance) than other types of backends. They allow changes, e.g., in spectral resolution, in software rather than requiring hardware modifications. They also expedite upgrades as computer technology advances, taking advantage of the Moore's law improvements in this area. As a result, they enable a number of sophisticated heterodyne instruments such as spatial arrays of receivers that required multiple backends. For example, this capability is central to the Jansky Very Large Array (JVLA), Atacama Large Millimeter/submillimeter Array (ALMA) and the LOw Frequency ARray (LOFAR) telescope, the latter being so dependent on these approaches to merge the outputs of ~ 7000 antennae that it is sometimes described as the first "software telescope." Further development of these techniques underlies plans to construct the "Square Kilometer Array," whose main instrument will be thousands of parabolic antennae with signals merged into the equivalent of a single telescope.

10.5 Performance Characteristics

10.5.1 Summary of Achieved Mixer Noise Temperatures

In discussing the noise in submm- and mm-wave receivers, we use the concept of noise temperature developed in Section 9.3.5. Achieved receiver noise temperatures are shown in Figure 10.19. In the mm-wave region, it has been possible for some time to produce mixers with performance close to the quantum limit. Most of the loss in performance has come from the difficulties in coupling the signal into the mixer. With sophisticated computer modeling of feed arrangements, these losses have largely been eliminated in the best devices. Existing receivers approach the quantum limit to within a factor of two to three up to ~ 700 GHz. The achieved noise temperatures increase above 700 GHz, largely because conventional niobium SIS junctions and striplines operate efficiently only up to this frequency. There are,

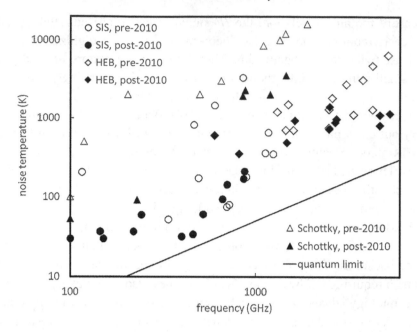

Figure 10.19 Achieved performance of high-frequency heterodyne receivers compared with the quantum limit.

however, other contributors to this behavior – increased difficulty in manufacturing waveguide feeds, for example.

10.5.2 Receiver Noise

The concept of noise temperature is readily extended to the entire data chain. It is often convenient to express the flux from a source as an antenna temperature, T_S, in analogy with the noise temperatures. This concept is particularly useful in the millimeter and radio regions, where the observations are virtually always at frequencies that are in the Rayleigh–Jeans regime ($h\nu \ll kT$). In this case, the antenna temperature is linearly related to the input flux density (see Problem 1.4):

$$\frac{P_S}{\Delta\nu} = 2k\,T_S, \tag{10.23}$$

where $\Delta\nu$ is the frequency bandwidth (for example, $2\Delta f_{IF}$ for a double sideband receiver) and we have assumed $A\Omega = \lambda^2$ from equation 9.18. T_S can be interpreted in terms of the thermodynamic temperature of a blackbody that fills the beam and produces the signal P_S. To maintain the simple formalism in terms of noise and antenna temperatures, it is conventional to use a Rayleigh–Jeans equivalent temperature such that equation 10.23 holds by definition whether the Rayleigh–Jeans approximation is valid or not (this is obviously a less useful assumption for optical and mid-infrared heterodyne systems). In this case, there is no direct correspondence with a thermodynamic temperature.

The concept of noise temperature provides a convenient means of quantifying the various sources of *LO*-independent noise, such as amplifier noise. The amplifier noise is usually expressed as the Johnson noise of an equivalent perfect resistor ("perfect" in the sense that it generates no excess noise) at a fictitious noise temperature; for example, assuming voltage noise can be ignored,

$$\langle I_A^2 \rangle = \frac{4k\, T_A\, \Delta f_{IF}}{R_A}, \tag{10.24}$$

where R_A and T_A are the amplifier input resistance and noise temperature, respectively, and $\langle I_A^2 \rangle^{1/2}$ is the amplifier noise current. A lower limit to the amplifier noise temperature is set by the quantum limit, equation 9.37, since this relation is grounded in fundamental physics that must apply generally. For example, an amplifier operating at 32 GHz has a quantum limit from equation 9.37 of 1.5 K; at this frequency the best HEMT amplifiers, even operated below 10 K, currently have noise temperatures about six times this limit (e.g., McCulloch et al. 2017).

To characterize the noise of receiver elements such as an amplifier, an imaginary resistor R_N having a resistance equal to the effective input resistance of the element is placed across its input. For example, refer to the simple equivalent circuit of the Schottky diode shown in Figure 10.8, where C_0 is from equation 10.9, R_S is the series resistance of the semiconductor substrate on which the diode is grown, and R_0 is the nonlinear equivalent resistance of the diode. Then $R_N = R_S + R_0$. The mixer noise temperature, T_M, is the temperature of R_N such that its Johnson noise (equation 2.25) would be equal to the observed mixer noise. We take R_0 from equation 10.7; we then have

$$\langle I_S^2 \rangle = 2q\, I_{LO}\, df = \langle I_J^2 \rangle = \frac{4kT_M\, df}{R} \approx \frac{4q\, I\, T_M\, df}{mT}, \tag{10.25}$$

or

$$T_M = \frac{\langle I_S^2 \rangle mT}{4q I\, df} = \frac{mT}{2}. \tag{10.26}$$

Assuming that the amplifier noise is dominant among the various *LO*-independent noise types, we can derive the *LO* power required to approach fundamental noise limits. We need $\langle I_A^2 \rangle < \langle I_{G-R}^2 \rangle + \langle I_B^2 \rangle$, or in the quantum limit (for which we take $\langle I_B^2 \rangle \to 0$),

$$P_{LO} > \frac{2k\, T_A\, L\, h\nu}{q^2 \eta\, R_A}. \tag{10.27}$$

Here, L represents the loss in transferring the output of the mixer to the amplifier.

The noise temperature of a receiver system contains contributions from the mixer, amplifiers, thermal background, and other sources. The total noise can be expressed as the sum of these temperatures all referred to the system input; hence, the temperatures of individual components must be multiplied by the appropriate losses (or divided by the gains) that occur as the signals lose energy (or are amplified) in being passed from one system component to another. In short, the system noise is

$$T_N{}^S = T_M + L_1 T_A + L_1 L_2 T_3 + \cdots , \tag{10.28}$$

where T_M is the equivalent input noise temperature of the mixer, T_A is the equivalent input noise temperature of the amplifier, and T_3 is the noise temperature of some other system component. The L's express the effective losses (or the inverses of the gains) for each noise contribution.

For example, L_1 is the loss or inverse of the gain in passing the signal from the mixer input to the first amplifier stage. Assume that the single sideband signal power into the mixer is P_S and the *IF* power out of the mixer from this input is P_{IF}. Then the conversion loss for the signal is $L_S = 1/\Gamma_{CS}$, where Γ_{CS} is the conversion gain:

$$L_S = \frac{1}{\Gamma_{CS}} = \frac{P_S}{P_{IF}}. \tag{10.29}$$

If the mixer and first-stage amplifier are coupled efficiently, L_S is equivalent to L_1 in equation 10.28. The conversion loss is often quoted in decibels (dB), where

$$L_S(\text{dB}) = 10 \, log(L_S) = 10 \, log\left(\frac{P_S}{P_{IF}}\right). \tag{10.30}$$

L_2 is the loss or inverse of the gain in passing the signal from the input of the amplifier (including the inverse of the amplifier gain) to the input of the next component, and so on. Comparing with equations 2.28 and 10.24, we see that this formalism in terms of temperatures adds noise powers, which we showed was the correct dependence in equation 2.28 and the following discussion.

The achievable signal-to-noise ratio for a coherent receiver is given in terms of antenna and system noise temperatures by the Dicke radiometer equation:

$$\left(\frac{S}{N}\right)_c = \mathbf{K}\frac{T_S}{T_N{}^S}(\Delta f_{IF}\Delta t)^{1/2}, \tag{10.31}$$

where Δt is the integration time of the observation. This relation follows from the definitions of the antenna and noise temperatures (and that the noise temperature is referred to the receiver input), and from equations 9.32 and 9.33. Details of the time response of the output stages may introduce minor corrections into the conversion to frequency bandwidth; for simplicity, these corrections are subsumed in the constant **K**, of order unity.

Although equation 10.31 seems straightforward, it is derived under the assumption that any receiver instabilities are small enough to be irrelevant. There are many potential sources of instability within the receiver: variations in the *LO* power, changes in the mixer temperature, microphonics (i.e., vibrations in receiver components), and so on. Particularly in the mm- and submm-wave regimes, where the atmosphere is far less than perfectly transparent, variations in the amount of water vapor along the line of sight also cause instabilities. The remedy for these effects is to nod the telescope periodically between the target and a nearby spot on the sky, with the largely fulfilled assumption that the causes of instability will be virtually the same for two closely adjacent positions. A metric for how often to nod is the Allan time. Equation 10.31 assumes that the noise will integrate down as the square root of the integration time. Instabilities will cause the noise to start to integrate down more slowly as time goes on, and the Allan time is when this effect is two standard deviations worse than the expectations for a perfectly stable system.

10.5.3 Test Procedures

The above discussion is all well and good, but how do we determine the performance numbers? The underlying principles for testing heterodyne receivers are similar to those discussed in Chapter 5. In those discussions, we showed that considerable care is required to specify and allow for the geometry of illumination, if we want to obtain meaningful results. In contrast, for heterodyne receivers the antenna theorem defines an appropriate geometry.

A particularly simple determination of the noise temperature of a receiver can be made with two blackbody emitters, or loads, at different temperatures that are alternately placed over the receiver input. It is assumed that the temperatures of these loads are well determined and that their emissivities are close to unity. A "*Y*-factor" is defined as

$$Y = \frac{V_{hot}}{V_{cold}}, \tag{10.32}$$

where *V* is the output voltage of the receiver and the subscripts indicate which load is over the input. Assuming that the receiver is linear, the output voltages should be proportional to the sum of the antenna temperatures of the loads and the noise temperature of the receiver, T_{rec}, or

$$Y = \frac{T_{hot} + T_{rec}}{T_{cold} + T_{rec}}. \tag{10.33}$$

Equation 10.33 can be solved for the receiver noise temperature,

$$T_{rec} = \frac{T_{hot} - Y T_{cold}}{Y - 1}. \tag{10.34}$$

Equation 10.34 assumes that the frequency is low enough that blackbody emission can be treated in the Rayleigh–Jeans regime. At high frequencies (i.e., $\gtrsim 2 \times 10^{12}$ Hz), the appropriate corrrections are embedded in the Callen–Welton temperatures (Callen and Welton 1951; Walker 2020), which should be used in place of the physical ones.

10.6 Improvements in Receiver Capabilities

10.6.1 IF bandwidth, Sideband Separating Backends

As shown in Figure 10.19, the very best noise temperatures achieved prior to 2010 were nearly as low as those achieved since, although these limiting performance levels are reached more often with the newer devices. Improvements in overall receiver performance have come in other areas. The most direct example is the improvement of the noise performance of broad-band HEMT-based amplifiers, with a resulting substantial increase in the *IF* frequency bandwidth. This by itself has revolutionized the performance of world-class radio- and mm-wave telescopes.

Another advance is the use of multiple mixers in a receiver to work around the limitations implied by the antenna theorem. To capture both polarizations in the signal, the incoming beam can be divided by a polarizer and the appropriate signals sent to two independent mixer chains, with their signals being combined after the full heterodyne process. By adding two more mixer chains, it is possible to separate the two sideband signals to eliminate the ambiguities they cause in double sideband operation. Sideband separation works by feeding the second mixer chain a local oscillator signal shifted in phase by 90° from that provided to the first mixer (or shifting the signal by 90° and leaving the *LO* unchanged), which results in complementary response to the sidebands. The standard for high-performance radio telescopes is now dual polarization, sideband separating receivers with very broad *IF* bandwidths (e.g., Perley et al. 2011; Kerr et al. 2014; Graf et al. 2015; Reddy et al. 2017; Yagoubov et al. 2019).

10.6.2 Spatial Arrays

Another important direction for performance improvements is the development of receiver arrays. We have seen that heterodyne receivers multiplex spectral information directly into their outputs. Although it would be relatively easy to construct a spatial array of mixers fed by a diplexer and local oscillator, every element in the array would require an *IF* amplifier and appropriate backend circuitry. These circuit elements are complex and not amenable to production in huge numbers. This situation is quite different from that of the infrared and CCD arrays discussed in Chapters 4 and 5, where the signals appear at distinct time intervals and the outputs of millions of detectors can be multiplexed to a simple set of output electronics. Because of their added complexities, heterodyne spatial arrays on the scale of CCDs or infrared hybrid arrays are not feasible. However, wideband digital correlators provide compact and relatively inexpensive backends, permitting the construction of modest sized spatial arrays that retain the advantage of spectral multiplexing. This topic has been reviewed by Graf et al. (2015).

10.7 Performance Comparison

Equations 10.23 and 10.31 give us the means to compare the performance of coherent and incoherent detectors, as long as we also keep in mind the bandwidth and single-mode detection restrictions that we have already discussed. From equation 10.23 and the definition of *NEP*, the signal-to-noise ratio with an incoherent detector system operating at the diffraction limit is

$$\left(\frac{S}{N}\right)_i = \frac{2kT_S\Delta\nu(\Delta t)^{1/2}}{NEP}. \tag{10.35}$$

Therefore, using equation 10.31, we obtain the ratios of signal to noise achievable with the two types of system under the same measurement conditions:

$$\frac{(S/N)_c}{(S/N)_i} = \frac{NEP\,(\Delta f_{IF})^{1/2}}{2kT_N{}^S\Delta\nu}. \tag{10.36}$$

To illustrate, consider a bolometer operating at the background limit and a heterodyne receiver in the thermal limit, both viewing a source through an etendue = λ^2 and against a background of unity emissivity at a temperature of T_B. Assume we are observing at a frequency $h\nu \ll kT_B$. Then it can be shown that

$$\frac{(S/N)_c}{(S/N)_i} = \left[\left(\frac{1}{\eta}\right)\left(\frac{\Delta f_{IF}}{\Delta\nu}\right)\left(\frac{h\nu}{kT_B}\right)\right]^{1/2}. \tag{10.37}$$

In this case, the incoherent detector yields better signal to noise unless $\Delta f_{IF}/(\eta\,\Delta\nu) \gg 1$, for example where measurements are being made at spectral resolution significantly higher than the *IF* bandwidth and the incoherent detector must be operated in a very narrow band.

On the other hand, where the bolometer is detector noise limited and the hetero-dyne receiver operates at the quantum limit,

$$\frac{(S/N)_c}{(S/N)_i} = \frac{NEP(\Delta f_{IF})^{1/2}}{2h\nu\Delta\nu}. \tag{10.38}$$

Particularly if the spectral resolution $\nu/\Delta\nu$ is kept constant (as is typical), the transition from the case favoring incoherent to that favoring coherent detectors occurs abruptly with decreasing frequency in this case because the figure of merit in equation 10.38 goes as $1/\nu^2$. Further gains in spectroscopy result from the ability of a heterodyne receiver to obtain more than one spectral measurement at a time by dividing its *IF* output into a bank of narrowband electronic filters or equivalent backend device (spectral multiplexing).

As an example, we compare the operation of a bolometer operating at the background limit and a heterodyne receiver operating at the quantum limit. We set the bolometer field of view at the diffraction limit, $A\Omega = \lambda^2$. Assume the background is in the Rayleigh–Jeans regime (e.g., thermal background at 270 K observed near 1 mm). The background limited *NEP* of the bolometer is

$$NEP = \frac{hc}{\lambda}\left(\frac{2\varphi}{\eta}\right)^{1/2}. \tag{10.39}$$

The photon incidence rate, φ, can be shown to be

$$\varphi = \frac{2\eta k\, T_B\, \Delta\nu}{h\nu}. \tag{10.40}$$

If we assume the bolometer is operated at 25% spectral bandwidth, that is $\Delta\nu = 0.25\,\nu$, then

$$\frac{(S/N)_c}{(S/N)_i} = \frac{2.4 \times 10^6}{\nu}\,(4\,\Delta f_{IF})^{1/2} \sim \frac{2.6 \times 10^{11}}{\nu}. \tag{10.41}$$

We have taken the *IF* bandwidth to be 3×10^9 Hz, a typical value.

Thus, the incoherent detector (bolometer) is more sensitive at frequencies above 2.6×10^{11} Hz (or at wavelengths shorter than about 1 mm). Actually, this comparison is slightly unfair to it (since, for example, it does not have to work at the diffraction limit, accepts all polarizations, and can more readily be constructed in small arrays), so it is usually the detector of choice for continuum detection to wavelengths as long as 2 or 3 mm. Of course, coherent detectors are preferred for high-resolution spectroscopy.

10.8 Example

Consider a heterodyne receiver operating between 10 and 11 μm, viewing a background of 290 K, and using a HdCdTe diode photomixer with the following properties:

cutoff wavelength = 15 μm,
quantum efficiency = 40% at wavelengths short of 13 μm,
size = 100 μm diameter,
depletion region width at operating bias = 1 μm,
reverse bias impedance = 150 Ω (due to electron–hole generation in the depletion region, there is no well-defined I_0, but a relatively constant effective resistance),
dielectric constant = 10.

The local oscillator is a CO_2 laser illuminating the photodiode by reflection off a diplexer with 10% reflectivity (the diplexer is built to transmit 90% of the incoming signal to maximize signal to noise). There are roughly 50 strong laser lines in the 10 to 11 μm region. Let the output of the photomixer go to an amplifier with an input impedance of 150 Ω, a gain of 100, a bandwidth of 3×10^9 Hz, and an output current noise of 33.2 μA $Hz^{-1/2}$. Viewing a 500 K source, the receiver output is 1.040 V, while viewing a 300 K source it is 1.000 V. Answer the following: (a) what is the *IF* bandwidth of the receiver; (b) will the receiver operate in the thermal or quantum limit; (c) what is the noise temperature of the amplifier; (d) what laser power is required to overcome amplifier noise; (e) what is the expected system noise temperature; and (f) what is the probability that the receiver can operate at an arbitrary frequency in the 10 to 11 μm band?

(a) The *IF* bandwidth will be limited by the amplifier or by the *RC* timeconstant of the mixer. To estimate the *RC* time constant, we use equation 3.34,

$$C = \frac{A\kappa_0\varepsilon_0}{w}, \tag{10.42}$$

where w is the depletion width and A the area of the diode, and κ_0 is the dielectric constant. Substituting, we find $C = 6.95 \times 10^{-13}$ F, $RC = 1.043 \times 10^{-10}$ s $= \tau$, and a cutoff frequency of

$$f_c = \frac{1}{2\pi\tau} = 1.5 \times 10^9 \text{Hz}. \tag{10.43}$$

Since this frequency is less than the bandwidth of the *IF* amplifier, the achievable *IF* band will be limited to the photomixer response. In the following, we assume the *IF* amplifier bandwidth is also limited to this value to avoid excess noise.

(b) At 10.5 μm, $v = 2.855 \times 10^{13}$ Hz. At the ambient background temperature of 290 K, $hv/kT = 4.7 > 1$, so we expect to operate in the quantum limit. A more precise determination can be made from equation 9.32 (with a little rearrangement); the quantum limit is obtained if

$$\frac{2\eta\varepsilon}{e^{hv/kT} - 1} \ll 1. \tag{10.44}$$

The left side of this expression is $\sim 0.007\varepsilon$, and since $\varepsilon \leq 1$, it is confirmed that we should be in the quantum limit.

(c) To convert the output noise of the amplifier to an input-referred noise, we divide by the gain to get the input-referred current noise $\langle I_A^2 \rangle^{1/2} = 3.32 \times 10^{-7}$ A Hz$^{-1/2}$. We also have $\Delta f_{IF} = 1.5 \times 10^9$ Hz and $R_A = 150$ Ω. From equation 10.24, we get

$$T_A = \frac{\langle I_A^2 \rangle R_A}{4k\Delta f_{IF}} = 200 \text{ K}. \tag{10.45}$$

(d) The *LO* power required for quantum-limited operation can be obtained from equation 10.27. Since the output impedance of the diode matches the input impedance of the amplifier, $L = 1$ and we have $\eta = 0.4$. Substituting, we need a local oscillator power of $\gg 0.068$ mW to overcome the amplifier noise. Since only 10% of the laser energy is reflected by the beamsplitter, the required laser power in the *LO* line is $\gg 0.68$ mW.

(e) For the mixer, we obtain the quantum limit noise temperature from equation 9.36 as

$$T_N = \frac{hv}{k \, ln(1 + 0.4)} = 2.97\frac{hv}{k} = 4073 \text{ K}. \tag{10.46}$$

Assume this mixer has a conversion gain of 2.34. Then L_1 in equation 10.28 $= 1/2.34 = 0.427$. From equation 10.28 with $L = 0.427$ and $T_A = 200$ K, the system noise temperature is expected to be 4158 K.

(f) The total frequency band between 10 and 11 μm is 2.73×10^{12} Hz. If each laser line allows a band of $\pm\Delta f_{IF}$ about the line, or a total of 3×10^9 Hz, the receiver can access a total frequency range of 1.5×10^{11} Hz or 5.5% of the 10 to 11 μm region. The probability of reaching an arbitrary frequency is 0.055.

10.9 Problems

10.1 Compare the signal to noise achievable at a wavelength of 300 μm on a continuum source with (1) a helium-3 cooled bolometer with an electrical *NEP* of 2×10^{-16} W Hz$^{-1/2}$ and a quantum efficiency of 0.53, operated through a spectral band of 30% of the center frequency; and (2) a heterodyne receiver with a single sideband noise temperature of 1500 K, an *IF* bandwidth of 3×10^9 Hz, and operated double sideband at the same frequency as the bolometer. At what spectral bandwidth would the two systems give equal S/N?

10.2 Assuming an insulator thickness of 2×10^{-9} m and an effective series resistance of $R_S = 50 \, \Omega$, taking the dielectric constant of Al_2O_3 to be 10, compute the frequency cutoff of a $Nb-Al_2O_3-Nb$ SIS mixer with an area of 1 µm on a side.

10.3 For the mixer in Problem 10.2 and a stripline feed as in Figure 10.13(b), determine the parameters of the feed to optimize the performance at 300 GHz.

10.4 Consider a Schottky diode mixer similar to that in Figure 10.4, with a diameter of 2 µm and a thickness for the epitaxial GaAs layer of 0.5 µm. Assume it is operated at room temperature and at a voltage of $(V_0 - V_b) = 5$ V. Take the mobility of the GaAs to be as in Table 2.3 and the saturation drift velocity of electrons in this material to be 10^7 cm s^{-1}. Determine the cutoff frequency and what mechanism determines it.

10.5 You have just taken delivery of a SIS receiver operating at 500 GHz, which the manufacturer claims is at the state of the art for noise temperature. You determine that its Y-factor with the hot load at room temperature and the cold load at liquid nitrogen temperature (77 K) is 2.5. Should you ask for your money back?

10.6 At one point there was a large effort to build a telescope with an aperture of 25 meters at the same site as ALMA, which has 54 12-meter telescopes, i.e., 13 times the area, combined via aperture synthesis into an equivalent 90-meter aperture. Both telescopes would operate at the same submillimeter wavelengths (e.g., 350 µm). Why was the 25-meter telescope of any interest?

10.10 Note

Sophisticated computer programs are now used for modeling mixer performance. Ward et al. (1999) describe the motivation for one such program: "Although there are many excellent software packages available to aid in the design of microwave circuits, none provide the specialized elements needed for the design of complete superconducting tunnel junction (SIS) receivers. For instance, thin film microstriplines, which are widely used in SIS mixers for impedance matching circuits, have characteristics which depend on the surface impedance if the normal or superconducting metal films. Surface impedance calculations usually involve nontrivial numerical computations, such as numerical integration or the solution of integral equations, and are not available in commercial microwave software packages. Furthermore, the calculation of the signal and noise properties of SIS mixers requires the use of Tucker [and Feldman]'s theory (1985) combined with a nonlinear harmonic-balance calculation of the local oscillator waveform. Again, the required calculations are numerically intensive and are not available in commercial packages. It is clear that a complete simulation of an SIS mixer is a substantial computational task. Because of this, SIS mixer design is usually performed using simplifying approximations."

Although the details can be found in Ward et al. (1999) and similar references, the basic issues are illustrated perhaps more clearly by considering the approach to such modeling prior to when computers made the job "easy." We therefore follow Torrey and Whitmer (1948) (hereafter TW) in calculating the conversion loss of a diode mixer. The reader is referred to TW for the full derivation.

TW consider the conversion loss for the single sideband case, L_S. They adopt the simplification that the mixer capacitance and other stray capacitances short-circuit the higher harmonics and hence that electrical signals need be considered at three frequencies, ω_S, ω_I, and ω_{IF}, corresponding respectively to the signal, the image, and the *IF*. They further assume that the signals at these frequencies are sufficiently small that the relevant voltages and currents can be related linearly, so the mixer can be regarded as a linear network with separate terminals for each of these three frequencies. The terminals then are connected to each other through the network by a set of complex impedances; for example, $Z_{S,I}$ is the impedance between the signal and image terminals.

If the complex current amplitudes at the three frequencies are designated by i_S, i_I, and i_{IF} and the complex voltage amplitudes by e_S, e_I, and e_{IF}, then by the assumption of linearity, the currents and voltages are related by three simultaneous linear equations:

$$i_S = y_{S,S}\, e_S + y_{S,IF}\, e_{IF} + y_{S,I}\, e_I^{\,*},$$
$$i_{IF} = y_{IF,S}\, e_S + y_{IF,IF}\, e_{IF} + y_{IF,I}\, e_I^{\,*}, \qquad (10.47)$$
$$i_I^{\,*} = y_{I,S}\, e_S + y_{I,IF}\, e_{IF} + y_{I,I}\, e_I^{\,*},$$

where the y's are the complex admittances, defined as the inverse of the impedance, $y = 1/Z$, and the * indicates complex conjugation (see TW for an explanation of why the image current and voltage appear as complex conjugates). It is often convenient to use the formalism of linear algebra for analysis of mixer performance, in which case equations 10.47 are expressed in matrix form:

$$\begin{vmatrix} i_S \\ i_{IF} \\ i_I^{\,*} \end{vmatrix} = Y \begin{vmatrix} e_S \\ e_{IF} \\ e_I^{\,*} \end{vmatrix}, \qquad (10.48)$$

where Y is the admittance matrix,

$$Y = \begin{vmatrix} y_{S,S} & y_{S,IF} & y_{S,I} \\ y_{IF,S} & y_{IF,IF} & y_{IF,I} \\ y_{I,S} & y_{I,IF} & y_{I,I} \end{vmatrix}. \qquad (10.49)$$

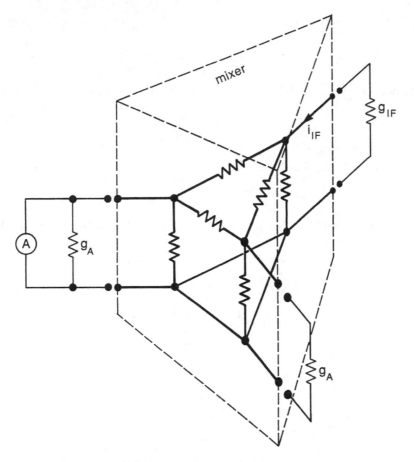

Figure 10.20 Linear network model of a diode mixer.

It is possible to simplify equations 10.47 and replace the nine complex admittances with five real conductances:

$$i_S = g_{S,S}e_S + g_{S,IF}e_{IF} + g_{S,I}e_I{}^*,$$
$$i_{IF} = g_{IF,S}e_S + g_{IF,IF}e_{IF} + g_{IF,S}e_I{}^*, \qquad (10.50)$$
$$i_I{}^* = g_{S,I}e_S + g_{S,IF}e_{IF} + g_{S,S}e_I{}^*.$$

According to equations 10.50, the mixer can be treated as a resistor network as illustrated in Figure 10.20. TW show how all the relevant mixer equivalent network resistances can be measured in the laboratory. If we attach a source of alternating current with amplitude A and conductance g_A to the signal terminals of the mixer, then the available signal power, defined as the power delivered into a matched load, is

$$P_S = \frac{A^2}{8g_A}. \qquad (10.51)$$

The power at the output will be

$$P_{IF} = \frac{i_{IF}^2}{8g_{IF}}, \tag{10.52}$$

and the conversion loss will be given as in equation 10.30 as P_S/P_{IF}. TW show that

$$g_{IF} = g_{IF,IF} - \frac{2g_{S,IF}g_{IF,S}}{g_{S,S} + g_{S,I} + g_A}, \tag{10.53}$$

and

$$i_{IF} = \frac{Ag_{IF,S}}{g_{S,S} + g_{S,I} + g_A}. \tag{10.54}$$

The conversion loss is the ratio of signal power into the mixer to *IF* power out of it:

$$L_S = \frac{P_S}{P_I F} = \frac{2g_{S,IF}}{g_{IF,S}} f(x), \tag{10.55}$$

where

$$f(x) = \frac{(x+a)(x+a-b)}{bx}, \tag{10.56}$$

and

$$a = 1 + \frac{g_{S,I}}{g_{S,S}},$$

$$b = \frac{2g_{S,IF}g_{IF,S}}{g_{S,S}g_{IF,IF}}, \tag{10.57}$$

$$x = \frac{g_A}{g_{S,S}}.$$

So long as $0 < b < a$, $f(x)$ has a unique minimum at

$$x_0 = [a(a-b)]^{1/2}, \tag{10.58}$$

and the minimum conversion loss is

$$L_S(min) = 2\frac{g_{S,IF}}{g_{IF,S}} \frac{1 + (1 - b/a)^{1/2}}{1 - (1 - b/a)^{1/2}}. \tag{10.59}$$

The smallest loss occurs when $a = b$. It is usually true that $g_{S,IF} = g_{IF,S}$, in which case the loss can never be less than 2:

$$L_S(min) \geq 2 \rightarrow 3\text{dB}. \tag{10.60}$$

This loss occurs because half the power is converted to the image frequency, along with half to the *IF*. Because of this behavior, useful Schottky diode mixers always deliver *less* power to the *IF* stage than the signal power into them. With large *LO* powers, Schottky diode mixers can be driven very close to the minimum 3 dB loss.

10.11 Further Reading

Goldsmith (1998) – thorough, definitive discussion of quasi-optical methods

Graf et al. (2015) – overview of heterodyne arrays

Maas (1993) – overview of high-frequency mixers and techniques in a variety of applications

Pozar (1998) – a standard textbook on microwave techniques

Siegel (2002) – terahertz technology applications, sensors, and sources are briefly reviewed

Sizov (2018) – review of modern terahertz detectors

Walker (2020) – an excellent introduction to terahertz astronomy and useful for easy-to-understand explanation of terahertz technology in general

Wild (2013) – approachable overview of far infrared and submillimeter heterodyne techniques

Wilson et al. (2014) – classical overview of radio astronomy techniques

Appendix A

Useful Constants and Conversions

Fundamental Constants

k	Boltzmann's constant	1.381×10^{-23} J K^{-1}
		8.617×10^{-5} eV K^{-1}
q	electronic charge	1.602×10^{-19} C
h	Planck's constant	6.626×10^{-34} J s
c	speed of light	2.998×10^{8} m s^{-1}
ε_0	permittivity of free space	8.854×10^{-12} F m^{-1}
m_e	rest mass of electron	9.109×10^{-31} kg
σ	Stefan–Boltzmann constant	5.669×10^{-8} W m^{-2} K^{-4}

Conversions and Relations

1 angstrom (1Å) $= 10^{-10}$ m

1 eV $= 1.602 \times 10^{-19}$ J

Mobility μ(m^2 V^{-1} s^{-1}) $= 10^{-4} \, \mu$(cm^2 V^{-1} s^{-1})

Density n(m^{-3}) $= 10^6 \, n$(cm^{-3})

Absorption coefficient a(m^{-1}) $= 100 \, a$(cm^{-1})

$\varepsilon = \kappa_0 \varepsilon_0$, where ε is the dielectric permittivity and κ_0 the dielectric constant.

$D = \mu kT/q$, where D is the diffusion coefficient and μ is the mobility.

D (m^2 s^{-1}) $= 10^{-4} \, D$ (cm^2 s^{-1})

$n_o p_o = n_i^2 = p_i^2$, where n_o and p_o are the free electron and hole concentrations in doped material, and n_i and p_i are the free electron and hole concentrations in the corresponding intrinsic material.

Appendix B

Answers to Selected Problems

1.1 (a) 1 μm: $L = 5.845 \times 10^{-28}$ W m^{-2} Hz^{-1} ster^{-1}
   10 μm: $L = 3.307 \times 10^{-12}$ W m^{-2} Hz^{-1} ster^{-1}
   (b) 1 μm: $E = 1.836 \times 10^{-35}$ W m^{-2} Hz^{-1}
   10 μm: $E = 1.039 \times 10^{-19}$ W m^{-2} Hz^{-1}
   (c) 1 μm: $P = 8.647 \times 10^{-27}$ W
   10 μm: $P = 4.894 \times 10^{-12}$ W
   (d) 1 μm: $N = 4.354 \times 10^{-27}$ s^{-1}
   10 μm: $N = 2.464 \times 10^{8}$ s^{-1}

1.2 At 2 μm, 1.10×10^{-10} W
   At 20 μm, 1.25×10^{-11} W

1.4 $\lambda_{max} = 30$ μm; Rayleigh–Jeans is within 20% for $\lambda > 40.7$ μm.

1.7 The noise correction is less than 10% for $\lambda < 11.5$ μm. The peak of the blackbody occurs at about 3 μm (from the relation derived in Problem 1.3).

1.8 $F(u) = (1/2)(\delta(u) - j/\pi u) + (\pi/10)sech(\pi^2 u/10)$

2.2 $S = 0.040$ A W^{-1}
   $\eta G = 0.050$

2.3 $DQE = 0.046$
   $G = 1.06$

2.7 For $T_d = 4$ K, $R_2 = 0.97 \times 10^6$ Ω, $S/N = 1$
   For $T_d = 77$ K, $R_2 = 0.73 \times 10^6$ Ω, $S/N = 0.91$
   For $T_d = 300$ K, $R_2 = 0.50 \times 10^6$ Ω, $S/N = 78$

2.7 (a) $\lambda_{max} = 0.52$ μm, or 2.39 eV:　　AlP ($E_g = 2.45$ eV)
　　　　　　　　　　　　　　　　　　　　　AlAs (2.16 eV)
　　　　　　　　　　　　　　　　　　　　　GaP (2.26 eV)
　　　　　　　　　　　　　　　　　　　　　CdS (2.42 eV)
　　　　　　　　　　　　　　　　　　　　　ZnTe (2.25 eV)
　　　　　　　　　　　　　　　　　　　　　AgBr (2.81 eV)
　　(b) $\lambda_{max} = 5.0$ μm, or 0.25 eV:　　InSb (0.18 eV)
　　　　　　　　　　　　　　　　　　　　　PbSe (0.27 eV)
　　　　　　　　　　　　　　　　　　　　　PnTe (0.29 eV)

(c) $\lambda_{max} = 21$ μm, or 0.06 eV: no intrinsic material, but extrinsic ones such as Si:B, Si:As, Si:P, or Si:Sb with λ_c from 23 to 29 μm.

3.3 $[(\phi\eta\, q)/(2I_0)]^{1/2} \geq 5$

3.8 (a) Detector thickness = 450 μm
(b) $G = 0.79$
(c) $S = 6.6$ A W^{-1}
(d) Assuming $\delta = 1$, $R_d = 4.3 \times 10^{11}$ Ω
(e) Detector time constant is 0.1 s

4.1 (a) Signal to noise of 1 with the TIA requires 723 photons s^{-1}
(b) S/N is larger by a factor of 8.45 compared with TIA
(c) S/N is larger by a factor of 17 compared with TIA; note, however, that it would be higher still if we sampled faster.

4.2 $V_b = 0$ V: 1.88×10^{-9} V electron^{-1}
$V_b = -1$ V: 3.11×10^{-9} V electron^{-1}
$V_b = -2$ V: 3.97×10^{-9} V electron^{-1}

4.4 Node capacitance = 0.5 pF
Amplifier read noise = 94 electrons rms

4.5 $\tau = 0.8\Delta t$

5.2 Need to use a read strategy that avoids kTC noise; (2) must slow readout down from 0.2 to 0.34 s.

5.3 0.4%

5.5 26%

6.2 The pulse counting will yield S/N of ten times that from current measurement.

6.3 $T < 190$ K, or $T < -83\,°C$

6.4 Gain = 3. It makes the light three times brighter, but also increases the noise by about a factor of 3; for most applications, it is not of much use.

7.1 (a) About 1000 cm s^{-1}; (b) 7 cm s^{-1}; and (c) 1.17×10^7 cm s^{-1} and 4770 cm s^{-1}

7.3 4.1 nH

7.4 0.85 pF, 2.7 GHz, 69,000, and 80 kHz

7.6 The sky brightness provides about 100 detections per second, and to avoid pileup one would need to read out ten times faster than this. so to maintain the clocking rate of the standard CMOS arrays would require about 10,000 outputs (!!!!!). In addition, the very high input impedance of the CMOS amplifiers would be difficult to match well to the STJs, and their power dissipation would pose a big problem in keeping them cold enough.

8.1 Bolometers do not show the dependence of performance on detector area that D^* is based upon. Their performance is based on heat capacity and thermal conductance.

8.4 Tunneling should become important for gallium concentrations of 10^{15} cm^{-3} and higher.

8.5 The semiconductor temperature sensors are much more forgiving of fluctuations in the temperature of the bolometer heat sink. In fact, tests of this behavior can be found in Holmes et al. (2008).

8.6 The time response will remain the same because both the heat capacity and the thermal link scale with temperature. However, the thermal limit to the *NEP* will go up more than an order of magnitude due to both the higher temperature and the stronger thermal link. It might be possible to bring the thermal link down to preserve the time response, with heroic efforts to reduce the heat capacity. However, this type of bolometer is clearly not capable with any plausible modifications of achieving a similar *NEP* at 0.3 K.

10.1 $(S/N)_c/(S/N)_I = 0.0033$. The S/N's are equal for $\Delta \nu = 1 \times 10^9$ Hz.

10.4 Frequency response is limited by transit time to $\sim 6 \times 10^{10}$ Hz.

10.6 The 25-meter telescope would use an array of incoherent detectors (e.g., MKIDs) for large field imaging. At 350 μm, each of these detectors would be more sensitive than the aperture synthesis array because the aperture synthesis is subject to the quantum noise limit. Also, of course, a large array is not technically feasible for the aperture synthesis telescopes.

References

Aadit, M. N. A., Kirtania, S. G., Afrin, F., Alam, Md. K., and Khosru, Q. D. M. (2017). *High electron mobility transistors: performance analysis, research trend, and applications*, London: IntechOpen.

Abergel, A., Miville-Deschênes, M. A., Désert, F. X., et al. (2000). 'The transient behaviour of the long wavelength channel of ISOCAM,' Exp. Ast., **10**, 353–368.

Adami, O.-A., Rodriguez, L., Reveret, V., et al. (2018). 'Characterization of doped silicon thermometers for very high sensitivity cryogenic bolometers,' J. Low Temp. Phys., **193**, 415–421.

American Institute of Physics (1972). *Handbook*, 3rd edn. New York: McGraw-Hill.

Andersson, J. Y., and Lundqvist, L. (1992). 'Grating-coupled quantum-well infrared detectors: theory and performance,' J. Appl. Phys, **71**, 3600–3610.

Annunziata, A. J., Santavicca, D. F., Frunzio, L., et al. (2010). 'Tunable superconducting nanoinductors,' Nanotech., **21**, 445202-1–6.

Aull, B. F., Duerr, E. K., Frechette, J. P., et al. (2018). 'Large-format geiger-mode avalanche photodiode arrays and readout circuits,' IEEE J. Sel. Top. Quant. Elec., **24**, 1–10.

Bahl, I. J. (1989). 'Transmission lines,' in *Handbook of microwave and optical components*, ed. K. Chang. New York: Wiley, **1**, 1–59.

Bai, Xiaogang, Yuan, Ping, McDonald, P., et al. (2012). 'Development of low excess noise SWIR APDs,' Proc. SPIE, **8353**, 83532H.

Bardeen, J., Cooper, L. N., and Schrieffer, J. R. (1957). 'Theory of superconductivity,' Phys. Rev. **108**, 1175–1204.

Baselmans, J. (2012). 'Kinetic inductance detectors,' J. Low Temp. Phys., **167**, 292–304. Excellent review of MKIDs.

Beck, J. D., Kinch, M., and Sun, Xiaoli (2014), 'Update on linear mode photon counting with the HgCdTe linear mode avalanche photodiode,' Opt. Eng., **53**, 081906-1–6.

Beeresha, R. S., Khan, A. M., and Manjunath Reddy, H. V. (2016). 'Design and optimization of interdigital capacitor,' Int. J. Res. Eng. Tech., **5**, 73–78.

Benford, D J., Allen, C. A., Chervenak, J. A., et al. (2000). 'Superconducting bolometer arrays for far infrared and submillimeter astronomy,' in *Imaging at Radio through Submillimeter Wavelengths*, ed. J. G. Mangum and S. J. E. Radford. ASP Conference Series, Vol. 217, 134–139.

Berggren, K. K., Dauler, E. A., Kerman, A. J., Sae-Woo, Nam, and Rosenberg, D. (2013). 'Detectors based on superconductors,' in *Single-Photon Generation and Detection*,

Vol. 45, ed. A. Migdall, S. V. Polykov, J. Fan, and J. C. Bienfang. Waltham, MA: Academic Press, 185–216.

Bielecki, Z. (2004). 'Readout electronics for optical detectors.' Opto-Elec. Rev., **12**, 129–137.

Billot, N., Agnése, P., Auguéres, J.-L., et al. (2006). 'The Herschel/PACS 2560 bolometers imaging camera,' Proc. SPIE, **6265**, 62650D - 62650D-12.

Blaney, T. G. (1975). 'Signal-to-noise ratio and other characteristics of heterodyne radiation receivers,' Space Sci. Rev., **17**, 691–702.

Boreman, Glenn D. (2001). *Modulation transfer function in optical and electro-optical systems*. Bellingham, WA: SPIE Publications. Relatively concise general reference on *MTF*, starting from first principles.

Born, M., Wolf, E., and Bhatia, A. B. (1999). *Principles of Optics*, 7th edn. Cambridge; New York: Cambridge University Press.

Bounissou, S., Revéret, V., Rodriguez, L., et al. (2018). 'Electromagnetic simulations of newly designed semiconductor bolometers for submillimeter observations,' J. Low Temp. Phys., **193**, 428–434.

Boussaha, F., Kawamura, J., Stern, J., et al. (2012). 'Terahertz-frequency waveguide HEB mixers for spectral line astronomy,' Proc SPIE, **8452**, 845211-1-7.

Boyd, R. W. (1983). *Radiometry and the Detection of Optical Radiation*. New York: Wiley. An extensive treatment of radiometry with general description of optical and infrared detectors and thorough discussion of noise mechanisms.

Bracewell, R. N. (2000). *The Fourier Transform and its Applications*, 3rd edn. Boston: McGraw-Hill. Comprehensive and standard reference on Fourier transforms.

Bratt, P. R. (1977). 'Impurity germanium and silicon infrared detectors,' in *Semiconductors and, Semimetals*, ed. R. K. Willardson and A. C. Beer. New York: Academic Press, **12**, 39–142. A classic review of the operation of photoconductors.

Brown, S. W., Eppeldauer, G. P., and Lykke, K. R. (2006). 'Facility for spectral irradiance and radiance responsivity calibrations using uniform sources,' App. Opt., **45**, 8218–8237.

Buckingham, M. J. (1983). *Noise in Electronic Devices and Systems*. New York: Ellis Horwood. A clear overview of noise mechanisms and behavior in a broad variety of electronic devices, ranging from linear networks through SQUIDs and gravitational wave detectors.

Budde, W. (1983). *Physical Detectors of Optical Radiation*, ed. F. Grum and C. J. Bartleson. New York: Academic Press.

Buil, C. (1991). *CCD Astronomy: Construction and Use of an Astronomical CCD Camera*. Richmond, VA: Willmann-Bell. Description of use of CCD detectors, from basics of operation through hardware and software and image reduction and analysis. Translated from the original French by E. and B. Davoust.

Bukshtab, M. (2019). *Photometry, Radiometry, and Measurements of Optical Losses*, 2nd edn. Heidelberg; New York: Springer. Massive, advanced treatment of photometry and radiometry.

Burke, B. E., Reich, R. K., Savoye, E. D., and Tonry, J. L. (1994). 'An orthogonal-transfer CCD imager,' IEEE Trans. Elec. Dev., **41**, 2482–2484.

Burke, B., Jorden, P., and Vu, P. (2005). 'CCD technology,' Exp. Ast., **19**, 69–102. Comprehensive review of CCDs, including comparison with CMOS detector arrays.

Burke, P. J., Schoelkopf, R. J., Prober, D. E., et al. (1996). 'Length scaling of bandwidth and noise in hot-electron superconducting mixers,' App. Phys. Let., **68**, 3344–3346.

Cabrera, M. S., McMurtry, C. W., Forrest, W. J., et al. (2020). 'Characterization of a 15 μm cutoff HgCdTe detector array for astronomy,' J. Ast. Tel., Inst., Sys., **6**, ID 011004.

Callen, H. B., and Welton, T. A. (1951). 'Irreversibility and generalized noise,' Phys. Rev., **83**, 34–40.

Calvo, M., Benoit, A., Catalano, A., et al. (2016). 'The NIKA2 instrument: a dual-band kilopixel KID array for millimetric astronomy,' J. Low Temp. Phys., **184**, 816–823.

Canali, C., Jacoboni, C., Nava, F., Ottaviani, G., and Alberigi-Quaranta, A. (1975). 'Electron drift velocity in silicon,' Phys. Rev. B, **12**, 2265–2284.

Capasso, F. (1985). 'Physics of avalanche diodes,' in *Semiconductors and Semimetals*, ed. W. T. Tsang. Orlando, FL: Academic Press, **22D**, 2–172.

Carlstrom, J. E., and Zmuidzinas, J. (1996). 'Millimeter and submillimeter techniques,' in *Reviews of Radio Science 1993–1996*, ed. W. R. Stone. Oxford: Oxford University Press, 839–882.

Carnes, J. E., and Kosonocky, W. F. (1972). 'Noise sources in charge coupled devices,' RCA Review, **33**, 327–343.

Carnes, J. E., Kosonocky, W. F., and Ramberg, E. G. (1972). 'Free charge transfer in charge-coupled devices,' IEEE Trans. Elec. Dev., **ED-19**, 798–808.

Chen, C. J., Choi, K. K., Tidrow, M. Z., and Tsui, D. C. (1996). 'Corrugated quantum well infrared photodetectors for normal incident light coupling,' App. Phys. Let., **68**, 1446–1448.

Chervenak, J. A., Irwin, K. D., Grossman, E. N., et al. (1999). 'Superconducting multiplexer for arrays of transition edge sensors,' App. Phys. Let., **74**, 4043–4045.

Choi, K. K. (1997). *The Physics of Quantum Well Infrared Photodetectors*. Singapore: World Scientific.

Christiansen, W. N., and Högbom, J. A. (1985). *Radiotelescopes*, 2nd edn. Cambridge; New York: Cambridge University Press.

Church, S. E., Price, M. C., Haegel, N. M., Griffin, M. J., and Ade, P. A. R. (1996). 'Transient response in doped germanium photoconductors under very low background operation,' App. Opt., **35**, 1597–1604.

Clarke, J., and Braginski, A. I. (2006). *The SQUID Handbook*, New York: Wiley.

Clarke, J., Hoffer, G. I., Richards, P. L., and Yeh, N.-H. (1977). 'Superconductive bolometers for submillimeter wavelengths,' J. Appl. Phys., **48**, 4865–4879.

Cova, S., Ghioni, M., Lacaita, A., Samori, C., and Zappa, F. (1996), 'Avalanche photodiodes and quenching circuits for single-photon detection,' App. Opt. LP, **35**, 1956–1976.

Coulais, A., and Abergel, A. (2000). 'Transient correction of the LW-ISOCAM data for low contrasted illumination,' A&AS, **141**, 533–544.

Coulais, A., Fouks, B. I., Giovanelli, J.-F., Abergel, A., and See, J. (2000). 'Transient response of IR detectors used in space astronomy: what we have learned from the ISO satellite,' Proc. SPIE, **4131**, 205–217.

Csorba, I. P. (1985). *Image Tubes*. Indianapolis, IN: Howard Sams. A comprehensive and readable discussion of image intensifiers, photomultipliers, and television tubes; includes both theoretical and practical aspects of these detectors. Unfortunately, may be difficult to obtain.

Cullum, M. (1988). 'Experience with the MAMA detector,' in *Instrumentation for Ground-based Optical Astronomy, Present and Future*, ed. L. B. Robinson. Heidelberg; New York: Springer, 568–581.

Davis, K. K., Kloosterman, J. L., Groppi, C., Kawamujra, J. H., and Underhill, M. (2017). 'Micromachined integrated waveguide transformers in THz Pickett-Potter feedhorn blocks,' IEEE Trans. THz Sci. Tech., **7**, 649–656.

Dieleman, P., Klapwijk, T. M., van de Stadt, H., et al. (1997). 'Performance limitations of NbN SIS junctions with Al striplines at 600–850 GHz,' in *Proceedings of the 8th International Symposium on Space Teraherztz Technology*, 291–300.

Doliber, D. L., Stock, J. M., Jelinsky, S. R., et al. (2000). 'Development challenges for the GALEX UV sealed tube detectors,' Proc. SPIE, **4013**, 402–410.

Donati, S. (2000). *Photodetectors: Devices, Circuits, and Applications*, Upper Saddle River, NJ: Prentice Hall. Overview of (mostly) visible light detectors, largely from an engineering viewpoint.

Driggers, R. G., Friedman, M. H., and Nichols, J. M. (2012). *Introduction to Infrared and Electro-Optical Systems*. Boston, MA: Artech House.

Drouet d'Aubigny, C. Y., Walker, C. K., and Jones, B. D. (2000). 'Laser micromachining of terahertz systems,' Proc. SPIE, **4015**, 584–588.

D'Souza, A., Wijewarnasuriya, P. S., and Poksheva, J. G. (2000). 'HgCdTe infrared detectors,' in *Handbook of Thin Film Devices, Vol. 2*, ed. A. G. U. Perera and H. C. Liu. San Diego, CA; London: Academic Press, 1–25.

Durini, D. (ed.) (2014). *High Performance Silicon Imaging: Fundamentals and Applications of CMOS and CCD Sensors*, Woodhead Publishing Series in Electronic and Optical Materials (Book 60). Cambridge: Woodhead Publishing. Useful collection of articles on CMOS arrays and CCDs that covers principles of operation, construction, and circuit architecture for applications from astronomy through cell phones.

Eastman Kodak Co. (1987). *Scientific Imaging with KODAK Films and Plates*. Rochester, NY: Eastman Kodak Co.

Eccles, M. J., Sim, M. E., and Tritton, K. P. (1983). *Low Light Level Detectors in Astronomy*. Cambridge; New York: Cambridge University Press.

Eisaman, M. D., Fan, J., Migdall, A., and Polyakov, S. V. (2011). 'Single-photon sources and detectors,' Rev. Sci. Inst., **82**, 071101–25. Review of single-photon detectors in general.

Eisenhauer, F., and Raab, W. (2015). 'Visible/infrared imaging spectroscopy and energy resolving detectors,' Ann. Rev. Ast. & Astrophys., **53**, 155–197.

Escher, J. S. (1981). 'NEA semiconductor photoemitters,' in *Semiconductors and Semimetals*, ed. P. K. Willardson and A. C. Beer. New York: Academic Press, **15**, 195–300.

Filipovic, D. F., Gearhart, S. S., and Rebeiz, G. M. (1993). 'Double-slot antennas on extended hemispherical and elliptical silicon dielectric lenses,' IEEE Trans. Microwave Theory, **41**, 1738–1749.

Finger, G., Beletic, J. W., Dorn, R., et al. (2005). 'Conversion gain and interpixel capacitance of CMOS hybrid focal plane arrays,' Exp. Ast., **19**, 135–147.

Finger, G., Baker, I., Alvarez, D., et al. (2012). 'Evaluation and optimization of NIR HgCdTe avalanche photodiode arrays for adaptive optics and interferometry,' Proc. SPIE, **8453**, ID 84530T-16.

Fossum, E. R., and Pain, B. (1993). 'Infrared readout electronics for space-science sensors: state of the art and future directions,' Proc. SPIE, **2020**, 262–285.

Fouks, B. I. (1992). 'Nonstationary behaviour of low background photon detectors,' in *Proceedings of an ESA Symposium on Photon Detectors for Space Instrumentation*, 167–174.

Garnett, J. D., Zandian, M., Dewames, R. E., et al. (2004). 'Performance of 5 micron, molecular beam epitaxy HgCdTe sensor chip assemblies (SCAs) for the NGST mission and ground-based astronomy,' in *Scientific Detectors for Astronomy: The Beginning of a New Era*, ed. P. Amico. Dordrecht, Netherlands: Kluwer, 59–79.

Geist, J., and Wang, C. S. (1983). 'New calculations of the quantum yield of silicon in the near ultraviolet,' Phys Rev. B, **27**, 4841–4847.

Gildemeister, J. M., Lee, A. T., and Richards, P. L. (1999). 'A fully lithographed voltage-biased superconducting spiderweb bolometer,' App. Phys. Let., **74**, 868–870.

Goldsmid, H. J. (1965). *The Thermal Properties of Solids*. New York: Dover Publications.

Goldsmith, P. F. (1998). *Quasioptical Systems: Gaussian Beam Quasioptical Propagation and Applications*. New York: Chapman and Hall. The ultimate source on quasi-optics.

Goldsmith, P. F., Itoh, T., and Stophan, K. D. (1989). 'Quasi-optical techniques,' in *Handbook of Microwave and Optical Components*, ed. K. Chang. New York: Wiley, **1**, 344–363.

Gonzalez, A. (2016). 'Frequency independent design of quasi-optical systems,' J. Infrared Mm THz Waves, **37**, 147–159.

Gordon, K. D., Engelbracht, C. W., Fadda, D., et al. (2007). 'Absolute calibration and characterization of the multiband imaging photometer for Spitzer. II. 70 μm imaging,' PASP, **119**, 1019–1037.

Graeme, J. G. (1996). *Photodiode Amplifiers: Op Amp Solutions*. Boston: McGraw Hill. Detailed discussion of TIA and related amplifiers, but virtually no discussion of integrating versions.

Graf, U. U., Honingh, C. E., Jacobs, K., and Stutzki, J. (2015). 'Terahertz heterodyne array receivers for astronomy,' J. Infrared Mm THz Waves, **36**, 896–921.

Grant, B. (2011). *Field Guide to Radiometry*, Bellingham, WA: SPIE Publications. Short and practical guide to practice of radiometry.

Grum, F. C., and Becherer, R. J. (1979). *Optical Radiation Measurement I. Radiometry*. New York: Academic Press. Classic description of radiometry.

Guériaux, V., de L'Isle, N. B., Berurier, A., et al. (2011). 'Quantum well infrared photodetectors: present and future,' Opt. Eng., **50**, 061013-1–19.

Gunapala, S. D., and Bandara, S. V. (2000). 'Quantum well infrared photodetectors (QWIP),' in *Handbook of Thin Film Devices, Vol. 2*, ed. A. G. U. Perera and H. C. Liu. San Diego; London: Academic Press, 63–99.

Hadfield, R. H. (2009). 'Single-photon detectors for optical quantum information applications,' Nature Phot., **3**, 696–705. Review of a broad variety of optical single-photon detectors, including APDs.

Haegel, N. M., Simoes, J. C., White, A. M., and Beeman, J. W. (1999). 'Transient behavior of infrared photoconductors: application of a numerical model,' App. Opt., **38**, 1910–1919.

Haegel, N. M., Schwartz, W. R., Zinter, J., White, A. M., and Beeman, J. W. (2001). 'Origin of the hook effect in extrinsic photoconductors,' App. Opt., **40**, 5748–5754.

Hall, D. N. B., Aikens, R. S., Joyce, R., and McCurnin, T. W. (1975). 'Johnson noise limited operation of photovoltaic InSb detectors.' App. Opt., **14**, 450–453.

Haller, E. E., Hueschen, M. R., and Richards, P. L. (1979). 'Ge:Ga photoconductors in low infrared backgrounds,' App. Phys. Let., **34**, 495–497.

Hamamatsu (2007). *Photomultiplier Tubes: Basics and Applications*, 3rd edn, Hamamatsu Corp., www.hamamatsu.com/resources/pdf/etd/PMT_handbook_v3aE.pdf. Up to date, thorough (theory and practice), and *free*!

Hamden, E. T., Jewell, A. D., Shapiro, C. A., et al. (2016). 'Charge-coupled device detectors with high quantum efficiency at UV wavelengths,' J. Ast. Tel., Inst. Sys., **2**, 036003-1–11.

Hanson, C. M. (1997). 'Hybrid pyroelectric-ferroelectric bolometer arrays,' in *Semiconductors and Semimetals*, ed. P. Kruse and D. Skatrud. San Diego, CA: Academic Press, **47**, 123–174.

Hawking, S. (1988). *A Brief History of Time: From the Big Bang to Black Holes*, London: Bantam Press.

Hartmann, R., Buttler, W., Gorky, H., et al. (2006). 'A high-speed pnCCD detector system for optical applications,' Nuc. Inst. Meth. Phys. Res., Sect. A, **568**, 118–123.

Hays, K. M., Laviolette, R. A., Stapelbroek, M. G., and Petroff, M. D. (1989). 'The solid state photomultiplier: status of photon counting beyond the near-infrared,' in *Proceedings of the Third Infrared Detector Technology Workshop,* ed. C. R. McCreight. NASA Technical Memorandum 102209, 59–80.

Hinken, J. H. (1989). *Superconductor Electronics: Fundamentals and Microwave Applications.* Heidelberg; New York: Springer. An excellent discussion of superconductivity, SIS junctions, and other devices as used for heterodyne receivers.

Hoffman, A., Loose, M., and Suntharalingam, V. (2005). 'CMOS detector technology,' Exp. Ast., **19**, 111–134.

Holland, A. (2013). 'X-ray CCDs,' in *Observing Photons in Space*, 2nd edn, ed. M. C. E., Huber, A., Pauluhn, J. L., Culhane, et al., Heidelberg; New York: Springer, 443–454.

Holland, W. S., Bintley, D., Chapin, E. L., et al. (2013). 'SCUBA-2: the 10 000 pixel bolometer camera on the James Clerk Maxwell Telescope,' MNRAS, **430**, 2513–2535.

Holloway, H. (1986). 'Collection efficiency and crosstalk in closely spaced photodiode arrays,' J. App. Phys., **60**, 1091–1096.

Holmes, W. A., Bock, J. J., Crill, B. P., et al. (2008). 'Initial test results on bolometers for the Planck high frequency instrument,' App. Opt., **47**, 5996–6008. Detailed discussion of the HFI bolometers, starting with theory and showing how the design and construction were developed.

Holst, G. C. (2008). *Testing and Evaluation of Infrared Imaging Systems*, 3nd edn. Bellingham, WA: SPIE Publications.

Horowitz, P., and Hill, W. (2015). *The Art of Electronics*, 3rd edn. Cambridge; New York: Cambridge University Press. An excellent practical discussion of electronic circuitry.

Howell, S. B. (2006). *Handbook of CCD Astronomy.* 2nd edn. Cambridge; New York: Cambridge University Press.

Hynecek, J. (2001). 'Impactron–a new solid state image intensifier,' IEEE Trans. Elec. Dev., **48**, 2238–2241.

Irwin, K. D., and Hilton, G. C. (2005). 'Transition-edge sensors,' in *Cryogenic Particle Detection*, ed. Ch. Enss, Topics in Applied Physics, 99. Heidelberg; New York: Springer, 63–150. Excellent and thorough review of this key type of bolometer thermometer.

Itsuno, A. M. (2012). 'Bandgap-engineered mercury cadmium telluride infrared detector structures for reduced cooling requirements,' Ph.D. thesis in electrical engineering, The University of Michigan.

Jackson, B. D., Baryshev, A. M., de Lange, G., et al. (2001). 'Low-noise 1 THz superconductor-insulator-superconductor mixer incorporating a NbTiN/SiO2/Al tuning circuit,' *Appl. Phys. Let.*, **79**, 436–438.

Jacoby, G. H. (1990). *CCDs in Astronomy.* ASP Conference Series, Vol. 8. San Francisco: Astronomical Society of the Pacific. Conference proceedings that give a summary of research in CCDs.

James, T. H. (ed.) (1977). *The Theory of the Photographic Process*, 4th edn. New York: Macmillan. The standard advanced treatment of the photographic process, consisting of individual articles prepared by members of the staff of Kodak. Not particularly suitable as an introduction.

James, T. H., and Higgins, G. C. (1960). *Fundamentals of Photographic Theory*, 2nd edn. Hastings-on-Hudson, NY: Morgan and Morgan.

Janesick, J. R. (2001). *Scientific Charge-Coupled Devices.* Bellingham, WA: SPIE Publications. The standard encyclopedia of CCDs – essential reading on this topic.

Janesick, J. R., Elliot, T., Dingizian, A., et al. (1990). 'New advances in charge-coupled technology – sub-electron noise and 4096×4096 pixel CCDs,' in *CCDs in Astronomy*, ed. G. H. Jacoby, ASP Conference Series, Vol. 8, 18–39.

Janesick, J. R., and Elliott, S. T. (1992). 'History and advancement of large area array scientific CCD imagers,' in *Astronomical CCD Observing and Reduction Techniques*, ed. S. B. Howell, ASP Conference Series, Vol. 23, 1–67. A detailed review of astronomical CCDs.

Jerram, P., Pool, P. J., Bell, R., et al. (2001). 'The LLCCD: low-light imaging without the need for an intensifier,' Proc. SPIE, **4306**, 178–186.

Joint, F., Gay, G., Vigneron, P.-B., et al. (2019). 'Compact and sensitive heterodyne receiver at 2.7 THz exploiting a quasi-optical HEB-QCL coupling scheme,' App. Phys. Let., **115**, 231104-1–5.

Jones, R. C. (1953). 'The general theory of bolometer performance,' JOSA, **43**, 1–14. A classic development of the theory of the bolometer and relevant figures of merit.

Joseph, C. L. (1995). 'UV image sensors and associated technologies,' Exp. Ast., **6**, 97–127.

Joseph, C. L., Argabright, V. S., Abraham, J., et al. (1995). 'Performance results of the STIS flight MAMA detectors,' Proc. SPIE, **2551**, 248–259.

Kaneda, T. (1985). 'Silicon and germanium avalanche photodiodes,' in *Semiconductors and Semimetals*, ed. W. T. Tsang. Orlando, FL: Academic Press, **22D**, 247–328.

Kaufmann, P., Marcon, R., Abrantes, A., et al. (2014). 'THz photometers for solar flare observations from space,' Exp. Ast., **37**, 579–598.

Kazanskii, A. G., Richards, P. L., and Haller, E. E. (1977). 'Far-infrared photoconductivity of uniaxially stressed germanium,' App. Phys. Let., **31**, 496–497.

Kemmer, J., and Lutz, G. (1987). 'New detector concepts,' Nucl. Inst. Meth. Phys. Res., A, **253**, 365–377.

Kenny, T. W. (1997). 'Tunneling infrared sensors,' in *Semiconductors and Semimetals*, ed. P. Kruse and D. Skatrud. San Diego, CA: Academic Press, **47**, 227–267.

Kerr, A. R., Pan, S.-K., Claude, S. M. X., et al. (2014). 'Development of the ALMA band-3 and band-6 sideband-separating SIS mixers.' IEEE Trans. THz Sci. Tech., **4**, 201–212.

Keyes, R. J., and Quist, T. M. (1970). 'Low-level coherent and incoherent detection in the infrared,' in *Semiconductors and Semimetals*, ed. R. K. Willardson and A. C. Beer. New York: Academic Press, **5**, 321–359.

Kinch, M. A. (2008). 'A theoretical model for the HgCdTe electron avalanche photodiode,' J. Elec. Mat., **37**, 1453–1459.

Kinch, M. A., Beck, J. D., Wan, C.-F., Ma, F., and Campbell, J. (2004). 'HgCdTe electron avalanche photodiodes,' J. Elec. Mat., **33**, 630–639.

Kinch, M. A., and Rollin, B. V. (1963). 'Detection of millimetre and submillimetre wave radiation by free carrier absorption in a semiconductor,' Brit. J. App. Phys., **14**, 672–676.

Kinch, M. A., and Yariv, A. (1989). 'Performance limitations of GaAs/AlGaAs infrared superlattices,' App. Phys. Let., **55**, 2093–2095.

Kocherov, V. F., Taubkin, I. I., and Zaletaev, N. B. (1995). 'Extrinsic silicon and germanium detectors,' in *Infrared Photon Detectors*, ed. A. Rogalski. Bellingham, WA: SPIE Publications, 189–297.

Kogan, Sh. (1996). *Electronic Noise and Fluctuations in Solids*. Cambridge; New York: Cambridge University Press.

Korneev, A., Matvienko, V., Minaeva, O., et al. (2005). 'Quantum efficiency and noise equivalent power of nanostructured, NbN, single photon detectors in the wavelength range from visible to infrared,' IEEE Trans. App. Supercond., **15**, 571–574.

Kraus, J. D. (1986). *Radio Astronomy,* 2nd edn. Powell, OH: Cygnus-Quasar Books. Dated, but still the classic reference for many principles of radio astronomy.

Krause, P. (1989). 'Color photography,' in *Imaging Processes and Materials, Neblette's Eighth Edition*, ed. J. Sturge, V. Walworth, and A. Shepp. New York: van Nostrand, 110–134.

Kruse, P. W. (1997). 'Principles of uncooled infrared focal plane arrays,' in *Semiconductors and Semimetals*, ed. P. Kruse and D. Skatrud. San Diego, CA: Academic Press, **47**, 17–44.

Lacaita, A. L., Zappa, F., Bigliardi, S., and Manfredi, M. (1993). 'On the bremsstrahlung origin of hot-carrier-induced photons in silicon devices,' IEEE Trans. Elec. Dev., **40**, 577–582.

Lampton, M. (1981). 'The microchannel image intensifier,' *Scientific American*, **245**, 62–71. A non-technical explanation of the manufacture and applications of microchannel plates in image intensifiers.

Lampton, M., Sigmund, O., and Raffanti, R. (1987). 'Delay line anodes for microchannel-plate spectrometers,' Rev. Sci. Inst., **58**, 2298–2305.

Lamsal, C. (2014). 'Electronic, thermoelectric and optical properties of vanadium oxides: VO2, V2O3 and V2O5,' Ph.D. thesis, Rutgers University. digitalcommons.njit.edu/dissertations/100.

Laviolette, R. A., and Stapelbroek, M. G. (1989). 'A non-Markovian model of avalanche gain statistics for a solid-state photomultiplier,' J. App. Phys., **65**, 830–836.

Lee, D., Carmody, M., Piquette, E., et al. (2016). 'High-operating temperature HgCdTe: a vision for the near future,' J. Elec. Mat., **45**, 4587–4595,

Lee, Y.-J., and Talghader, J. J. (2018). 'Observational limitations of Bose-Einstein photon statistics and radiation noise in thermal emission,' Phys. Rev. A, **97**, ID 013844, 1–11.

Lei, Wen, Antoszewski, J., and Faraone, L. (2015). 'Progress, challenges, and opportunities for HgCdTe infrared materials and detectors,' App. Phys. Rev., 2, ID 041303-1–34. Broad review of the status of this important detector type, with emphasis on detector architectures and processing.

Leitz, C., Rabe, S., Prigozhin I., et al. (2017). 'Germanium CCDs for large-format SWIR and X-ray imaging,' J. Instrum, **12**, C05014.

Lesser, M. (2015). 'A summary of charge-coupled devices for astronomy,' PASP, **127**, 1097–1104. A short and readable review of the use of modern CCDs in astronomy.

Lesser, M. P., and Iyer, V. (1998). 'Enhancing back-illuminated performance of astronomical CCDs,' Proc. SPIE, **3355**, 446–456.

Levine, B. F. (1990). 'Comment on "Performance limitations of GaAs/AlGaAs infrared superlattices",' App. Phys. Let., **56**, 2354–2356. Followed by a response from M. A. Kinch and A. Yariv (above).

Levine, B. F. (1993). 'Quantum-well infrared photodetectors,' J. App. Phys., **74**, R1–R81.

Lim, B. W., Chen, Q. C., Yang, J. Y., and Asif Khan, M. (1996). 'High responsivity intrinsic photoconductor based on $Al_x Ga_{1-x}N$,' App. Phys. Let., **68**, 3761–3762.

Liu, H. C. (2000). 'An introduction to the physics of quantum well infrared photodetectors and other related new devices,' in *Handbook of Thin Film Devices, Vol. 2*, ed. A. G. U. Perera and H. C. Liu. San Diego, CA; London: Academic Press, 101–134.

London, F., and London, H. (1935). 'The electromagnetic equations of the supraconductor,' Proc. Roy. Soc. London, Ser. A, **149**, 71–88.

Low, F. J. (1961). 'Low-temperature germanium bolometer,' JOSA, **51**, 1300–1304. A classic paper that describes the first high performance semiconductor bolometer.

Lutz, G., Porro, M., Aschauer, S., Wölfel, S., and Strüder, L. (2016). 'The DEPFET sensor-amplifier structure: a method to beat 1/f noise and reach sub-electron noise in pixel detectors,' Sensors, **16**, 608–621.

Maas, S. A. (1993). *Microwave Mixers*, 2nd edn. Boston, MA: Artech. Excellent, thorough discussion of Schottky diode misers and supporting devices such as HEMTs.

Martin, D. D. E., and Verhoeve, P. (2013). 'Superconducting tunnel junctions,' in *Observing Photons in Space*, 2nd edn, ed. M. C. E. Huber, A. Pauluhn, J. L. Curhane, et al. Heidelberg; New York: Springer, 479–496.

Martin, D. D. E., Verhoeve, R., Oosterbroek, T., et al. (2006). 'Accurate time-resolved optical photospectroscopy with superconducting tunnel junction arrays,' Proc. SPIE, **6269**, 626900–11.

Mather, J. C. (1982). 'Bolometer noise: nonequilibrium theory,' App. Opt., **21**, 1125–1129. The complete development of modern theory of bolometer operation in five pages – not for the faint of heart.

Mauskopf, P. D. (2018). 'Transition edge sensors and kinetic inductance detectors in astronomical instruments,' PASP, **130**, 082001–082028. This provides a quite advanced discussion of these two detector types. Some preliminary material might be advisable before tackling this review.

McCammon, D. (2005). 'Semiconductor thermistors,' in *Cryogenic Particle Detection, Topics in Applied Physics*. Heidelberg; New York: Springer, **99**, 35–61. A very thorough and definitive review of this type of bolometer thermometer.

McCluney, W. R. (2014). *Introduction to Radiometry and Photometry*, 2nd edn. Boston, MA: Artech House. Thorough introduction to radiometry.

McCulloch, M. A., Grahn, J., Melhuish, S. J., et al. (2017). 'Dependence of noise temperature on physical temperature for cryogenic low-noise amplifiers,' J. Ast. Tel. Inst. Syst., **3**, 014003-1–014003-4.

McGrath, W. R., Karasik, B. S., Skalare, A., et al. (1999). 'Hot-electron superconductive mixers for THz frequencies,' Proc. SPIE, **3617**, 80–88.

McIntyre, R. J. (1972). 'The distribution of gains in uniformly multiplying avalanche photodiodes: theory,' IEEE Trans. Elec. Dev., **ED-19**, 703–713.

McMurtry, C., Lee, D., Beletic, J., et al. (2013). 'Development of sensitive long-wave infrared detector arrays for passively cooled space missions,' Opt. Eng., **52**, 091804-1–9.

McMurtry, C., Dorn, M. L., Cabrera, M. S., et al. (2016a). 'Candidate 10 micron HgCdTe arrays for the NEOCam space mission,' Proc. SPIE, **9915**, 99150D–8.

McMurtry, C., Cabrera, M. S., Dorn, M. L., Pipher, J. L., and Forrest, W. J. (2016b). '13 micron cutoff HgCdTe detector arrays for space and ground-based astronomy,' Proc. SPIE, **9915**, 99150E–10.

Meeker, S. R., Mazin, B. A., Walter, A. B., et al. (2018). 'DARKNESS: a microwave kinetic inductance detector integral field spectrograph for high-contrast astronomy,' PASP, **130**, 065001–65017.

Melen, R., and Buss, D. (eds.) (1977). *Charge Coupled Devices: Technology and Applications*. New York: IEEE Press.

Mizrahi, U., Argaman, N., Elkind, S., et al. (2013). 'Large-format 17μm high-end VOx μ-bolometer infrared detector,' Proc. SPIE, **8704**, 87041H-8.

Monroy, E., Omnès, F., and Calle, F. (2003). 'Wide-bandgap semiconductor ultraviolet photodetectors,' Semicond. Sci. Tech., **18**, R33–R51. Review of semiconductor materials useful for ultraviolet detectors.

Moroni, G. F., Estrada, J., Paolini, E. E., et al. (2011). 'Achieving sub-electron readout noise in skipper CCDs,' arXiv 1106.1839.

Müller-Seidlitz, J., Bähr, A., Meidinger, N., and Treverspurg, W. (2018). 'Recent improvements on high-speed DEPFET detectors for x-ray astronomy,' Proc. SPIE, **10709**, 107090F–8.

Muth, J. F., Brown, J. D., Johnson, M. A. L., et al. (1999). 'Absorption coefficient and refractive index of GaN, AIN, and AlGaN alloys,' MRS Internet J. Nitride Semicond. Res. **4**, 502–507.

Nahum, M., Richards, P. L., and Mears, C. A. (1993). 'Design analysis of a novel hot-elecron microbolometer,' IEEE Trans. Appl. Supercond., **3**, 2124–2127.

Nakagawa, H., Aoki, S., Sagawa, H., et al. (2016). 'IR heterodyne spectrometer MILAHI for continuous monitoring observatory of Martian and Venusian atmospheres at Mt. Haleakala, Hawaii,' Planetary Space Sci., **126**, 34–48.

National Institute of Standards and Technology (2020). physics.nist.gov/PhysRefData/FFast/html/form.html.

National Research Council (2014). 'Laser radar: progress and opportunities in active electro-optical sensing,' Washington, DC: National Academies Press.

Newhall, B. (1983). *Latent Image: The Discovery of Photography*. Albuquerque, NM: University of New Mexico Press.

Niigaki, M., Hirohata, T., Suzuki, H. K., and Hiruma, T. (1997). 'Field-assisted photoemission from InP/InGaAsP photocathode with p/n junction,' App. Phys. Let., **71**, 2493–2495.

Nikzad, S., Hoenk, M. E., Greer, F., et al. (2012). 'Delta-doped electron-multiplied CCD with absolute quantum efficiency over 50% in the near to far ultraviolet range for single photon counting applications,' App. Opt., **51**, 365–369.

Norton, P. R., Braggins, T., and Levinstein, H. (1973). 'Impurity and lattice scattering parameters as determined from Hall and mobility analysis in n-type silicon,' Phys. Rev. B, **8**, 5632–5653.

Novoselov, E., and Cherednichenko, S. (2017). 'Low noise terahertz MgB_2 hot electron bolometer mixers with an 11 GHz bandwidth,' App. Phys. Let., **110**, 032601-1–5.

Palaio, N. P., Rodder, M., Haller, E. E., and Kreysa, E. (1983). 'Neutron-transmutation-doped germanium bolometers,' Int. J. IR Mm Waves, **4**, 933–943.

Palmer, J. M., and Grant, B. G. (2009). *The Art of Radiometry*, Bellingham, WA: SPIE Publications. A thorough introduction, starting from first principles.

Pearsall, T. P., and Pollack, M. A. (1985). 'Compound semiconductor photodiodes,' in *Semiconductors and Semimetals*, ed. W. T. Tsang. Orlando, FL: Academic Press, **22D**, 173–245.

Perera, A. G. U., Shen, W. Z., Matsik, S. G., et al. (1998). 'GaAs/AlGaAs quantum well photodetectors with a cutoff wavelength at 28μm,' App. Phys. Let., **72**, 1596–1598.

Perkin Elmer Corp. (2003). 'Avalanche Photodiodes: A User Guide,' www.perkinelmer.com/CMSResources/Images/44-6538APP_AvalanchePhotodiodesUsersGuide.pdf.

Perley, R. A., Chandler, C. J., Butler, B. J., and Wrobel, J. A. (2011). 'The expanded Very Large Array: a new telescope for new science,' ApJL, **739**, 1–5.

Petroff, M. D., Stapelbroek, M. G., and Kleinhans, W. A. (1987). 'Detection of individual 0.4–28μm wavelength photons via impurity-impact ionization in a solid-state photomultiplier,' App. Phys. Let., **51**, 406–408.

Phillips, J. D., Edwall, D. D., and Lee, D. L. (2002). 'Control of very-long-wavelength infrared HgCdTe detector-cutoff wavelength,' J. Elec. Mat., **31**, 664–668.

Pierret, R. F. (1996). *Semiconductor Device Fundamentals*, 2nd edn., Reading, MA: Addison-Wesley. A highy recommended text on semiconductor devices.

Pinkie, B., Schuster, J., and Bellotti, E. (2013). 'Physics-based simulation of the modulation transfer function in HgCdTe infrared detector arrays,' Opt. Let., **38**, 2546–2549.

Pobell, F., and Luth, S. (1996). *Matter and Methods at Low Temperatures,* Heidelberg; New York: Springer. A modern overview of methods for achieving low temperatures.

Pollock, D. D. (1985). *Thermoelectricity: Theory, Thermometry, Tool.* Philadelphia: ASTM Special Publication no. 852.

Polyakov, S. V. (2013). 'Photomultiplier tubes,' in *Single-Photon Generation and Detection*, Vol. 45, ed. A. Migdall, S. V. Polykov, J. Fan, and J. C. Bienfang. Waltham, MA: Academic Press, 69–82.

Pozar, D. M. (1998). *Microwave Engineering*, 2nd edn. New York: Wiley.

Press, W. H., Teukolsky, S. A, Vetterling, W. T., and Flannery, B. P. (2007). *Numerical Recipes: The Art of Scientific Computing*, 3rd edn, Cambridge; New York: Cambridge University Press. A thorough and practical general description of numerical methods, including Fourier transformation.

Putley, E. H. (1970). 'The pyroelectric detector,' in *Semiconductors and Semimetals*, ed. R. K. Willardson and A. C. Beer. New York: Academic Press, **5**, 259–285.

Putley, E. H. (1977). 'InSb submillimeter photoconductive devices,' in *Semiconductors and Semimetals*, ed. R. K. Willardson and A. C. Beer. New York: Academic Press, **12**, 143–168.

Rana, V. R., Cook, W. R., III, Harrison, F. A., Mao, P. H., and Miyasaka, H. (2009). 'Development of focal plane detectors for the Nuclear Spectroscopic Telescope Array (NuSTAR) mission,' Proc. SPIE, **7435**, 743503–743508.

Razeghi, M., and Nguyen, Binh-Minh (2014). 'Advances in mid-infrared detection and imaging: a key issues review,' Rep. Prog. Phys, **77**, 082401–17.

Reddy, S. H., Kudale, S., Gokhale, U., et al. (2017). 'A wideband digital back-end for the upgraded GMRT,' J. Astr. Inst., **6**, 1641011-336.

Ressler, M. E., Cho, Hyung, Lee, R. A. M., et al. (2008). 'Performance of the JWST/MIRI Si:As detectors,' Proc. SPIE, **7021**, 70210O-1-12.

Richards, P. L. (1994). 'Bolometers for infrared and millimeter waves,' J. Appl. Phys., **76**, 1–24.

Rieke, F. F., DeVaux, L. H., and Tuzzolino, A. J. (1959). 'Single-crystal infrared detectors based upon intrinsic absorption,' Proc. IRE, **47**, 1475–1478.

Rieke, G. H. (2007). 'Infrared detector arrays for astronomy,' Ann. Rev. A&A, **45**, 77–115. General review of high-performance detector arrays operating from 1 to 1000 μm, reasonably current except for far infrared, submillimeter ranges.

Rieke, G. H., Montgomery, E. F., Lebofsky, M. J., and Eisenhardt, P. R. M. (1981). 'High sensitivity operation of discrete solid state detectors at 4K,' App. Opt., **20**, 814–818.

Rieke, G. H., Ressler, M. E., Morrison, J. E., et al. (2015). 'The mid-infrared instrument for the James Webb Space Telescope, VII: the MIRI detectors,' PASP, **127**, 665–674.

Robinson, F. N. H. (1962). *Noise in Electrical Circuits.* London: Oxford University Press.

Rogalski, A. (2010). *Infrared Detectors*, 2nd edn, Boca Raton, FL: CRC Press. Comprehensive discussion of virtually all infrared detector types.

Rogalski, A. (2012). 'Progress in focal plane array technologies,' Prog. Quant. Elec., **36**, 342–473. Useful general review of infrared detector arrays.

Rogalski, A., and Piotrowski, J. (1988). 'Intrinsic infrared detectors,' Prog. Quant. Elec., **12**, 87–289.

Rosfjord, K. M., Yang, J. K., Dauler, E. A., et al. (2006). 'Nanowire single-photon detector with an integrated optical cavity and anti-reflection coating,' Opt. Exp., **14**, 527–534.

Ruggiero, S. T. and Rudman, D. A. (2013). *Superconducting Devices.* New York: Academic Press.

Saito, Terubumi (2012). 'Optical properties of semiconductor photodiodes/solar cells,' Metrologia, **49**, S118–S123.

Schlaerth, J., Golwala, S., Zmuidzinas, J., et al. (2009). 'Sensitivity optimization of millimeter/submillimeter MKID camera pixel device design,' AIP Conf. Proc., **1185**, 180–183.

Schmit, J. L., and Stelzer, E. L. (1969). 'Temperature and alloy compositional dependences of the energy gap of $Hg_{1-x}Cd_xTe$,' J. App. Phys., **40**, 4865–4869.

Schmülling, F., Klumb, B., Harter, M., et al. (1998). 'High-sensitivity mid-infrared heterodyne spectrometer with a tunable diode laser as a local oscillator,' App. Opt., **37**, 5771–5776.

Schoelkopf, R. J., Wahlgren, P., Kozhevnikov, A. A., Delsing, P., and Prober, D. E. (1998). 'The radio-frequency single-electron transistor (RF-SET): a fast and ultrasensitive electrometer,' Science, **280**, 1238–1242.

Scholze, F., Rabus, H., and Ulm, G. (1998). 'Mean energy required to produce an electron-hole pair in silicon for photons of energies between 50 and 1500 eV,' J. App. Phys., **84**, 2926–2939.

Schubert, J., Fouks, B. I., Lemke, D., and Wolf, J. (1995). 'Transient response of ISOPHOT Si:Ga infrared photodetectors: experimental results and application of the theory of nonstationary processes,' Proc. SPIE, **2553**, 461–469.

Schühle, U., and Hochedez, J.-F. (2013). 'Solar-blind UV detectors based on wide band gap semiconductors,' in *Observing Photons in Space*, 2nd edn., ed. M. C. E. Huber, A. Pauluhn, J. L. Curhane, et al. Heidelberg; New York: Springer, 467–478.

Schulman, T. (2006). 'Si, CdTe and CdZnTe radiation detectors for imaging applications,' Ph.D. thesis in physics, University of Helsinki, Finland.

Sclar, N. (1984). 'Properties of doped silicon and germanium infrared detectors,' Prog. Quant. Elec., **9**, 149–257. An advanced review addressed specifically to extrinsic photoconductivity with extensive details on behavior with different dopants and on nonlinear effects.

Semenov, A. D., Gol'tsman, G. N., and Sobolewski, R. (2002). 'Hot-electron effect in superconductors and its applications for radiation sensors,' Supercond. Sci. Tech., **15**, R1–R16.

Séquin, C. H., and Tompsett, M. F. (1975). 'Charge transfer devices,' in *Advances in Electronics and Electron Physics, Supplement 8*. New York: Academic Press. A dated but very thorough review of the operational principles of CCDs and related devices.

Shaw, M. D., Bueno, J., Day, P., Bradford, C. M., and Echternach, P. M. (2009). 'Quantum capacitance detector: a pair-breaking radiation detector based on the single Cooper-pair box,' Phys. Rev. B, **79**, 144511.

Shi, S.-C., and Noguchi, T. (1998). 'Low-noise superconducting receivers for millimeter and submillimeter wavelengths,' IEICE Trans. Electron., **E81-C**, 1584–1594.

Shockley, W. (1961). 'Problems related to p-n junctions in silicon,' Sol.-St. Elec., **2**, 35–67.

Shu, S., Calvo, M., Leclercq, S., et al. (2018). 'Prototype high angular resolution LEKIDs for NIKA2,' J. Low Temp. Phys., **193**, 141–148.

Siegel, P. H. (2012). 'Terahertz technology: an overview,' Int. J. High Speed Elec. Sys., **13**, 351–394.

Simoens, F. (2013). 'THz bolometer detectors,' in *Physics and Applications of Terahertz Radiation*, ed. M. Perenzoni and D. Paul, Springer Series in Optical Sciences, Vol. 173. Heidelberg; New York: Springer. 35–75.

Singh, A., Srivastav, V., and Pal, R. (2011). 'HgCdTe avalanche photodiodes: a review.' Opt. Laser Tech., **43**, 1358–1370.

Sizov, F. (2018). 'Terahertz radiation detectors: the state-of-the-art,' Semicond. Sci. Tech., **33**, 123001–123026.

Smith, A. G., and Hoag, A. A. (1979). 'Advances in astronomical photography at low light levels,' Ann. Rev. A&A, **17**, 43–71.

Solymar, L. (1972). *Superconducting Tunneling and Applications*. London: Chapman and Hall. A readable yet extensive account of the physical principles behind SIS and Josephson junction mixers.

Sonnabend, G., Sornig, M., Krötz, P., Stupar, D., and Schieder, R. (2008). 'Ultra high spectral resolution observations of planetary atmospheres using the Cologne tuneable heterodyne infrared spectrometer,' J. Quan. Spec. Rad. Transfer, **109**, 1016–1029.

Spicer, W. E. (1977). 'Negative affinity 3-5 photocathodes: their physics and technology,' App. Phys., **12**, 115–130.

Stevens, N. B. (1970). 'Radiation thermopiles,' in *Semiconductors and Semimetals*, ed. R. K. Willardson and A. C. Beer. New York: Academic Press, **5**, 287–318.

Stillman, G. E., and Wolfe, C. M. (1977). 'Avalanche photodiodes,' in *Semiconductors and Semimetals*, ed. R. K. Willardson and A. C. Beer. New York: Academic Press, **12**, 291–393.

Stratton, J. A. (1941). *Electromagnetic Theory*. New York; London: McGraw-Hill.

Streetman, B. G., and Banerjee, S. (2014). *Solid State Electronic Devices*, 7th edn. London: Pearson. A standard text on the operation and construction of solid state electronic devices.

Strüder, L., and Meidinger (2008). 'CCD detectors,' in *The Universe in X-rays*, ed. J. E. Trümper and G. Hasinger. Heidelberg; New York: Springer, 51–71.

Strüder, L., Kanbach, G., Meidinger, N., et al. (2008) 'The development of avalanche amplifying pnCCDs: a status report,' in *High Time Resolution Astrophysics*, Astrophysics and Space Science Library, Vol. 351, ed. D. Phelan, O. Ryan, and A. Shearer. Heidelberg; New York: Springer, 281–289.

Sturmer, D. M., and Marchetti, A. P. (1989). 'Silver halide imaging,' in *Imaging Processes and Materials, Neblette's Eighth Edition*, ed. J. Sturge, V. Walworth, and A. Shepp. New York: van Nostrand, 71–109.

Sze, S. M. (2000). *Semiconductor Devices, Physics and Technology*. 2nd edn. New York: Wiley. A standard and widely used text on solid state physics and the construction and operation of solid state electronic devices.

Szmulowicz, F., and Madarsz, F. L. (1987). 'Blocked impurity band detectors - an analytical model: figures of merit.' J. App. Phys., **62**, 2533–2540. One of the few descriptions of silicon IBC detectors available in the general literature.

Szypryt, P., Mazin, B. A., Ulbricht, G., et al. (2016). 'High quality factor platinum silicide microwave kinetic inductance detectors,' App. Phys. Let., **109**, 151102-1–151102-4.

Szypryt, P., Meeker, S. R., Coiffard, G., et al. (2017). 'Large-format platinum silicide microwave kinetic inductance detectors for optical to near-IR astronomy,' Optics Exp., **25**, 25894–25909.

Tabbert, B., and Goushcha, A. (2012). 'Optical detectors,' in *Springer Handook of Lasers and Optics*. Heidelberg/New York: Springer, 543–619.

Takahashi, T., and Watanabe, S. (2001). 'Recent progress in CdTe and CdZnTe detectors,' IEEE Trans. Nuc. Sci., **48**, 950–959.

Teich, M. C. (1970). 'Coherent detection in the infrared,' in *Semiconductors and Semimetals*, ed. R. K. Willardson and A. C. Beer. New York: Academic Press, **5**, 361–407.

Teranishi, N. (1997). 'Thermoelectric uncooled infrared focal plane arrays,' in *Semiconductors and Semimetals,* ed. P. Kruse and D. Skatrud. San Diego, CA: Academic Press, **47,** 203–218.

Theuwissen, A. J. P. (1995). *Solid-State Imaging with Charge-Coupled Devices*. Dordrecht, Boston: Kluwer. A very thorough discussion of CCD principles, operation, and construction.

Timothy, J. G. (1988). 'Photon-counting detector systems,' in *Instrumentation for Groundbased Optical Astronomy, Present and Future*, ed. L. B. Robinson. Heidelberg; New York: Springer, 516–527.

Timothy, J. G., and Morgan, J. S. (1988). 'Status of the MAMA detector development program,' in *Instrumentation for Groundbased Optical Astronomy, Present and Future*, ed. L. B. Robinson. Heidelberg; New York: Springer, 557–567.

Timothy, J. G. (2013). 'Microchannel plates for photon detection and imaging in space,' in *Observing Photons in Space*, 2nd edn, ed. M. C. E. Huber, A. Pauluhn, J. L. Curhane, J. G. Timothy, K. Wilhelm, and A. Zehnder. Heidelberg; New York: Springer, 391–421.

Tinkham, M. (1996). *Introduction to Superconductivity*, 2nd edn. New York: McGraw-Hill. Classic textbook on superconductivity, brought up to date in a second edition.

Tong, C.-Y. E., Blundell, R., Bumble, B., Stern, J. A., and LeDuc, H. G. (1995). 'Quantum-limited heterodyne detection in superconducting nonlinear transmission lines at submillimeter wavelengths,' *Appl. Phys. Let.*, **67**, 1304–1306.

Tonry, J., and Burke, B. E. (1998). 'The orthogonal transfer CCD,' Exp. Ast., **8**, 77–87.

Torrey, H. C., and Whitmer, C. A. (1948). *Crystal Rectifiers*. Massachusetts Institute of Technology Radiation Laboratory Series, Vol. 15. New York: McGraw-Hill.

Tucker, J. R. (1979). 'Quantum limited detection in tunnel junction mixers,' IEEE J. Quant. Elec., **QE-15**, 1234–1258.

Tucker, J. R. (1980). 'Predicted conversion gain in superconductor-insulator-superconductor quasiparticle mixers,' App. Phys. Let., **36**, 477–479.

Tucker, J. R., and Feldman, M. J. (1985). 'Quantum detection at millimeter wavelengths,' Rev. Mod. Phys., **57**, 1055–1113.

Tulloch, S. M., and Dhillon, V. S. (2011). 'On the use of electron-multiplying CCDs for astronomical spectroscopy,' MNRAS, **411**, 211–225.

Ullom, J. N., and Bennett, D. A. (2015). 'Review of superconducting transition-edge sensors for x-ray and gamma-ray spectroscopy,' Supercond. Sci. Tech., **28**, 084003–084039.

Vampola, J. L. (1993). 'Readout electronics for infrared sensors,' in *The Infrared & Electro-Optical Systems Handbook, Vol. 3*, ed. J. S. Accetta and D. L. Shumaker. Bellingham, WA: SPIE Publications, 285–342.

Van Cleve, J. E., Herter, T., Pirger, B., et al. (1994). 'The first large format Si:Sb BIB arrays,' Exp. Ast., **3**, 177–178.

Van Cleve, J. E., Herter, T. L., Butturini, R., et al. (1995). 'Evaluation of Si:As and Si:Sb blocked-impurity-band detectors for SIRTF and WIRE,' Proc. SPIE, **2553**, 502–513.

Van der Ziel, A. (1976). *Noise in Measurements*. New York: Wiley.

Van Duzer, T., and Turner, T. W. (1998). *Superconductive Devices and Circuits*, 2nd edn. Upper Saddle River, NJ: Prentice-Hall.

Van Vliet, K. M. (1967). 'Noise limitations in solid state photodetectors,' App. Opt., **6**, 1145–1169.

Vincent, J. D., Hodges, S., Vampola, J., Stegall, N., and Pierce, G. (2015). *Fundamentals of Infrared and Visible Detector Operation and Testing*, 2nd edn. New York: Wiley.

Von Zanthier, C., Braeuninger, H., Dennerl, K., et al. (1998). 'A fully depleted pn-junction CCD for infrared-, UV-, and X-ray detection,' Exp. Ast., **8**, 89–96.

Walker, C. K. (2020). *Terahertz Astronomy*, Boca Raton, FL: CRC Press. An excellent, highly approachable overview of submillimeter astronomy covering both the instrumentation and how it is used.

Walker, C. K., Kooi, J. W., Chan, M., et al. (1992). 'A low noise 492 GHz SIS waveguide receiver,' Int. J. IR Mm Waves, **13**, 785–798.

Walker, D., Zhang, X., Saxler, A., et al. (1997). '$Al_xGa_{1-x}N$ ($0 \leq x \leq 1$) ultraviolet photodetectors grown on sapphire by metal-organic chemical-vapor deposition,' App. Phys. Let., **70**, 949–951.

Wang, J.-Q., Richards, P. L., Beeman, J. W., Haegel, N. M., and Haller, E. E. (1986). 'Optical efficiency of far-infrared photoconductors,' App. Opt., **25**, 4127–4134.

Ward, J., Rice, F., Chattopadhyay, G., and Zmuidzinas, J. (1999). 'SuperMix: a flexible software library for high-frequency circuit simulation, including SIS mixers and superconducting electronics,' Proceedings of the 10th International Symposium on Space THz Technology, 268–281.

White, D., McGenn, W., George, D., et al. (2019). '125–211 GHz low noise MMIC amplifier design for radio astronomy,' Exp. Ast., **48**, 137–143.

Wild, W. (2013). 'Coherent far-infrared/submillimetre detectors,' in *Observing Photons in Space*, 2nd edn. ed. M. C. E. Huber, A. Pauluhn, J. L. Curhane, et al. Heidelberg; New York: Springer, 503–523.

Wilson, T. L. (2018). 'Introduction to millimeter/sub-millimeter astronomy,' in *Millimeter Astronomy*, ed. M. Dessauger-Zavadsky and D. Pfeniger. Heidelberg; New York: Springer, 1–110. A companion piece to '*Tools...*' with more emphasis on the high frequencies.

Wilson, T. L., Rohlfs, K., and Hüttemeister, S. (2014). *Tools of Radio Astronomy*, 6th edn. Heidelberg; New York: Springer. The standard introduction to the tools and methods of radio astronomy; emphasizes cm-waves, but there is some discussion of underlying principles as well as mm- and submm-wave techniques.

Woodcraft, A. I., Sudiwala, R. V., Wakui, E., et al. (2000). 'Thermal conductance measurements of a silicon nitride membrane at low temperatures,' Physica B: Cond. Mat., **284–288**, 1968–1969.

Woodward, J. T., Shaw, P.-S., Yoon, H. W., et al. (2018). 'Invited article: advances in tunable laser-based radiometric calibration applications at the National Institute of Standards and Technology, USA,' Rev. Sci. Inst., **89**, 0913011-1–25.

Wyatt, C. L., Baker, D. J., and Frodsham, D. G. (1974). 'A direct coupled low noise preamplifier for cryogenically cooled photoconductive IR detectors,' IR Phys., **14**, 165–176.

Yagoubov, P., Mroczkowski, T., Belitsky, V., et al. (2019). 'Wideband 67–116 GHz receiver development for ALMA Band 2,' A&A, **634**, A46-1–22.

Young, A. T. (1974). 'Photomultipliers, their cause and cure,' in *Methods of Experimental Physics, Vol. 12: Astrophysics Part A*, ed. N. P. Carleton, New York; London: Academic Press, 1–94. An extensive discussion of the theory, construction, and operation of photomultipliers, with emphasis on their use in accurate photometry.

Yun, Minhee, Beeman, J., Bhatia, R., et al. (2003). 'Bolometric detectors for the Planck Surveyor,' Proc. SPIE, **4855**, 136–147.

Zhang, J., Slysz, W., Verevkin, A., et al. (2003). 'Response time characterization of NbN superconducting single-photon detector,' IEEE Trans. App. Supercond., **13**, 180–183.

Zhang, Yijun, and Jiao, Gangcheng (2019). 'Energy bandgap engineering of transmission-mode AlGaAs/GaAs photocathode,' in *Advances in Photodetectors - Research and Applications*, ed. Chee, Kuan, London: IntechOpen.

Zmuidzinas, J. (2012). 'Superconducting microresonators: physics and applications,' Ann. Rev. Cond. Mat. Phys., **3**, 169–214. Excellent in-depth review of MKIDs and related technologies.

Zmuidzinas, J., Kooi, J. W., Kawamura, J., et al. (1998). 'Development of SIS mixers for 1 THz,' Proc. SPIE, **3357**, 53–62.

Zmuidzinas, J., and Richards, P. L. (2004). 'Superconducting detectors and mixers for millimeter and submillimeter astrophysics,' Proc. IEEE, **92**, 1597–1616.

Zwicker, H. R. (1977). 'Photoemissive detectors,' in *Optical and Infrared Detectors*, 2nd edn., ed. R. J. Keyes. Heidelberg; New York: Springer, 149–196. An extensive discussion of photocathodes, with an emphasis on negative electron affinity materials.

Index

CPSIA information can be obtained
at www.ICGtesting.com
Printed in the USA
LVHW060852030821
694401LV00007B/448